Food Safety and Toxicology

Food Safety and Toxicology

Edited by Freddy Lawrence

SYRAWOOD
PUBLISHING HOUSE

New York

Published by Syrawood Publishing House,
750 Third Avenue, 9th Floor,
New York, NY 10017, USA
www.syrawoodpublishinghouse.com

Food Safety and Toxicology
Edited by Freddy Lawrence

International Standard Book Number: 978-1-64740-439-0 (Hardback)

Trademark Notice: Registered trademark of products or corporate names are used only for explanation and identification without intent to infringe.

Cataloging-in-publication Data

Food safety and toxicology / edited by Freddy Lawrence.
    p. cm.
Includes bibliographical references and index.
ISBN 978-1-64740-439-0
1. Food--Safety measures. 2. Food security. 3. Food. 4. Food supply. I. Lawrence, Freddy.
TX531 .F66 2023
363.192--dc23

TABLE OF CONTENTS

PREFACE

The main aim of this book is to educate learners and enhance their research focus by presenting diverse topics covering this vast field. This is an advanced book which compiles significant studies by distinguished experts in the area of analysis. This book addresses successive solutions to the challenges arising in the area of application, along with it; the book provides scope for future developments.

Food contains nutritional substances that are necessary for the development, upkeep, and repair of body tissues as well as the control of vital processes. Therefore, it is essential to consume safe foods, which can be ensured by practicing food safety. The practice of preparing, handling, and storing food in such a way that minimizes the risk of individuals becoming ill from foodborne illnesses is called food safety. Issues related to food safety arise from pathogenic microorganisms such as bacteria, which cause food intoxication or poisoning. Toxicology is the scientific study of the harmful effects of chemicals on living organisms. It also involves diagnosing and treating conditions that occur as a result of exposure to toxins and toxicants. In the field of toxicology, the main focus remains on determining the safety of the food by carrying out a detailed research on the effects of toxins. This book explores all the important aspects of food safety and toxicology in the present day scenario. It consists of contributions made by international experts. The book will serve as a reference to a broad spectrum of readers.

It was a great honour to edit this book, though there were challenges, as it involved a lot of communication and networking between me and the editorial team. However, the end result was this all-inclusive book covering diverse themes in the field.

Finally, it is important to acknowledge the efforts of the contributors for their excellent chapters, through which a wide variety of issues have been addressed. I would also like to thank my colleagues for their valuable feedback during the making of this book.

Editor

Screen-Printed Electrode-Based Sensors for Food Spoilage Control: Bacteria and Biogenic Amines Detection

Ricarda Torre [1], Estefanía Costa-Rama [1,2,*], Henri P. A. Nouws [1] and
Cristina Delerue-Matos [1,*]

[1] REQUIMTE/LAQV, Instituto Superior de Engenharia do Porto, Instituto Politécnico do Porto, Dr. António
Bernardino de Almeida 431, 4200-072 Porto, Portugal; rdvdt@isep.ipp.pt (R.T.); han@isep.ipp.pt (H.P.A.N.)
[2] Departamento de Química Física y Analítica, Universidad de Oviedo, Av. Julián Clavería 8,
33006 Oviedo, Spain
* Correspondence: costaestefania@uniovi.es (E.C.-R.); cmm@isep.ipp.pt (C.D.-M.)
† This article is dedicated to the memory of Professor Agustín Costa-García.

Abstract: Food spoilage is caused by the development of microorganisms, biogenic amines, and other harmful substances, which, when consumed, can lead to different health problems. Foodborne diseases can be avoided by assessing the safety and freshness of food along the production and supply chains. The routine methods for food analysis usually involve long analysis times and complex instrumentation and are performed in centralized laboratories. In this context, sensors based on screen-printed electrodes (SPEs) have gained increasing importance because of their advantageous characteristics, such as ease of use and portability, which allow fast analysis in point-of-need scenarios. This review provides a comprehensive overview of SPE-based sensors for the evaluation of food safety and freshness, focusing on the determination of bacteria and biogenic amines. After discussing the characteristics of SPEs as transducers, the main bacteria, and biogenic amines responsible for important and common foodborne diseases are described. Then, SPE-based sensors for the analysis of these bacteria and biogenic amines in food samples are discussed, comparing several parameters, such as limit of detection, analysis time, and sample type.

Keywords: screen-printed electrode; electroanalysis; electrochemical sensor; biosensor; immunosensor; food analysis; bacteria; biogenic amines; histamine

1. Introduction

The impact of food contamination by microorganisms and other poisonous substances is considered a major public health and safety concern. According to the World Health Organization (WHO), each year 600 million people (almost 1 in 10) fall ill because of contaminated food [1]. Pathogens have the ability to adapt to various environments, causing contaminations in different stages of the food production and supply chains. Thus, they can appear in raw food but also at any point of the food production process and even after the consumer acquires the food if the necessary precaution to transport and store is not taken. Many microorganisms are affected by heat and can be destroyed or inactivated after cooking [2,3]. However, some of them, and substances such as histamine (the main biogenic amine), are not affected by cooking, freezing, or canning processes [3,4]. Taking this into account, the importance of the control of contamination along the whole food chain is clear. Analytical methods and devices for real-time control of food safety and quality provide immediate information that allows corrective actions to be taken before the food products are made available for consumption.

Among the microorganisms that cause foodborne illnesses, bacteria are the most important because of their high occurrence [1,5]. These bacteria can be detected by appropriate techniques and methods such as cell culture and colony counting, polymerase chain reaction (PCR) and immunological assays [3,6]. Biogenic amines are nitrogenous species usually present in different foods that, at normal levels, do not entail health risks. However, their levels increase when food, especially fish, is stored for a long time and/or at an inadequate temperature (>4 °C) [7–9]. Therefore, the quantification of biogenic amines, especially histamine, is included in the routine analysis of many food industries. The analysis of biogenic amines is often performed through chromatographic methods (mainly liquid) coupled to different detectors [10]. Enzymatic kits are also employed since they are simpler and require cheaper instrumentation [11].

Although the above-mentioned methods (i.e., cell culture and colony counting, PCR, chromatography) are very useful, robust and provide accurate results, they are time consuming, involve complex processing steps and require highly trained analysts and expensive/complex instrumentation (Figure 1A). Therefore, the analyses have to be performed in centralized laboratories and the results are not available in real-time. Taking into account the time the different steps of the analytical process take (sampling, sample preparation, analysis, results interpretation, and communication) and the short shelf life of food products, the development of analytical methods that allow rapid screening of pathogens and spoilage indicators is critical to ensure food safety.

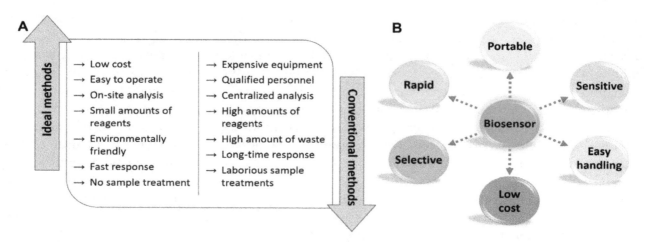

Figure 1. Schematic representation of (**A**) the advantageous features of ideal analytical methods vs. disadvantageous features of conventional ones and (**B**) the advantages of screen-printed based biosensors. The last one adapted from [12] with permission from Elsevier.

In this context, electrochemical (bio)sensors based on screen-printed electrodes (SPEs) have gained increasing interest as analytical tools for food analysis since SPEs provide great advantages that make these kind of sensors have the important characteristics of ideal biosensors (Figure 1B) [12]: ease of use, low-cost, and portability [13,14]. So, the screen-printed technology has significantly contributed to the transition from the traditional unwieldy electrochemical cells to miniaturized and portable electrodes that meet the needs for on-site analysis [12,15]. Although a screen-printed electrode (SPE) is not as robust as a conventional electrode, such as glassy carbon or gold disk, and the surface of its working electrode is not as perfect as the one of a mirror-like polished solid electrode, the advantages of SPEs regarding cost and size led to their increasing use in the last years as transducers in (bio)sensing. The use of SPE-based sensors in the control of food spoilage as complementary analytical tools to the conventional methods allows a rapid screening at any point of the food production chain, preventing the occurrence of foodborne illnesses and the reduction of food waste.

The purpose of this article is to review SPE-based biosensors for the analysis of bacteria and biogenic amines related with food spoilage, focusing on the analyte, and discussing the different approaches and trends in the development of these sensors. The main characteristics of SPEs as transducers and the main challenges on improved SPE-based biosensors are also highlighted.

2. Screen-Printed Electrodes as Transducers

2.1. Production and Design of Screen-Printed Electrodes

The screen-printing technology was adapted from the microelectronics industry and is used, among others, to produce screen-printed electrodes (SPEs) (Figure 2A,B). These electrodes offer the main characteristics required to obtain electrochemical sensing platforms for on-site analysis. Although this technology exists in its present form since the 20th century [16], it began to be used for the fabrication of electrochemical cells in the 1990's. Since then, the use of SPEs as transducers for many different electrochemical sensors has steadily increased (Figure 2C). Nowadays, the screen-printing technology is a common and well-established technique for the conception of electroanalytical devices with assorted applications: from point-of-care (POC) devices for biomedical applications [17–19] to portable sensors for food analysis [13,14] and detection of environmental contaminants [12,20,21]. SPEs usually contain an electrochemical cell composed of three electrodes (working-WE-, reference-RE-, and counter-CE-electrodes) printed on a solid substrate (Figure 2A). Different inks (the most common are carbon and metallic inks) to print the electrodes [22] and different substrates (often ceramic or plastic) can be used. The SPE's fabrication process is fast and allows large scale and highly reproducible production of small-sized, cheap, and disposable electrodes. Therefore, it is not necessary to clean and/or polish them, avoiding tedious pretreatment steps, saving a lot of time. In contrast, the robustness of the printed electrodes and their electrochemical features are not as good as those of conventional electrodes (e.g., glassy carbon, gold disk, etc.). However, SPEs show adequate electroanalytical features for sensing applications and this, together with their low-cost and ease of use (which avoids the need of highly skilled analysts) make SPEs clearly advantageous as transducers for applications in which on-site one-point measurements are required. Moreover, the miniaturized design of SPEs not only allows to transport them to perform on-site measurements for real-time analysis, but also avoids the use of high amounts of reagents and samples. All these characteristics are in accordance with the principles of Green Analytical Chemistry [23,24].

Briefly, the fabrication of SPEs consists of the following steps: (i) design of the screen or mesh that will define the geometry and size of the SPE; (ii) selection and preparation of the conductive inks and selection of the substrate material; (iii) layer-by-layer deposition of the chosen inks on the solid substrate and (iv) drying and curing [17,20]. By covering the electrical circuits with an insulating material it is possible to perform the analytical measurement by depositing a single drop of the reagent/sample solution onto the SPE, by immersing it into a solution or by including it in a flow system. Regarding the inks for the WE, as mentioned before, the most popular ones are based on carbon (graphite, graphene, fullerene, carbon nanomaterials, etc.) because of their suitable features for electroanalysis (i.e., good conductivity, chemical inertness, ease of modification, low background currents, and a wide potential range) and their low costs [17,25]. Besides carbon inks, conductive metallic inks have increasingly been used; among them gold ink is the most common due to its high affinity with thiol moieties that allows easy surface modification with proteins by the formation of self-assembled monolayers (SAMs). SPEs with a WE made of other metallic inks such as silver, platinum or palladium are also available on the market [26] but their use is scarce and limited to specific applications. The use of SPEs with an optically transparent WE, made of indium tin oxide (ITO), PEDOT or even gold (obtained by sputtering process) or carbon (made of carbon nanotubes), is increasing because of the growing interest in spectroelectrochemistry [26–31]. The RE is often made of silver or silver/silver chloride ink. This is considered a pseudo-reference or quasi-reference electrode since its potential is not as stable as that of an ideal reference electrode (e.g., conventional silver/silver chloride RE, which is the most common). Therefore, the applied potential is not as exact and reproducible as when an Ag/AgCl electrode is used. This can be problematic for electrochemical studies in which the control of the potential is essential;

however, for sensing applications, this is rarely a problem. The CE is usually made of the same ink as the WE. Because the composition of the inks defines the electrochemical characteristics of the electrode, SPEs are highly versatile. However, the versatility of SPEs is not only due to the use of different inks, but also because of the ease of modification of the WE. The purpose of these modifications is to enhance the electroanalytical characteristics of the SPEs (such as sensitivity, precision, operational stability) and to improve the immobilization of the recognition element (which are often biological elements (e.g., proteins, DNA, etc.), but can also be synthetic (e.g., molecularly imprinted polymers (MIPs), see Section below)) [17]. For example, great enhancements of the analytical features have been achieved by using carbon nanomaterials (nanotubes, nanofibers, graphene, among others) and metallic nanoparticles (primarily gold nanoparticles, since they are cheaper than a WE made of gold ink) [18,32]. Besides these nanomaterials many other materials can be used: redox mediators, polymers, complexing agents, metallic oxides, etc. The simplest procedure to modify SPEs is by deposition of the modifying agent onto the WE; this procedure is facilitated because of the planar nature of the SPE, so it can be performed through an automatic dispenser in a mass-producible way. However, the WE of an SPE can also be modified by adding the modifier to the ink before printing, by chemical adsorption or by electrochemical deposition (a good example is the in-situ generation of metallic nanoparticles) [32–34].

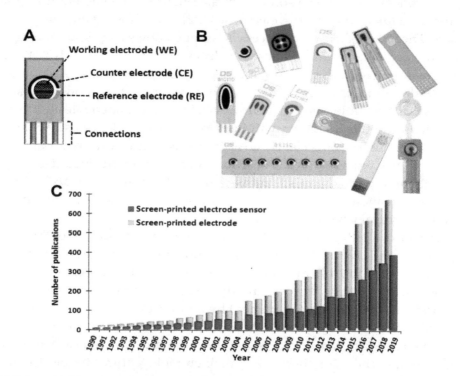

Figure 2. (**A**) Scheme of the most common configuration of a screen-printed electrode. (**B**) Examples of commercial screen-printed electrodes with different configurations and designs. Reproduced from [19] with permission from Wiley. (**C**) Number of publications per year when searching "screen-printed electrode" and "screen-printed electrode sensor" in Scopus database for the last 30 years (1990–2019).

Another source of the SPE's versatility is the possibility of printing the electrochemical cell on a wide variety of substrates. The choice of the substrate will determine the stability, robustness, disposability, and applicability of the SPE. As mentioned before, the most common are rigid substrates such as ceramics. However, although printing the electrodes on non-planar and non-rigid surfaces is not so easy as on rigid ones, there are several works describing SPEs that were fabricated using paper sheets, cloths, stretch and foldable films, and even epidermis [18,20,25,35–38]. To choose the correct substrate, it is important to keep the final application in mind: for example, ceramics are easy to print on and are highly robust but are more expensive than paper. Paper is light and easy to transport but its flexibility and moisture tolerance is limited. Polymeric substrates, especially flexible ones, are interesting for wearable sensors; in these cases, the limitation is related to the bending endurance of the printed electrodes.

So, SPEs offer numerous advantages, but the most important one is their high adaptability. This adaptability is not confined to the materials to fabricate them (inks and substrates); it also covers their design. As said before, the most common option is printing one electrochemical cell (with three electrodes) on the substrate, but many others configurations are possible: SPEs with more than one WE sharing the same RE and CE, platforms with several complete electrochemical cells, 96-well SPE plate or even SPEs with an integrated micro-well/reactor [20,26,39,40] (Figure 2B). Thus, their high versatility together with their ease of use and portability make SPEs one of the main transducers for the development of electroanalytical devices.

2.2. (Bio)Sensors Based on SPEs

As mentioned before, there are a great amount and variety of (bio)sensors based on SPEs with applications in very different fields. A biosensor is a type of chemical sensor; it can be defined as an analytical device able to provide (bio)chemical information, usually the concentration of a substance in a complex matrix, which consists of two main parts: a biological recognition element that selectively identifies the analyte of interest, and a transducer that transforms that recognition event into an measurable signal (Figure 3A) [41,42]. A biosensor should therefore contain biological elements that can be, mainly, (i) enzymes (catalytic biosensors) and (ii) proteins (antibody or antigen), or DNA or RNA strands (affinity biosensors) (Figure 3C). However, because of the advantages of artificial biomimetic receptors, such as MIPs and aptamers, regarding physical and chemical stability, it is increasingly accepted to include them in the "biosensor" category [25].

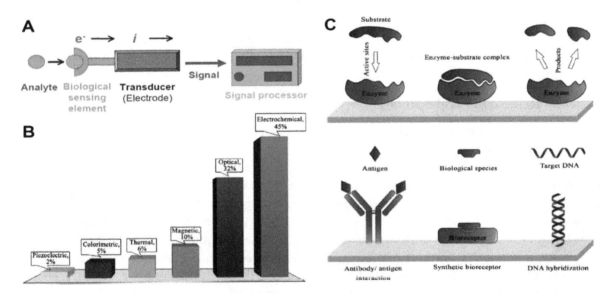

Figure 3. (**A**) A schematic representation of a biosensor with electrochemical transduction. Reproduced from [43] with permission from The Royal Society of Chemistry. (**B**) Distribution of different types of techniques for signal transduction using biosensors (data retrieved from the Scopus database from 2017 to August 2019). Reproduced from [25] with permission from Springer 2019. (**C**) Schematic illustration of enzymatic reaction on catalytic-based biosensors (top) and three different types of affinity-based biosensors (bottom).

The most common SPE-based biosensors for food analysis are enzymatic- and immunosensors. Enzymatic biosensors are based on the highly selective interaction of the target analyte with an enzyme through its active sites, forming a complex that transforms the analyte into a (or more than one) product(s) [25,44]. The determination of the analyte is usually carried out by measuring the amount of generated product. Nevertheless, since co-factors or other co-reagents are sometimes needed, their consumption can be also used to monitor the analyte–enzyme interaction. Immunosensors are based on antibody–antigen interactions and take advantage of the high specificity of an antibody towards the corresponding antigen. In these sensors, the target analyte can either be the antibody or the antigen. Briefly, there are two main ways of following the immunoreaction: (i) using a label attached to a one of the immunoreagents, e.g., an enzyme or a nanoparticle that produces a detectable signal, and (ii) through label-free detection; in this case, the formation of the immunocomplex (antibody–antigen) produces a detectable physical/chemical change [14,19]. Immunosensors are highly specific and can be applied to a wide variety of analytes provided that an antibody that interacts with the analyte is available. Moreover, different strategies (e.g., the use of different labels or nanomaterials) can be used to improve their sensitivity. However, compared to enzymatic sensors, immunosensors are usually more labour intensive and less robust since several steps with long incubation times are required.

Independent of the type of recognition element, they have to be immobilized on the surface of the WE. The versatility of SPEs allows to choose between many different immobilization procedures: from the simplest one, the direct adsorption of the receptor by incubating it on the WE surface, to others that require more steps such as crosslinking, SAM formation, covalent binding, entrapment, or affinity binding (e.g., using the avidin-biotin system). By taking advantage of the transducer, immobilization of the recognition element through electrodeposition is also possible (a good example is the case of electrogenerated MIPs [45]). These immobilization methods are extensively described in several previous reviews [32,44,46–48].

When using biosensors, mainly electrochemical techniques are used for signal transduction, but colorimetric (without instrumentation), optical, magnetic, piezoelectric, and thermal techniques can also be employed (Figure 3B) [25,49]. In electrochemical biosensors, the analytical signal can be provided by different techniques: amperometry and voltammetry (based on current measurement), electrochemical impedance spectroscopy (EIS), potentiometry or conductometry [43,50,51]. The amperometric and voltammetric sensors are the most widely used because of their simplicity and applicability. Nevertheless, EIS sensors are gaining interest since there is no need for labels (especially used in immunosensing), but their sensitivities are often lower than the amperometric and voltammetric sensors [43].

3. Parameters Related to Food Spoilage

The production of safe and high-quality foodstuffs requires the control of several parameters at different points of the food production and supply chain: the quality of raw ingredients, the hygienic conditions of food production, the suitability of storage conditions, and the nutritional properties of the finished products [2]. Inadequate conditions at any stage of this chain often lead to food spoilage, involving chemical and physical changes (e.g., oxidation, colour changes, nasty smells, physical damages, etc.). Although food spoilage can be originated by various causes, the growth of microorganisms is the most common [1,5]. Many factors can contribute or accelerate food deterioration, such as exposure to inadequate levels of oxygen, moisture, light, or temperature. The microorganisms responsible for food spoilage include several bacteria, viruses, moulds, and yeasts. Among these, the bacteria *Salmonella*, *Escherichia coli*, *Campylobacter*, and *Listeria* are the most common foodborne pathogens [1,34]. Besides microorganisms, mycotoxins, which are toxic metabolites produced by fungi, are also important causes of foodborne illnesses [1]. Another common parameter to evaluate food safety and freshness is the level of biogenic amines, which are produced by the microbial decarboxylation of amino acids present in some foodstuffs such as fish, meat, and fermented foods [10,52,53].

3.1. Bacteria

Bacteria are the most common cause of foodborne illnesses. It is often difficult to notice their presence at low but harmful levels since visual or olfactory changes are not always easy to observe. Consequently, the consumer can ingest food contaminated with bacteria without realising it, causing illnesses with important implications. Although there are a great variety of bacteria responsible for foodborne illnesses, *Campylobacter*, *Escherichia coli*, *Salmonella*, and *Listeria* are the most common causes [1,5,42,54,55]. Therefore, these bacteria will be focussed on in this review.

Campylobacter are Gram-negative bacteria that live as commensals organisms in the gastrointestinal tract of humans and many domestic animals [56,57]. Campylobacteriosis is the most commonly reported gastrointestinal infection in humans in the EU [54,55]. It normally produces symptoms such as diarrhoea and vomiting that can last from 2 to 10 days. Its main food sources are undercooked meat (specially poultry), unpasteurized milk and vegetables [3,58]. Within the genus *Campylobacter*, the species *C. jejuni* is responsible for more than 80% of *Campylobacter* infections [3].

Escherichia coli (*E. coli*) are Gram-negative bacteria belonging to the *Enterobacteriaceae* family that inhabit the gastrointestinal tract of humans and warm-blooded animals. As commensal microorganism, *E. coli* live in mutually beneficial association with its host without causing diseases [59,60]. However, there are several *E. coli* strains with virulent attributes associated mainly with three clinical syndromes: diarrhoea, urinary tract infections or meningitis. [3,60]. Its ease of handling and the availability of its complete genome sequence makes *E. coli* an important microorganism in biotechnological, medical and industrial applications [59]. Among the intestinal pathogenic *E. coli* there are six well-describes variants known as pathovars or pathotypes: Enteropathogenic *E. coli* (EPEC), Enterotoxigenic *E. coli* (ETEC), Enteroaggregative *E. coli* (EAEC), Enteroinvasive *E. coli* (EIEC) and Diffusely adherent *E. coli* (DAEC) and Shiga toxin-producing *E. coli* (STEC, which includes the Enterohemorrhagic *E. coli*-EHEC-) [3,60]. Foodborne illness outbreaks related with *E. coli* can be associated with many types of food: from meats and unpasteurized milk or fruit juice to vegetables such as lettuce and spinach [3]. Symptoms of *E. coli* infections can be minor for some people; however, sometimes the infection may become a life-threatening illness causing serious problems such as kidney failure. STEC is the third most common cause of foodborne zoonotic illness [54]. *E. coli* O157:H7 currently accounts for most of the EHEC infections worldwide [3].

Salmonella are Gram-negative bacteria that belong to the *Enterobacteriaceae* family and is classified into two species: *S. bongori* and *S. enterica*. The latter are associated with the main public health concern [3,61]. Based on the Kaufmann-White scheme, *Salmonella* spp. are subdivided into serotypes and consequently, they are usually referred to by their serotype names [61]. Within *S. enterica*, which includes more than 2500 serotypes, *S.* Typhi and *S.* Paratyphi are responsible for typhoid illness (typhoid and paratyphoid fever, respectively) characterized by fever, headache, abdominal pain, and diarrhoea which can be fatal if suitable treatment is not provided [3,42,61,62]. Besides typhoidal illness, the other *Salmonella* serotypes can cause gastrointestinal illness (salmonellosis) that is less serious and its symptoms normally last for a few days [3]. In 2018, nearly 30% of the total foodborne illness outbreaks reported in the EU (5146 outbreaks affecting 48,365 people) were caused by *Salmonella* [54]. These outbreaks were mainly linked to eggs [54], however salmonellosis can also occur by the ingestion of other animal-derived contaminated foods such as milk, meat, or poultry, or even of contaminated fruits or raw vegetables [3,63].

Listeria are Gram-positive bacteria that comprise seventeen species, including *Listeria monocytogenes*, which is responsible for Listeriosis that, although presenting a low incidence, leads to high hospitalizations and mortality rates [50,55,64]. *L. monocytogenes* is highly persistent: it is salt-tolerant and can survive, and even grow, at temperatures below 1°C unlike many other pathogens [3]. *Listeria* can grow in several kinds of foods: raw milk, smoked fish, meats, and raw vegetables [3,50].

The main methods for the detection of these pathogens in foods are based on culturing and colony counting, which are characterized by laborious and time-consuming procedures, consumption of high amounts of reagents and the need for highly-trained personnel [3,5]. Alternative methods are those based on polymerase chain reaction (PCR) or real-time (quantitative) PCR that considerably reduce the analysis time (24 h or 3–6 h, respectively) but also involve laborious procedures [3,5,6,42]. Other detection methods are those based on immunoassays. Among them, enzyme-linked immunosorbent assays (ELISA) are the most common since their commercialization as kits facilitate their use and large-scale application [5,50,51].

3.2. Biogenic Amines

Biogenic amines (BAs) are nitrogenous low-molecular weight compounds that are mainly produced by the microbial decarboxylation of amino acids. There are eight BAs commonly present in animals, plants, and foods and can be classified in three groups based on their structure: (i) aliphatic (putrescine, cadaverine, spermine, and spermidine); (ii) aromatic (tyramine and phenylethylamine); and (iii) heterocyclic (histamine and tryptamine) (Table 1) [7–9,52]. BAs are important in several physiological processes, such as neuromodulating functions, and each one of them has key roles in organisms [7]. For example, histamine acts as a neurotransmitter, is related with intestinal physiological functions, and is involved in allergic reactions; tyramine has antioxidant effects, and putrescine is an important constituent of all mammalian cells [52,65].

Table 1. Classification and basic information of the eight most common biogenic amines. Adapted from [52] with permission from Elsevier.

Classification	Name	Molecular Formula	Structure	Molecular Weight (g/mol)
Heterocyclic	Histamine (HIS)	$C_5H_9N_3$		111.15
	Tryptamine (TRYP)	$C_{10}H_{12}N_2$		160.21
Aromatic	Phenylethylamine (PHEN)	$C_8H_{11}N$		121.18
	Tyramine (TYR)	$C_8H_{11}NO$		137.18
Aliphatic	Spermidine (SPD)	$C_{10}H_{26}N_4$		145.25
	Spermine (SPM)	$C_7H_{19}N_3$		202.34
	Cadaverine (CAD)	$C_5H_{14}N_2$		102.18
	Putrescine (PUT)	$C_4H_{12}N_2$		88.15

In suitable levels BAs have beneficial effects, but the consumption of an excessive amount of BAs can be toxic to humans. The most common example is histamine fish poisoning (also known as scombroid poisoning) which is generally caused by the consumption of fish with high levels of histamine. The symptoms are headache, gastrointestinal, and skin problems, and their severity depends on the dosage [4,7,9]. The rapid increase of the concentration of histamine in fish is induced by unsuitable storage conditions (mainly temperatures >4 °C and long storage times) [4,7]. Fish with high levels of histidine (such as sardine or tuna) are more prone to develop histamine than histidine-poor fish. Moreover, histamine shows a high temperature stability, so it is not affected by cooking or freezing nor by sterilization or canning processes [4,7]. Thus, the concentration of histamine is a common parameter that is used in the fish industry as a quality and freshness indicator.

Although histamine is the main BA of concern due to its toxicity, the other BAs can also induce harmful effects on human health; for example, tyramine, phenylethylamine, and tryptamine cause hypertension, and putrescine and cadaverine can cause hypotension and bradycardia, and potentiate the toxicity of other amines, especially of histamine [7,8,66,67].

Besides fish and sea-food, BAs are found in several daily-life foodstuffs (wine, beer, cheese, other fermented foods and meat) [9,10,66]. Because of their microbiological origin, the concentration of BAs has been used for the assessment of the freshness of certain foods. With this aim, the biogenic amine index (BAI) has been proposed; this parameter can include different BAs depending on the type of food to be evaluated [10]. The most widely used BAI includes histamine, cadaverine, putrescine, and tyramine. A BAI lower than 5–10 mg/kg indicates a good quality and fresh food [8,52,66].

Hence, the concentration of BAs in food is an important parameter to control. The main methods for their quantification are based on chromatographic techniques combined with different extraction techniques such as solid phase extraction, ultrasound-assisted extraction or dispersive liquid–liquid microextraction [10,52,66,67]. Regarding chromatographic techniques, the most common is liquid chromatography (LC) combined with ultraviolet or fluorescence detectors (in which the BAs need to be derivatized since they exhibit neither UV absorption nor fluorescence emission), or tandem mass spectrometry [8,10,52,65–67].

4. SPE-Based (Bio)Sensors for the Determination of Food Spoilage Parameters

When a bibliographic search for articles about SPE-based sensors for determining parameters related to food spoilage is performed, a considerable number of works is found. However, this number is lower than for clinical or biomedical applications. Therefore, the development of these kind of sensors in this area will surely continue to be explored in the next years. Among the SPE-based sensors, there are several enzymatic- and immunosensors, but aptasensors were also described. The electroactivity of some of the analytes is also explored in some studies, avoiding the use of a recognition element. In the following section, different SPE-based sensors for the determination of important food spoilage parameters are discussed. The classification of these sensors is based on the analytes and mainly focuses on those with applications in food analysis.

4.1. SPE-Based (Bio)Sensors for Bacteria Detection

As mentioned before, among the bacteria responsible for foodborne diseases the main contributors, because of their incidence and the illnesses they cause, are *Campylobacter*, *Salmonella*, *E. coli*, and *Listeria*. Because of this, numerous biosensors have been developed for the determination of these microorganisms in food, as a whole or through target indicators of their presence (for example, specific cell membrane proteins or toxins). The wide incidence of salmonellosis has led to the development of many biosensors for the determination of *Salmonella* in foods such as milk or chicken meat (Table 2): most of them are immunosensors, for the serotype *Salmonella* Typhimurium, and based on SPEs with a carbon WE, both unmodified [68,69] or modified with nanomaterials [70–72], polymers [70], or an ionic liquid [71]. Although immunosensors are the main type of sensors, an aptasensor for *Salmonella* detection in apple juice is also reported [73] (Figure 4B). This is a label-free impedimetric sensor that used an SPCE modified with diazonium salt through chemical grafting on which the aminated-aptamer is immobilized. With this approach a concentration range between 10^1 and 10^8 CFU/mL and a limit of detection (LOD) of 6 CFU/mL was achieved. Several sensors use magnetic beads on which either the capture or the detection antibody is immobilized. A good example of this is the one developed for Ngoensawat et al. [74] in which the monoclonal capture antibody is immobilized on carboxylic acid-modified Fe_2O_3 magnetic particles on which multiwalled carbon nanotubes (MWCNT) modified with Methylene blue (the detection label) are immobilized. Once the immunomagnetic separation of *Salmonella* from the sample is performed, the detection is carried out through a sandwich type assay on an avidin-modified SPCE on which a biotin-labelled polyclonal antibody is immobilized. Using DPV as detection technique, a good LOD in milk samples is obtained: 17.3 CFU/mL. As in this case, using labels for monitoring the immunoaffinity event is the most common: the enzyme horseradish peroxidase is widely used. However, the use of nanomaterials such as gold nanoparticles (AuNP) [69,75,76] and CdS nanocrystals [68] is also frequent. A remarkable label is the one based on a polymeric dendron modified with CdTe Quantum Dots (QD) which was recently developed by Murasova et al. [77]. Using a specific anti-*Salmonella* antibody modified with this label, a sandwich immunoassay is performed using an antibody attached to magnetic beads. The detection is carried out through square-wave anodic stripping voltammetry (SWASV) on an SPCE modified with an on-site generated bismuth film obtaining a LOD of 4 CFU/mL. Viswanathan et al. explored the fact that metals show different redox potentials by using different metallic nanoparticles (CuS, CdS and PbS) to construct a multiplexed immunosensor for the simultaneous determination of *Salmonella*, *E. coli*, and *Campylobacter* [78]. The sensor consisted of a MWCNT-polyallylamine-modified SPCE on which specific antibodies for each one of the bacteria are immobilized. The sandwich is formed with detection antibodies specifically labelled with each one of the three different nanocrystals (Figure 4A). Using SWASV as technique detection, calibration curves in the range 10^3–5×10^5 cell/mL, and LODs of 400 cell/mL for *Salmonella* and *Campylobacter* and 800 cell/mL for *E. coli* are obtained. Another approach to develop sensors able to perform simultaneous measurements is the use of SPEs with more than one WE or more than one electrochemical cell. Examples of these sensors for *Salmonella* detection are also reported: from SPEs with two WEs [79], to a microfluidic system with eight WEs [75] (Figure 4C) or a 96-well-SPE plate [80].

Table 2. SPE-based sensors for *Salmonella* and *Listeria* detection in foods.

Serotype	Sensor Construction	Detect. Tech.	Conc. Range	LOD	Analysis Time	Sample	Ref.
Salmonella Pullorum and *Salmonella* Gallinarum	Immunoassay; HRP as indirect label; ERGO/PVA-PDMS/SPCE	CV	10–10^9 CFU/mL	1.61 CFU/mL	≈31 min	Chicken, eggs	[70]
Salmonella Pullorum and *Salmonella* Gallinarum	Immunoassay (sandwich); HRP as label; IL/Ab/AuNP/SPCE	CV	10^4–10^9 CFU/mL	3×10^3 CFU/mL	≈81 min	Chicken, eggs	[71]
Salmonella Typhimurium	Immunoassay (sandwich); Ab-coated MB; Ag measurement; Avidin-SPCE	DPASV	10–10^6 CFU/mL	12.6 CFU/mL	≈105 min	Milk, green bean sprouts, eggs	[81]
Salmonella Typhimurium	Immunoassay; Au-coated MB/SAM/Ab; CdSNP as label; SPCE	SWASV	10–10^6 cell/mL	13 cells/mL	≈40 min	Milk	[68]
Salmonella Typhimurium	Immunoassay (sandwich); Ab on MB-MWCNT-Methylene Blue (which is the label); Avidin-SPCE	DPV	10–10^6 CFU/mL in buffer and milk	7.9 CFU/mL in buffer; 17.3 CFU/mL in milk	≈55 min	Milk	[74]
Salmonella Typhimurium	Immunoassay (sandwich); Capture Ab on MB; AuNP as label; SPCE	DPV	10^3–10^6 cell/mL	143 cells/mL	≈95 min	Milk	[69]
Salmonella Typhimurium	Immunoassay; Label free; Ferrocyanide measurement; rG-GO/SPCE	EIS	–	10^1 CFU/mL in samples	≈15 min	Water, orange juice	[72]
Salmonella Typhimurium	Aptasensor; Label free; diazonium salt- modified SPCE	EIS	10–10^8 CFU/mL	6 CFU/mL	≈45 min	Apple juice	[73]
Salmonella Typhimurium	Paper-based immunoassay (sandwich); AuNP as label; SPCE	C	10–10^8 CFU/mL	10 CFU/mL	≈35 min	Water	[76]

Table 2. *Cont.*

Serotype	Sensor Construction	Detect. Tech.	Conc. Range	LOD	Analysis Time	Sample	Ref.
Salmonella Typhimurium	Immunoassay (sandwich); HRP as label; SPAuE	CA	$10-10^7$ CFU/mL	≈20 CFU/mL	≈150 min	Chicken	[82]
Salmonella Typhimurium	Immunoassay (sandwich); HRP as label; SAM/Protein A/SPAuE	CA	–	10 CFU/mL	≈125 min	Milk	[62]
Salmonella (no serotype)	Immunoassay (sandwich); Capture Ab on MB; QD (CdTe) dendron as label; BiSPCE	SWASV	–	4 CFU/mL	≈80 min	Milk	[77]
Salmonella Typhimurium	Immunoassay; Label free; SAM/GA/Ab/2-SPAuE	EIS	10^3-10^8 CFU/mL	10^3 CFU/mL	≈20 min	Milk	[79]
Salmonella. Typhimurium (and *E. coli* O157:H7)	Immunoassay (sandwich); Capture biotinylated Ab on stretavidin-MB; GOX-as label; SP-IDME (gold)	EIS	10^2-10^6 CFU/mL for both	1.66×10^3 CFU/mL (3.90×10^2 CFU/mL for *E. coli*)	≈180 min	Chicken carcass (ground beef for *E. coli*)	[83]
Salmonella Typhimurium and *Salmonella aureus* (and *E. coli*)	Antimicrobial petide melittin on MB; SP-IDME (silver)	EIS	$10-10^4$ CFU/mL; $10-10^6$ CFU/mL ($1-10^6$ CFU/mL for *E. coli*)	10 CFU/mL for both (1 CFU/mL for *E. coli*)	≈30 min	Water, apple juice	[84]
Salmonella Pullorum and *Salmonella* Gallinarum	Immunoassay (sandwich); Capture Ab on AuNP-modified MB (SiO_2/Fe_3O_4); HRP as label; 4-SPCE	CV	10^2-10^6 CFU/mL	32 CFU/mL	≈70 min	Chicken	[85]
Salmonella Typhimurium	Immunoassay (sandwich); Capture Ab on MB; AuNP as label; μFD-8-SPCE	DPV	10.0–100.0 cell/mL in milk	7.7 cells/mL	≈75 min	Milk	[75]
Salmonella (no serotype)	Immunoassay (sandwich); HRP as label; 96-well SPCE plate	IPA	$5 \times 10^6-5 \times 10^8$ CFU/mL	2×10^6 CFU/mL	≈100 min	Pork, chicken, beef	[80]
Salmonella (no serotype) (Multiplexed: *E. coli*, *Campylobacter*)	Immunoassay (sandwich); specific nanolabel for each specie (CuS, CdS, PbS); MWCNT-PAH/SPCE	SWASV	$10^3-5 \times 10^5$ cell/mL	400 cells/mL for *Salmonella* and *Campylobacter*; 800 cells/mL for *E. coli*	≈70 min	Milk	[78]
Listeria monocytogenes	Immunoassay (sandwich); Ab capture on MB; Ab detection/urease (as label) modified AuNP; SP-IDE (gold)	EIS	$1.9 \times 10^3-1.9 \times 10^6$ CFU/mL	1.6×10^3 CFU/mL	≈115 min	Lettuce	[86]
Listeria innocua Serovar 6b	Label-free; Bacteriophage endolysin CBD500 covalent immobilized on SPAuE	EIS	10^4-10^9 CFU/mL	1.1×10^4 CFU/mL	≈25 min	Milk	[87]

4-SPCE: screen-printed carbon electrode with 4 working electrodes; 2-SPE: screen-printed electrode with 2 working electrodes; 8-SPCE: screen-printed carbon electrode with 8 working electrodes; μFD: microfluidic device; Ab: antibody; AP: alkaline phosphatase; AuNP: gold nanoparticles; BiSPCE: Bi film-modified screen-printed carbon electrode; C: conductometry; CA: chronoamperometry; CdSNP: CdS nanoparticles; CV: cyclic voltammetry; DPASV: differential pulse anodic stripping voltammetry; DPV: differential pulse voltammetry; EIS: electrochemical impedance spectroscopy; ERGO: electrochemically reduced graphene oxide; GA: Glutaraldehyde; HRP: horseradish peroxidase; GOX: glucose oxidase; IL: ionic liquid; IPA: intermittent pulse amperometry; MB: magnetic beads; MWCNT: multiwalled carbon nanotube; n.r: not reported; LSV: linear sweep voltammetry; PAH: polyallylamine; PDMS: polydimethylsiloxane; PVA: polyvinyl alcohol; QD: Quantum Dot; rG-GO: reduced graphene-graphene oxide; SPAuE: screen-printed gold electrode; SPCE: screen-printed carbon electrode; SP-IDME: screen-printed interdigitated electrode; SWASV: square wave anodic stripping voltammetry.

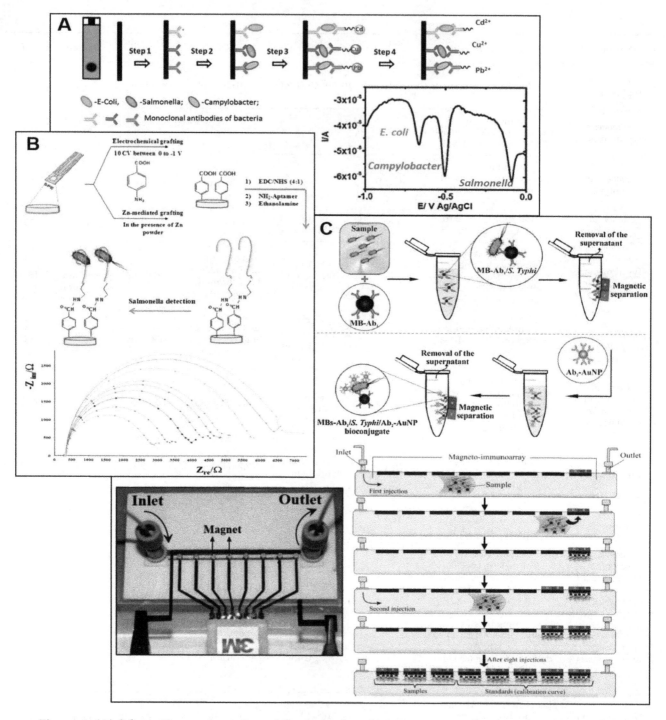

Figure 4. (**A**) Schematic representation of the multiplexed immunosensor developed by Viswanathan et al. for *E. coli*, *Salmonella*, and *Campylobacter* detection and the analytical signals obtained for the three bacteria by square-wave anodic stripping voltammetry (SWASV). Reproduced from [78] with permission from Elsevier. (**B**) Schematic representation of the aptasensor developed by Bagheryan et al. for *Salmonella* detection and the electrochemical impedance spectroscopy (EIS) signals obtained for different *Salmonella* concentrations included in the calibration curve. Reproduced from [73] with permission from Elsevier. (**C**) Schematic representation of the immunoassay based on magnetic beads developed by T.R. de Oliveira et al. using the microfluidic multiplex system shown in the picture. Reproduced from [75] with permission from Elsevier.

For *Listeria* detection in food samples scarce works were found. A noteworthy example is the one developed by Tolba et al. [87] that used the cell wall binding domain (CBD) of bacteriophage-encoded peptidoglycan hydrolases (endolysin) as biorecognition element. CBD500 was immobilized by covalent binding on an SPE with a gold WE. After the reaction with the *Listeria* present in the sample, the analytical signal was obtained by EIS using $[Fe(CN)_6]^{3-/4-}$ as redox probe.

As in the case of *Salmonella*, the SPE-based sensors for *E. coli* detection in foods are mostly based on immunoassays (Table 3). A simple biosensor was developed by Yueh-Hui Lui et al. that consisted of a sandwich immunoassay. The capture antibody was immobilized on an SPCE modified with AuNP and ferrocene dicarboxylic acid (FeDC) [88]. The detection antibody was labelled with HRP and H_2O_2 was used as substrate. The combination of AuNP and FeDC resulted in a significant improvement of the current intensity (studied by CV) when compared with an SPCE that was modified only with AuNP or FeDC. The SPCE contained two electrodes, both made of carbon ink: one acting as working and the other acting as both reference and counter electrodes. Using chronoamperometry (at 300 mV vs. carbon counter/reference electrode) as detection technique, the obtained immunosensor showed a quite good LOD of 600 CFU/mL. A lower LOD (309 CFU/mL in tap water and 457 CFU/mL in minced beef) was obtained by Hassan et al. [89] who immobilized the capture antibody on magnetic beads and carried out a sandwich immunoassay using a AuNP-modified detection antibody (Figure 5A). The quantification of the bacteria was performed through the Hydrogen Evolution Reaction (HER) catalysed by the AuNP using chronoamperometry (applying +1.35 V for 60 s and then, −1.00 V for 100 s) and an SPCE as transducer. Wenchao Dou et al. used a nanocomposite consisting of gold-platinum core/shell nanoparticles, neutral red, and reduced graphene oxide (rGO-NR-Au@Pt) to develop different sensors for *E. coli* determination [90–92]. Using this nanocomposite to label the detection antibody, a sandwich-type immunosensor was developed by immobilizing the capture antibody on an AuNP/polyaniline-SPCE [90]. *E. coli* was quantified by taking advantage of the catalytic effect of the Au@Pt particles on the reduction of H_2O_2, achieving a high LOD of 2840 CFU/mL. Using a similar immunoassay and detection system (measurement of the reduction of H_2O_2 by CV on SPCE), Wenchao Dou et al. achieved a much better LOD (450 CFU/mL) by immobilizing the capture antibody on magnetic beads and using thionine as electron mediator [91]. They achieved an even lower LOD (91 CFU/mL), introducing HRP as a label (to catalyse the H_2O_2 reaction) in the rGO-NR-Au@Pt-detection antibody composite [92] (Figure 5B).

Table 3. SPE-based sensors for *E. coli* O157:H7 detection in foods.

Sensor Construction	Detect. Tech.	Conc. Range	LOD	Analysis Time	Sample	Ref.
Immunoassay (sandwich); HRP as label; AuNP/FeDC-SPCE	CA	10^2 to 10^7 CFU/mL	600 CFU/mL	≈35 min	Milk	[88]
Immunoassay (sandwich); Ab capture on MB; AuNP as label (catalysing HER); SPCE	CA	10^2–10^5 CFU/mL in samples	309 CFU/mL in tap water, 457 CFU/mL in minced beef	≈70 min	Water, minced beef	[89]
Immunoassay (sandwich); rGO-NR-Au@Pt nanocomposite-detection Ab (measurement of H_2O_2 reduction); AuNP/PANI-SPCE	CV	8.9×10^3–8.9×10^9 CFU/mL	2840 CFU/mL	≈110 min	Milk, pork	[90]
Immunoassay (sandwich); Capture Ab on MB; rGO-NR-Au@Pt nanocomposite-detection Ab (measurement of H_2O_2 reduction); Thionine as mediator; SPCE	CV	4×10^3–4×10^8 CFU/mL	450 CFU/mL	≈115 min	Milk, pork	[91]

Table 3. *Cont.*

Sensor Construction	Detect. Tech.	Conc. Range	LOD	Analysis Time	Sample	Ref.
Immunoassay (sandwich); Capture Ab on MB; rGO-NR-Au@Pt nanocomposite HRP-modified detection-Ab; HRP as label; Thionine as mediator; 4-SPCE	CV	4×10^2–4×10^8 CFU/mL	91 CFU/mL	≈135 min	Milk, pork	[92]
Immunoassay; Label-free (measurement of $Fe(CN)_6^{3-/4-}$); AuNP-SPCE	CV	1.19×10^3–1.19×10^9 CFU/mL	594 CFU/mL	≈55 min	Milk powder	[93]
Immunoassay; Label-free (measurement of $Fe(CN)_6^{3-/4-}$); AuNP/PANI-SPCE	DPV	4×10^4–4×10^9 CFU/mL	7980 CFU/mL	≈45 min	Milk	[94]
Immunoassay (sandwich); Ab photochemical immobilization; Label free; SPAuE	EIS	10^2–10^3 CFU/mL in drinking water	30 CFU/mL	≈70 min	Drinking water	[95]
Immunoassay; Capture Ab on MB; Label free; SP-IDME of gold	EIS	10^4–10^7 CFU/mL	$10^{4.45}$ CFU/mL	≈60 min	Ground beef	[96]
Immunoassay; Ab on AuNP/MB-GOX@PDA; Filtration step; GOX as label; Prussian Blue-modified SP-IDME of gold	A	10^3–10^6 CFU/g in ground beef	190 CFU/g	≈75 min	Ground beef	[97]

4-SPCE: screen-printed carbon electrode with 4 working electrodes; A: amperometry; Ab: antibody; AuNP: gold nanoparticles; CA: chronoamperometry; CV: cyclic voltammetry; DPV: differential pulse voltammetry; EIS: electrochemical impedance spectroscopy; FeDC: ferrocene dicarboxylic acid; HER: hydrogen evolution reaction; HRP: horseradish peroxidase; ITO: indium tin oxide; MB: magnetic beads; NP: nanoparticles; NR: neutral red; PANI: polyaniline; PDA: polydopamine; rGO: reduced graphene oxide; SPAuE: screen-printed gold electrode; SPE: screen-printed electrode; SP-IDME: screen-printed interdigitated microelectrode.

Among the label-free sensors for *E. coli* detection, it is worthy to note the one recently developed by Cimafonte et al. for drinking water [95]. It consisted of a sandwich-type immunosensor in which the capture antibody was immobilized with a suitable orientation on a SPAuE by a photochemical technique. The determination was performed by EIS using the $[Fe(CN)_6]^{3-/4-}$ redox probe, achieving a very low LOD of 30 CFU/mL.

Studies using screen-printed interdigitated electrodes were also found. An interesting one is the sensor developed by Xu et al. [97] that used a Prussian blue (PB)-modified screen-printed interdigitated gold microelectrode achieving a LOD of 190 CFU/g. An anti-E.coli antibody was immobilized on magnetic beads that were coated with polydopamine and modified with glucose oxidase (GOX) and AuNP. After the immunoreaction with the bacteria, a filtration was performed (through a paper with 0.8 μm pores) to separate the immunocomplex formed with the bacteria from the free nanocomposite-Ab. The analytical signal was recorded by amperometry dropping the filtered solution onto the PB-modified electrode together with a glucose solution (enzymatic substrate for GOX).

4.2. SPE-Based (Bio)Sensors for Biogenic Amines Detection

As mentioned before, histamine is the main BA. This explains the large number of sensors developed for its determination when compared to those developed for the other BAs. The most frequently reported SPE-based sensors for the determination of BAs in food samples (mainly fish) are enzymatic, although some immunosensors [98,99] or sensors based on the electroactivity of the amines [100,101] can also be found (Table 4). An interesting example among the immunosensors for histamine determination is the one recently developed by Shkodra et al. [98] using a flexible SPE with a WE made of a silver polymeric paste. The three-electrode cell is screen-printed on a polyethylene terephthalate (PET) flexible substrate to obtain a sensor that withstands frequent bending without signal loss (Figure 6A). To perform the immunoassay, an anti-histamine antibody is immobilized on the WE, previously modified with oxygen plasma-treated carbon nanotubes. Then, the competitive immunoassay is carried out using HRP-labelled histamine to compete with the histamine of the sample. Using 3,3′,5,5′-tetramethylbenzidine (TMB) as enzymatic substrate and chronoamperometry as detection technique, a sensor with a very low LOD (0.022 nM) and a high selectivity (tested using other BAs (cadaverine, putrescine, and tyramine) was obtained.

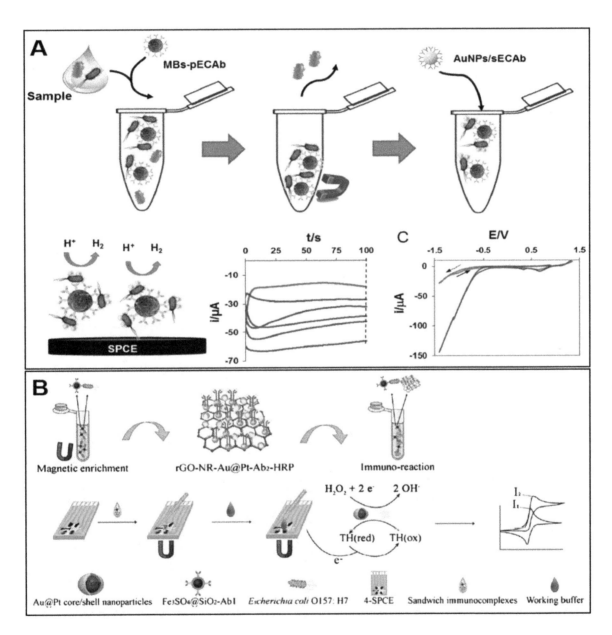

Figure 5. (**A**) Schematic representation of the magneto immunoassay developed by Hassan et al. for *E.coli* O157:H7 detection based on the Hydrogen Evolution Reaction electrocatalyzed by AuNP; chronoamperograms for different bacteria concentration; cyclic voltammograms in absence (red line) and presence of bacteria (blue line). Reproduced from [89] with permission from Elsevier. (**B**) Schematic representation of the magneto immunoassay, using rGO-NR-Au@Pt nanocomposite and HRP as label, developed by Wenchao Dou et al. *E. coli* O157:H7 detection.

Table 4. SPE-based sensors for biogenic amines detection in food samples.

Biogenic Amines	Sensor Construction	Detect. Tech.	Conc. Range	LOD	Analysis Time	Sample	Ref.
HIS	Rhenium (IV) oxide-SPCE	A	4.5–90 μM	1.8 μM	≈3 min	Fish sauce	[100]
HIS	Nafion/$Cu_3(PO_4)_2$NP/SPCE	A	0.045–4.5 mM	0.027 mM	≈3 min	Fish	[101]
HIS	Immunoassay (competitive); Histamine labelled with HRP; Capture Ab on SWCNT/SPE (flexible with a silver WE)	CA	0.045–450 nM	0.022 nM	≈140 min	Fish	[98]
HIS	Immunoassay (competitive); HRP-labelled detection Ab; Histamine-ovalbumin conjugate on PB/chitosan/AuNP/SPCE	CV	0.09–900 μM	0.01 nM	≈130 min	Fish	[99]
HIS	DAO on SPCE	CA	9–675 μM	4.5 μM	≈1 min	Fish (hake, mackerel)	[102]
HIS	DAO on SPCE; $[Fe(CN)_6]^{3-}$ in solution as mediator	CA	45–675 μM	8.7 μM	≈7 min	Fish (tuna, mackerel)	[103]
HIS	DAO and HRP on polysulfone/MWCNT/ferrocene membrane/SPCE; SPCE with two WE, ferrocene as mediator	A	0.3–20 μM	0.17 μM	≈2 min	Fish (anchovy, tuna, sardine, mackerel, shrimp, grater weever)	[104]
HIS	DAO on PtNP/rGO/chitosan/SPCE	A	0.1–300 μM	25.4 nM	≈2 min	Fish (carp, tench, catfish, perch)	[105]
PUT	MAO on TTF-SPCE; TTF as mediator	A	16–101 μM	17.2 μM	≈2 min	Anchovy, Courgette	[106]
PUT	PUO on TTF-SPCE; TTF as mediator	A	10–74 μM	10.1 μM	≈2 min	Octopus, courgette	[107]
TYR	DAO on GO/PVF-modified SPCE / MAO on GO/PVF-modified SPCE	A	0.99–120 μM / 0.9–110 μM	0.41 μM / 0.61 μM	≈2 min	Cheese	[108]
TYR	Ty on SWCNT/SPCE	A	5–180 μM	0.62 μM	≈2 min	Fish	[109]
TYR	1-methyl-4-mercaptopyridine/AuNP/PEDOT:PSS/SPCE	DPV	5–100 nM	2.31 nM	≈6 min	Milk	[110]
TYR	Nafion/Ty/Fe_3O_4-chitosan/poly-L-lysine/SPCE	A	0.49–63 μM	0.075 μM	≈2 min	Cheese	[111]
TYR	PAO on SPCE (hydroxymethylferrocene in cell solution as mediator)	A	2–164 μM	2.0 μM	≈2 min	Cheese	[112]
TYR	HRP on SPCE	A	2–456 μM	2.1 μM	≈2 min	Cheese	[113]

Table 4. *Cont.*

Biogenic Amines	Sensor Construction	Detect. Tech.	Conc. Range	LOD	Analysis Time	Sample	Ref.
HIS CAD	DAO on PB/ITO nanoparticles/SPCE MAO on PB/ITO nanoparticles/SPCE	A	6.0-690 µM 3-1000 µM	1.9 µM 0.9 µM	≈2 min	Cheese	[114]
HIS PUT	HMD and PUO respectively on TTF-SPCE (with 4 WE); TTF as mediator	A	– –	8.1 µM 10 µM	≈2 min	Octopus	[115]
PUT CAD	MAO (for PUT) or MAO/AuNPs (for PUT and CAD) on TTF-SPCE (with two WE); TTF as mediator	A	9.9-74.1 µM 19.6-107.1 µM	9.9 µM 19.9 µM	≈2 min	Octopus	[116]
Total biogenic amines (calibration with HIS, PUT, CAD)	DAO on MB; PB-SPCE	CA	0.01-1 mM for HIS, PUT, CAD	4.8 µM for HIS; 0.9 µM for PUT; 0.67 µM for CAD	≈15 min	Fish (sea bass)	[117]
Total biogenic amines (calibration with HIS)	DAO and HRP on aryl diazonium salt/SPCE	A	0.2-1.6 µM	0.18 µM	≈2 min	Fish (anchovy)	[118]
Total biogenic amines (calibration with PUT)	DAO on polyazetidine prepolimer/SPE (with two WE of gold)	A	8-227 µM	2.3 µM	≈2 min	Wine, beer	[119]
Total biogenic amines (calibration with CAD, PUT, TYR, HIS)	Nafion/DAO/MrO$_2$-SPCE (MnO$_2$ as mediator)	A	1-50 µM for CAD and PUT; 10-300 µM for TYR and HIS	0.3 µM for CAD and PUT; 3.0 µM for TYR and HIS	≈5 min	Chicken meat	[120]

A: amperometry; AuNP: gold nanoparticles; BSA: bovine serum albumin; CA: chronoamperometry; CAD: cadaverine; CV: cyclic voltammetry; DAO: diamine oxidase; DPV: differential pulse voltammetry; GO: graphene oxide; HIS: histamine; HMD: histamine dehydrogenase; HRP: horseradish peroxidase; ITO: indium tin oxide; MAO: monoamine oxidase; MB: magnetic beads; MWCNT: multi-walled carbon nanotubes; NP: nanoparticles; PAO: plasma amine oxidase; PB: Prussian blue; PEDOT:PSS: poly(3,4-ethylenedioxythiophene):poly-styrene sulfonate; PtNP: platinum nanoparticles; PUO: putrescine oxidase; PUT: putrescine; PVF: polyvinylferrocene; rGO: reduced graphene oxide; SPCE: screen-printed carbon electrode; SPE: screen-printed electrode; SWCNT: single-walled carbon nanotubes; TTF: tetrathiafulvalene; Ty: tyrosinase; TYR: tyramine; WE: working electrode.

Among the large number of enzymatic sensors reported for the determination of BAs in food, most are based on the use of the enzymes monoamine oxidase (MAO) or diamine oxidase (DAO). These enzymes catalyse the oxidation of BAs, producing hydrogen peroxide [102,105,106,116–118]. The detection in these sensors is usually carried out by amperometric techniques and the use of redox mediators, such as $[Fe(CN)_6]^{3-/4-}$ [103], ferrocene [104,112], or tetrathiafulvalene (TTF) [106,107,115,116] is very common. The use of these mediators decreases the detection potential, improving the selectivity of the sensor. An interesting work was reported by S. Leonardo et al. in which different mono- (DAO) and bienzymatic (DAO and HRP) sensors using magnetic beads and different mediators (Co(II)-phthalocyanine (CoPh), Prussian Blue (PB), and Os-polyvinylpyridine (Os-PVP)) were developed and compared [117] (Figure 6C). Although calibration curves for histamine, putrescine and cadaverine were obtained for each one of the sensors (DAO-MB/CoPh-SPCE, DAO-MB/PB-SPCE, and DAO-MB/Os-PVP-HRP/SPCE), obtaining LODs from 0.47 µM to 5.13 µM, the one that included Prussian Blue as mediator was chosen for the determination of BAs in sea bass.

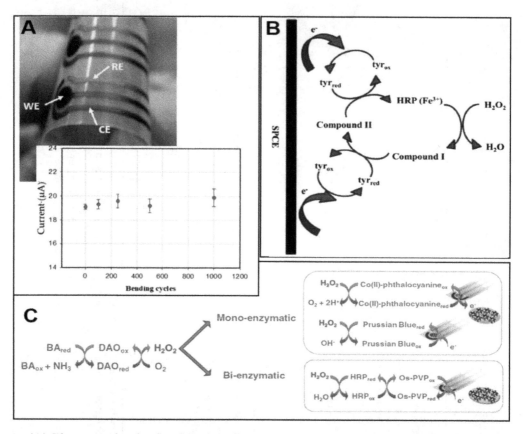

Figure 6. (**A**) Photograph of a flexible three-electrode SPE with silver working electrode used as transducer of an immunosensor for histamine, and flexibility test of that sensor (current intensity obtained after bending it). Reproduced from [98]). (**B**) Schematic representation of the enzymatic reaction occurring at the surface of the HRP/SPCE sensor for tyramine detection; Compound I and compound II are reaction intermediates (compound I (oxidation state +5) comprising a ferryl species ($Fe^{4+}=O$) and a porphyrin radical cation; compound II (oxidation state +4) is formed by the first reduction of the porphyrin radical cation). Reproduced from [113] with permission from Wiley. (**C**) Scheme of the enzymatic and electrochemical reaction occurring on DAO-MB mono- and bi-enzymatic sensors for biogenic amines (BAs) detection.

The use of nanomaterials is also frequent: single or multi-walled carbon nanotubes [104,109], graphene [108], nanoparticles [110,111,114,116], or the combination of different kind of nanomaterials [105]. An example is the sensor developed by Pérez et al. [104] that combines the use of two enzymes, DAO and horseradish peroxidase (HRP), with MWCNT and ferrocene as mediator. In this case, an SPCE with two

WEs was used: one contained the enzymes immobilized on a polysulfone/MWCNT/ferrocene membrane and the other only contained the membrane. With this strategy, the response towards any electroactive species present in the samples that could interfere in the determination is eliminated, improving the selectivity of the sensor.

Sensors based on DAO and MAO or oxidase enzymes such as putrescine oxidase or plasma amine oxidase sometimes show problems regarding selectivity when just one biogenic amine is the target analyte [102,106,107,112,121]. Since for many food applications the objective is obtain the BAI (biogenic amine index), this fact could be not a problem and indeed, several sensors based on DAO and MAO are focused on the quantification of the total amount of BAs [117–120]. With the aim of obtaining more selective enzymatic sensors, the use of other enzymes as recognition element has been reported: e.g., tyrosinase [109] and histamine dehydrogenase [115]. A noteworthy example of a selective SPE-based enzymatic sensor for tyramine is the one developed by Calvo-Pérez et al. [113]. In this work HRP was used as recognition element for tyramine, which is not among the common substrates for this enzyme. The recognition of tyramine through HRP is based on the oxidation of the –OH group present in the molecular structure of tyramine (Figure 6B). Two immobilization procedures were assessed: (i) cross-linking with glutaraldehyde and bovine serum albumin and (ii) mixing the carbon ink used for screen-printing the WE with HRP. Since the second procedure was easier and provided better reproducibility, the sensor obtained in that way was the chosen for its application in real samples. A high selectivity of this HRP-sensor was demonstrated; no response was observed when calibration curves for other BAs (putrescine, cadaverine, histamine, tryptamine, spermine and spermidine were evaluated) in the same concentration range than for tyramine were performed. Another approach to greatly improve the selectivity is to add a separation step before the measurement with the sensor as reported by Li et al. [110]. In this work, a sensor based on an SPCE modified with a conductive polymer (PEDOT.PSS), AuNP and 1-methyl-4-mercaptopyridine (1m-4-MP) was developed to detect tyramine using DPV as technique detection. Before the electrochemical sensing, a sample was treated through a solid-phase extraction based on MIP technology (using a MIP synthetized with methacrylic acid as monomer). The combination of the MIP-based solid phase extraction with the 1-m-4-MP/AuNP/PEDOT:PSS/SPCE provided a LOD of 2.31 nM.

5. Conclusions

Nowadays, food safety is a key concern because it is directly related to public health. Therefore, the development of methods that allow rapid and on-site analysis has gained special relevance in food safety and quality assurance. Disposable electrodes, such as screen-printed electrodes (SPEs), have attracted attention worldwide since they allow the development of easy-to-handle and cost-efficient biosensors. The easy mass-production of reproducible SPEs allows the use of SPE-based sensors as one-shot devices. Besides the concern from the food industry and public-health-related administration about food safety and quality, the growing consumer concern about the security and healthiness of the food they eat enormously increases the interest in point-of-need sensors that can be used by untrained people.

Although there are many published papers on biosensors for food applications, the number of those commercially available is scare since the knowledge transfer from research laboratories to the market is hard. The main challenge for the commercialization of biosensors (for any kind of application) is often the low stability of their recognition element since they are biological compounds that requires special storage conditions. In the case of biosensors for food applications, another important difficulty is related to the sample since it is usually solid, and the measurements normally have to be performed in aqueous medium. This is an important limitation when compared with biosensors for clinical application that are typically applied to bodily fluids. Although the development of multiplex biosensors is increasing, multi-analyte detection is still a big challenge. In the case of food sensors, this is a key issue since, for example, a biosensor can be able to selectively determine a single bacteria serotype (i.e., *S. typhimurium*) but does not provide any information about the presence of other serotypes that can also be harmful. In those cases, the selection of a biological recognition able to detect several serotypes (or kinds of analytes) or the design of a multiplex devices is of paramount importance.

Therefore, it is obvious that biosensors cannot replace the conventional methods (e.g., PCR or HPLC-MS), since these show better features in terms of accuracy, selectively, sensitivity, or multi-analyte detection ability. However, the advantages of SPE-based biosensors mentioned in this review make them exceptional devices for on-site screening. The hard work of electroanalytical researchers to make portable sensors a suitable alternative to the centralized analysis, together the great advances in digital communication networks is leading to promising tools for food control and analysis. Nowadays, in a growing number of situations, it is much more advantageous to have simple tools for fast and on-site screening than sophisticated instrumentation in centralized laboratories.

Funding: This work was supported by UIDB/50006/2020 and UIDP/50006/2020 with funding from the Fundação para a Ciência e a Tecnologia (FCT)/the Ministério da Ciência, Tecnologia e Ensino Superior (MCTES) through national funds. The authors also thank FCT and the EU for funding through the projects: FishBioSensing—Portable electrochemical (bio)sensing devices for safety and quality assessment of fishery products (02/SAICT/2016, POCI-01-0145-FEDER-023817), PTDC/QUI-QAN/30735/2017—TracAllerSens—Electrochemical sensors for the detection and quantification of trace amounts of allergens in food products (POCI-01-0145-FEDER-030735), and PTDC/ASP-PES/29547/2017—CECs(Bio)Sensing—(Bio)sensors for assessment of contaminants of emerging concern in fishery commodities (POCI-01-0145-FEDER-029547), supported by national funds by FCT/MCTES and co-supported by Fundo Europeu de Desenvolvimento Regional (FEDER) through COMPETE 2020—Programa Operacional Competitividade e Internacionalização. E. Costa-Rama thanks the Government of Principado de Asturias and Marie Curie-Cofund Actions for the post-doctoral grant "Clarín-Cofund" ACA17-20. R. Torre is grateful to FCT for her PhD grant (SFRH/BD/143753/2019), financed by POPH–QREN–Tipologia 4.1–Formação Avançada, subsidized by FSE and MCTES.

References

1.   World Health Organization. Available online: https://www.who.int/news-room/fact-sheets/detail/food-safety (accessed on 29 September 2020).
2.   Den Besten, H.M.W.; Wells-Bennik, M.H.J.; Zwietering, M.H. Natural Diversity in Heat Resistance of Bacteria and Bacterial Spores: Impact on Food Safety and Quality. *Annu. Rev. Food Sci. Technol.* **2018**, *9*, 383–410. [CrossRef] [PubMed]
3.   Abraham, A.; Al-Khaldi, S.; Assimon, S.A.; Beuadry, C.; Benner, R.A.; Bennett, R.; Binet, R.; Cahill, S.M.; Burkhardt, W., III. Bad Bud Book. In *Handbook of Foodborne Pathogenic Microorganisms and Natural Toxins Introduction*, 2nd ed.; Food and Drug Administration: Silver Spring, MD, USA, 2012; ISBN 9780323401814.
4.   Naila, A.; Flint, S.; Fletcher, G.; Bremer, P.; Meerdink, G. Control of biogenic amines in food—Existing and emerging approaches. *J. Food Sci.* **2010**, *75*, R139–R150. [CrossRef] [PubMed]
5.   Velusamy, V.; Arshak, K.; Korostynska, O.; Oliwa, K.; Adley, C. An overview of foodborne pathogen detection: In the perspective of biosensors. *Biotechnol. Adv.* **2010**, *28*, 232–254. [CrossRef] [PubMed]
6.   Zhao, X.; Lin, C.W.; Wang, J.; Oh, D.H. Advances in rapid detection methods for foodborne pathogens. *J. Microbiol. Biotechnol.* **2014**, *24*, 297–312. [CrossRef] [PubMed]
7.   European Food Safety Authority (EFSA). BIOHAZ Scientific Opinion on risk based control of biogenic amine formation in fermented foods. *EFSA J.* **2011**, *9*, 2393. [CrossRef]
8.   Biji, K.B.; Ravishankar, C.N.; Venkateswarlu, R.; Mohan, C.O.; Gopal, T.K.S. Biogenic amines in seafood: A review. *J. Food Sci. Technol.* **2016**, *53*, 2210–2218. [CrossRef] [PubMed]
9.   Jairath, G.; Singh, P.K.; Dabur, R.S.; Rani, M.; Chaudhari, M. Biogenic amines in meat and meat products and its public health significance: A review. *J. Food Sci. Technol.* **2015**, *52*, 6835–6846. [CrossRef]
10.   Papageorgiou, M.; Lambropoulou, D.; Morrison, C.; Kłodzińska, E.; Namieśnik, J.; Płotka-Wasylka, J. Literature update of analytical methods for biogenic amines determination in food and beverages. *TrAC Trends Anal. Chem.* **2018**, *98*, 128–142. [CrossRef]
11.   Köse, S.; Kaklikkaya, N.; Koral, S.; Tufan, B.; Buruk, K.C.; Aydin, F. Commercial test kits and the determination of histamine in traditional (ethnic) fish products-evaluation against an EU accepted HPLC method. *Food Chem.* **2011**, *125*, 1490–1497. [CrossRef]

12.   Mishra, R.K.; Nunes, G.S.; Souto, L.; Marty, J.L. Screen printed technology—An application towards biosensor development. In *Encyclopedia of Interfacial Chemistry*; Elsevier: Amsterdam, The Netherlands, 2018; pp. 487–498.

13.   Smart, A.; Crew, A.; Pemberton, R.; Hughes, G.; Doran, O.; Hart, J.P. Screen-printed carbon based biosensors and their applications in agri-food safety. *TrAC Trends Anal. Chem.* **2020**, *127*, 115898. [CrossRef]

14.   Vasilescu, A.; Nunes, G.; Hayat, A.; Latif, U.; Marty, J.L. Electrochemical affinity biosensors based on disposable screen-printed electrodes for detection of food allergens. *Sensors* **2016**, *16*, 1863. [CrossRef] [PubMed]

15.   Díaz-Cruz, J.M.; Serrano, N.; Pérez-Ràfols, C.; Ariño, C.; Esteban, M. Electroanalysis from the past to the twenty-first century: Challenges and perspectives. *J. Solid State Electrochem.* **2020**. [CrossRef] [PubMed]

16.   Roberts, G.; Age, S.; Simon, S. History's Influence on Screen Printing's Future Explore How Screenprinting's Past Will Shape Its Future. *Screen Print.* February 2006. Available online: https://www.screenweb.com/content/historys-influence-screen-printings-future (accessed on 30 September 2020).

17.   Couto, R.A.S.; Lima, J.L.F.C.; Quinaz, M.B. Recent developments, characteristics and potential applications of screen-printed electrodes in pharmaceutical and biological analysis. *Talanta* **2016**, *146*, 801–814. [CrossRef] [PubMed]

18.   Arduini, F.; Micheli, L.; Moscone, D.; Palleschi, G.; Piermarini, S.; Ricci, F.; Volpe, G. Electrochemical biosensors based on nanomodified screen-printed electrodes: Recent applications in clinical analysis. *Trends Anal. Chem.* **2016**, *79*, 114–126. [CrossRef]

19.   Rama, E.C.; Costa-García, A. Screen-printed Electrochemical Immunosensors for the Detection of Cancer and Cardiovascular Biomarkers. *Electroanalysis* **2016**, *28*, 1700–1715. [CrossRef]

20.   Li, M.; Li, Y.T.; Li, D.W.; Long, Y.T. Recent developments and applications of screen-printed electrodes in environmental assays-A review. *Anal. Chim. Acta* **2012**, *734*, 31–44. [CrossRef]

21.   Hayat, A.; Marty, J.L. Disposable screen printed electrochemical sensors: Tools for environmental monitoring. *Sensors* **2014**, *14*, 10432–10453. [CrossRef]

22.   Cano-Raya, C.; Denchev, Z.Z.; Cruz, S.F.; Viana, J.C. Chemistry of solid metal-based inks and pastes for printed electronics–A review. *Appl. Mater. Today* **2019**, *15*, 416–430. [CrossRef]

23.   Yáñez-Sedeño, P.; Campuzano, S.; Pingarrón, J.M. Electrochemical (bio)sensors: Promising tools for green analytical chemistry. *Curr. Opin. Green Sustain. Chem.* **2019**, *19*, 1–7. [CrossRef]

24.   Gałuszka, A.; Migaszewski, Z.; Namieśnik, J. The 12 principles of green analytical chemistry and the SIGNIFICANCE mnemonic of green analytical practices. *TrAC Trends Anal. Chem.* **2013**, *50*, 78–84. [CrossRef]

25.   Sanati, A.; Jalali, M.; Raeissi, K.; Karimzadeh, F.; Kharaziha, M.; Mahshid, S.S.; Mahshid, S. A review on recent advancements in electrochemical biosensing using carbonaceous nanomaterials. *Microchim. Acta* **2019**, *186*, 773. [CrossRef] [PubMed]

26.   Metrohm DropSens. Available online: http://www.dropsens.com/ (accessed on 29 September 2020).

27.   Micrux Technologies. Available online: http://www.micruxfluidic.com/ (accessed on 29 September 2020).

28.   Pine Research. Available online: https://pineresearch.com/ (accessed on 29 September 2020).

29.   Gwent Group. Available online: http://www.gwent.org/ (accessed on 29 September 2020).

30.   PalmSens. Available online: https://www.palmsens.com/ (accessed on 29 September 2020).

31.   Rusens. Available online: http://www.rusens.com/indexeng.html (accessed on 29 September 2020).

32.   Putzbach, W.; Ronkainen, N.J. Immobilization techniques in the fabrication of nanomaterial-based electrochemical biosensors: A review. *Sensors (Basel)* **2013**, *13*, 4811–4840. [CrossRef] [PubMed]

33.   Antuña-Jiménez, D.; González-García, M.B.; Hernández-Santos, D.; Fanjul-Bolado, P. Screen-printed electrodes modified with metal nanoparticles for small molecule sensing. *Biosensors* **2020**, *10*, 9. [CrossRef]

34.   Duffy, G.F.; Moore, E.J. Electrochemical Immunosensors for Food Analysis: A Review of Recent Developments. *Anal. Lett.* **2017**, *50*, 1–32. [CrossRef]

35.   Windmiller, J.R.; Bandodkar, A.J.; Parkhomovsky, S.; Wang, J. Stamp transfer electrodes for electrochemical sensing on non-planar and oversized surfaces. *Analyst* **2012**, *137*, 1570–1575. [CrossRef]

36.   Mishra, R.K.; Hubble, L.J.; Martín, A.; Kumar, R.; Barfidokht, A.; Kim, J.; Musameh, M.M.; Kyratzis, I.L.; Wang, J. Wearable flexible and stretchable glove biosensor for on-site detection of organophosphorus chemical threats. *ACS Sens.* **2017**, *2*, 553–561. [CrossRef]

37.   Desmet, C.; Marquette, C.A.; Blum, L.J.; Doumèche, B. Paper electrodes for bioelectrochemistry: Biosensors and biofuel cells. *Biosens. Bioelectron.* **2016**, *76*, 145–163. [CrossRef]

38. Moro, G.; Bottari, F.; Van Loon, J.; Du Bois, E.; De Wael, K.; Moretto, L.M. Disposable electrodes from waste materials and renewable sources for (bio)electroanalytical applications. *Biosens. Bioelectron.* **2019**, *146*. [CrossRef]

39. Neves, M.M.P.S.; González-García, M.B.; Hernández-Santos, D.; Fanjul-Bolado, P. Screen-Printed Electrochemical 96-Well Plate: A High-Throughput Platform for Multiple Analytical Applications. *Electroanalysis* **2014**, *26*, 2764–2772. [CrossRef]

40. Piermarini, S.; Micheli, L.; Ammida, N.H.S.; Palleschi, G.; Moscone, D. Electrochemical immunosensor array using a 96-well screen-printed microplate for aflatoxin B1 detection. *Biosens. Bioelectron.* **2007**, *22*, 1434–1440. [CrossRef]

41. Thévenot, D.R.; Toth, K.; Durst, R.A.; Wilson, G.S. Electrochemical biosensors: Recommended definitions and classification1International Union of Pure and Applied Chemistry: Physical Chemistry Division, Commission I.7 (Biophysical Chemistry); Analytical Chemistry Division, Commission V.5 (Electroanalytical). *Biosens. Bioelectron.* **2001**, *16*, 121–131. [CrossRef]

42. Sharma, H.; Mutharasan, R. Review of biosensors for foodborne pathogens and toxins. *Sens. Actuators B Chem.* **2013**, *183*, 535–549. [CrossRef]

43. Ronkainen, N.J.; Halsall, H.B.; Heineman, W.R. Electrochemical biosensors. *Chem. Soc. Rev.* **2010**, *39*, 1747–1763. [CrossRef] [PubMed]

44. Sassolas, A.; Blum, L.J.; Leca-Bouvier, B.D. Immobilization strategies to develop enzymatic biosensors. *Biotechnol. Adv.* **2012**, *30*, 489–511. [CrossRef] [PubMed]

45. Crapnell, R.D.; Hudson, A.; Foster, C.W.; Eersels, K.; van Grinsven, B.; Cleij, T.J.; Banks, C.E.; Peeters, M. Recent advances in electrosynthesized molecularly imprinted polymer sensing platforms for bioanalyte detection. *Sensors (Switzerland)* **2019**, *19*, 1204. [CrossRef]

46. Tudorache, M.; Bala, C. Biosensors based on screen-printing technology, and their applications in environmental and food analysis. *Anal. Bioanal. Chem.* **2007**, *388*, 565–578. [CrossRef]

47. Ricci, F.; Adornetto, G.; Palleschi, G. A review of experimental aspects of electrochemical immunosensors. *Electrochim. Acta* **2012**, *84*, 74–83. [CrossRef]

48. Cesewski, E.; Johnson, B.N. Electrochemical biosensors for pathogen detection. *Biosens. Bioelectron.* **2020**, *159*, 112214. [CrossRef]

49. Wang, Y.; Duncan, T.V. Nanoscale sensors for assuring the safety of food products. *Curr. Opin. Biotechnol.* **2017**, *44*, 74–86. [CrossRef]

50. Silva, N.F.D.; Neves, M.M.P.S.; Magalhães, J.M.C.S.; Freire, C.; Delerue-Matos, C. Emerging electrochemical biosensing approaches for detection of *Listeria monocytogenes* in food samples: An overview. *Trends Food Sci. Technol.* **2020**, *99*, 621–633. [CrossRef]

51. Silva, N.F.D.; Magalhães, J.M.C.S.; Freire, C.; Delerue-Matos, C. Electrochemical biosensors for *Salmonella*: State of the art and challenges in food safety assessment. *Biosens. Bioelectron.* **2018**, *99*, 667–682. [CrossRef] [PubMed]

52. Zhang, Y.-J.; Zhang, Y.; Zhou, Y.; Li, G.-H.; Yang, W.-Z.; Feng, X.-S. A review of pretreatment and analytical methods of biogenic amines in food and biological samples since 2010. *J. Chromatogr. A* **2019**, *1605*, 360361. [CrossRef] [PubMed]

53. Prabhakar, P.K.; Vatsa, S.; Srivastav, P.P.; Pathak, S.S. A comprehensive review on freshness of fish and assessment: Analytical methods and recent innovations. *Food Res. Int.* **2020**, *133*, 109157. [CrossRef] [PubMed]

54. European Food Safety Authority (EFSA). Available online: http://www.efsa.europa.eu/en/news/salmonella-most-common-cause-foodborne-outbreaks-european-union (accessed on 29 September 2020).

55. European Food Safety Authority (EFSA). ECDC The European Union summary report on trends and sources of zoonoses, zoonotic agents and food-borne outbreaks in 2015. *EFSA J.* **2016**, *14*, 4364. [CrossRef]

56. Bolton, D.J. *Campylobacter* virulence and survival factors. *Food Microbiol.* **2015**, *48*, 99–108. [CrossRef]

57. Silva, J.; Leite, D.; Fernandes, M.; Mena, C.; Gibbs, P.A.; Teixeira, P. *Campylobacter* spp. As a foodborne pathogen: A review. *Front. Microbiol.* **2011**, *2*, 1–12. [CrossRef]

58. Fabiani, L.; Delibato, E.; Volpe, G.; Piermarini, S.; De Medici, D.; Palleschi, G. Development of a sandwich ELIME assay exploiting different antibody combinations as sensing strategy for an early detection of *Campylobacter*. *Sens. Actuators B Chem.* **2019**, *290*, 318–325. [CrossRef]

59. Allocati, N.; Masulli, M.; Alexeyev, M.F.; Di Ilio, C. *Escherichia coli* in Europe: An overview. *Int. J. Environ. Res. Public Health* **2013**, *10*, 6235–6254. [CrossRef]

60. Kaper, J.B.; Nataro, J.P.; Mobley, H.L.T. Pathogenic *Escherichia coli*. *Nat. Rev. Microbiol.* **2004**, *2*, 123–140. [CrossRef]

61. Eng, S.K.; Pusparajah, P.; Ab Mutalib, N.S.; Ser, H.L.; Chan, K.G.; Lee, L.H. *Salmonella*: A review on pathogenesis, epidemiology and antibiotic resistance. *Front. Life Sci.* **2015**, *8*, 284–293. [CrossRef]

62. Alexandre, D.L.; Melo, A.M.A.; Furtado, R.F.; Borges, M.F.; Figueiredo, E.A.T.; Biswas, A.; Cheng, H.N.; Alves, C.R. A Rapid and Specific Biosensor for *Salmonella* Typhimurium Detection in Milk. *Food Bioprocess Technol.* **2018**, *11*, 748–756. [CrossRef]

63. Cinti, S.; Volpe, G.; Piermarini, S.; Delibato, E.; Palleschi, G. Electrochemical biosensors for rapid detection of foodborne *Salmonella*: A critical overview. *Sensors (Switzerland)* **2017**, *17*, 1910. [CrossRef] [PubMed]

64. Buchanan, R.L.; Gorris, L.G.M.; Hayman, M.M.; Jackson, T.C.; Whiting, R.C. A review of *Listeria monocytogenes*: An update on outbreaks, virulence, dose-response, ecology, and risk assessments. *Food Control* **2017**, *75*, 1–13. [CrossRef]

65. De Jong, W.H.A.; De Vries, E.G.E.; Kema, I.P. Current status and future developments of LC-MS/MS in clinical chemistry for quantification of biogenic amines. *Clin. Biochem.* **2011**, *44*, 95–103. [CrossRef]

66. Ahmad, W.; Mohammed, G.I.; Al-Eryani, D.A.; Saigl, Z.M.; Alyoubi, A.O.; Alwael, H.; Bashammakh, A.S.; O'Sullivan, C.K.; El-Shahawi, M.S. Biogenic Amines Formation Mechanism and Determination Strategies: Future Challenges and Limitations. *Crit. Rev. Anal. Chem.* **2019**, *0*, 1–16. [CrossRef]

67. Ordóñez, J.L.; Troncoso, A.M.; García-Parrilla, M.D.C.; Callejón, R.M. Recent trends in the determination of biogenic amines in fermented beverages—A review. *Anal. Chim. Acta* **2016**, *939*, 10–25. [CrossRef]

68. Freitas, M.; Viswanathan, S.; Nouws, H.P.A.; Oliveira, M.B.P.P.; Delerue-Matos, C. Iron oxide/gold core/shell nanomagnetic probes and CdS biolabels for amplified electrochemical immunosensing of *Salmonella typhimurium*. *Biosens. Bioelectron.* **2014**, *51*, 195–200. [CrossRef]

69. Afonso, A.S.; Pérez-López, B.; Faria, R.C.; Mattoso, L.H.C.; Hernández-Herrero, M.; Roig-Sagués, A.X.; Maltez-da Costa, M.; Merkoçi, A. Electrochemical detection of *Salmonella* using gold nanoparticles. *Biosens. Bioelectron.* **2013**, *40*, 121–126. [CrossRef]

70. Wang, D.; Dou, W.; Chen, Y.; Zhao, G. Enzyme-functionalized electrochemical immunosensor based on electrochemically reduced graphene oxide and polyvinyl alcohol-polydimethylsiloxane for the detection of *Salmonella* pullorum & *Salmonella* gallinarum. *RSC Adv.* **2014**, *4*, 57733–57742. [CrossRef]

71. Fei, J.; Dou, W.; Zhao, G. A sandwich electrochemical immunosensor for *Salmonella* pullorum and *Salmonella* gallinarum based on a screen-printed carbon electrode modified with an ionic liquid and electrodeposited gold nanoparticles. *Microchim. Acta* **2015**, *182*, 2267–2275. [CrossRef]

72. Mutreja, R.; Jariyal, M.; Pathania, P.; Sharma, A.; Sahoo, D.K.; Suri, C.R. Novel surface antigen based impedimetric immunosensor for detection of *Salmonella typhimurium* in water and juice samples. *Biosens. Bioelectron.* **2016**, *85*, 707–713. [CrossRef] [PubMed]

73. Bagheryan, Z.; Raoof, J.B.; Golabi, M.; Turner, A.P.F.; Beni, V. Diazonium-based impedimetric aptasensor for the rapid label-free detection of *Salmonella typhimurium* in food sample. *Biosens. Bioelectron.* **2016**, *80*, 566–573. [CrossRef] [PubMed]

74. Ngoensawat, U.; Rijiravanich, P.; Surareungchai, W.; Somasundrum, M. Electrochemical Immunoassay for *Salmonella* Typhimurium Based on an Immuno-magnetic Redox Label. *Electroanalysis* **2018**, *30*, 146–153. [CrossRef]

75. De Oliveira, T.R.; Martucci, D.H.; Faria, R.C. Simple disposable microfluidic device for *Salmonella typhimurium* detection by magneto-immunoassay. *Sens. Actuators B Chem.* **2018**, *255*, 684–691. [CrossRef]

76. Wonsawat, W.; Limvongjaroen, S.; Supromma, S.; Panphut, W.; Ruecha, N.; Ratnarathorn, N.; Dungchai, W. A paper-based conductive immunosensor for the determination of *Salmonella* Typhimurium. *Analyst* **2020**, *145*, 4637–4645. [CrossRef] [PubMed]

77. Murasova, P.; Kovarova, A.; Kasparova, J.; Brozkova, I.; Hamiot, A.; Pekarkova, J.; Dupuy, B.; Drbohlavova, J.; Bilkova, Z.; Korecka, L. Direct culture-free electrochemical detection of *Salmonella* cells in milk based on quantum dots-modified nanostructured dendrons. *J. Electroanal. Chem.* **2020**, *863*, 114051. [CrossRef]

78. Viswanathan, S.; Rani, C.; Ho, J.A.H.A. Electrochemical immunosensor for multiplexed detection of food-borne pathogens using nanocrystal bioconjugates and MWCNT screen-printed electrode. *Talanta* **2012**, *94*, 315–319. [CrossRef]

79. Farka, Z.; Juřík, T.; Pastucha, M.; Kovář, D.; Lacina, K.; Skládal, P. Rapid Immunosensing of *Salmonella* Typhimurium Using Electrochemical Impedance Spectroscopy: The Effect of Sample Treatment. *Electroanalysis* **2016**, *28*, 1803–1809. [CrossRef]

80. Delibato, E.; Volpe, G.; Stangalini, D.; De Medici, D.; Moscone, D.; Palleschi, G. Development of SYBR-green real-time PCR and a multichannel electrochemical immunosensor for specific detection of *Salmonella* enterica. *Anal. Lett.* **2006**, *39*, 1611–1625. [CrossRef]

81. Pratiwi, F.W.; Rijiravanich, P.; Somasundrum, M.; Surareungchai, W. Electrochemical immunoassay for *Salmonella* Typhimurium based on magnetically collected Ag-enhanced DNA biobarcode labels. *Analyst* **2013**, *138*, 5011–5018. [CrossRef]

82. Salam, F.; Tothill, I.E. Detection of *Salmonella* typhimurium using an electrochemical immunosensor. *Biosens. Bioelectron.* **2009**, *24*, 2630–2636. [CrossRef] [PubMed]

83. Xu, M.; Wang, R.; Li, Y. Rapid detection of *Escherichia coli* O157:H7 and *Salmonella* Typhimurium in foods using an electrochemical immunosensor based on screen-printed interdigitated microelectrode and immunomagnetic separation. *Talanta* **2016**, *148*, 200–208. [CrossRef]

84. Wilson, D.; Materón, E.M.; Ibáñez-Redín, G.; Faria, R.C.; Correa, D.S.; Oliveira, O.N. Electrical detection of pathogenic bacteria in food samples using information visualization methods with a sensor based on magnetic nanoparticles functionalized with antimicrobial peptides. *Talanta* **2019**, *194*, 611–618. [CrossRef] [PubMed]

85. Fei, J.; Dou, W.; Zhao, G. A sandwich electrochemical immunoassay for Salmonella pullorum and *Salmonella* gallinarum based on a AuNPs/SiO$_2$/Fe$_3$O$_4$ adsorbing antibody and 4 channel screen printed carbon electrode electrodeposited gold nanoparticles. *RSC Adv.* **2015**, *5*, 74548–74556. [CrossRef]

86. Wang, D.; Chen, Q.; Huo, H.; Bai, S.; Cai, G.; Lai, W.; Lin, J. Efficient separation and quantitative detection of *Listeria monocytogenes* based on screen-printed interdigitated electrode, urease and magnetic nanoparticles. *Food Control* **2017**, *73*, 555–561. [CrossRef]

87. Tolba, M.; Ahmed, M.U.; Tlili, C.; Eichenseher, F.; Loessner, M.J.; Zourob, M. A bacteriophage endolysin-based electrochemical impedance biosensor for the rapid detection of *Listeria* cells. *Analyst* **2012**, *137*, 5749–5756. [CrossRef]

88. Lin, Y.H.; Chen, S.H.; Chuang, Y.C.; Lu, Y.C.; Shen, T.Y.; Chang, C.A.; Lin, C.S. Disposable amperometric immunosensing strips fabricated by Au nanoparticles-modified screen-printed carbon electrodes for the detection of foodborne pathogen *Escherichia coli* O157:H7. *Biosens. Bioelectron.* **2008**, *23*, 1832–1837. [CrossRef]

89. Hassan, A.R.H.A.A.; de la Escosura-Muñiz, A.; Merkoçi, A. Highly sensitive and rapid determination of *Escherichia coli* O157:H7 in minced beef and water using electrocatalytic gold nanoparticle tags. *Biosens. Bioelectron.* **2015**, *67*, 511–515. [CrossRef]

90. Mo, X.; Wu, Z.; Huang, J.; Zhao, G.; Dou, W. A sensitive and regenerative electrochemical immunosensor for quantitative detection of: *Escherichia coli* O157:H7 based on stable polyaniline coated screen-printed carbon electrode and rGO-NR-Au@Pt. *Anal. Methods* **2019**, *11*, 1475–1482. [CrossRef]

91. Zhu, F.; Zhao, G.; Dou, W. A non-enzymatic electrochemical immunoassay for quantitative detection of *Escherichia coli* O157:H7 using Au@Pt and graphene. *Anal. Biochem.* **2018**, *559*, 34–43. [CrossRef]

92. Zhu, F.; Zhao, G.; Dou, W. Electrochemical sandwich immunoassay for *Escherichia coli* O157:H7 based on the use of magnetic nanoparticles and graphene functionalized with electrocatalytically active Au@Pt core/shell nanoparticles. *Microchim. Acta* **2018**, *185*. [CrossRef] [PubMed]

93. Huang, Y.; Wu, Z.; Zhao, G.; Dou, W. A Label-Free Electrochemical Immunosensor Modified with AuNPs for Quantitative Detection of *Escherichia coli* O157:H7. *J. Electron. Mater.* **2019**, *48*, 7960–7969. [CrossRef]

94. Mo, X.; Zhao, G.; Dou, W. Electropolymerization of Stable Leucoemeraldine Base Polyaniline Film and Application for Quantitative Detection of *Escherichia coli* O157:H7. *J. Electron. Mater.* **2018**, *47*, 6507–6517. [CrossRef]

95. Cimafonte, M.; Fulgione, A.; Gaglione, R.; Papaianni, M.; Capparelli, R.; Arciello, A.; Censi, S.B.; Borriello, G.; Velotta, R.; Ventura, B. Della Screen printed based impedimetric immunosensor for rapid detection of *Escherichia coli* in drinking water. *Sensors* **2020**, *20*, 274. [CrossRef] [PubMed]

96. Wang, R.; Lum, J.; Callaway, Z.; Lin, J.; Bottje, W.; Li, Y. A label-free impedance immunosensor using screen-printed interdigitated electrodes and magnetic nanobeads for the detection of *E. coli* O157:H7. *Biosensors* **2015**, *5*, 791–803. [CrossRef] [PubMed]

97. Xu, M.; Wang, R.; Li, Y. An electrochemical biosensor for rapid detection of: *E. coli* O157:H7 with highly efficient bi-functional glucose oxidase-polydopamine nanocomposites and Prussian blue modified screen-printed interdigitated electrodes. *Analyst* **2016**, *141*, 5441–5449. [CrossRef]

98. Shkodra, B.; Abera, B.D.; Cantarella, G.; Douaki, A.; Avancini, E.; Petti, L.; Lugli, P. Flexible and printed electrochemical immunosensor coated with oxygen plasma treated SWCNTs for histamine detection. *Biosensors* **2020**, *10*, 35. [CrossRef]

99. Dong, X.X.; Yang, J.Y.; Luo, L.; Zhang, Y.F.; Mao, C.; Sun, Y.M.; Lei, H.T.; Shen, Y.D.; Beier, R.C.; Xu, Z.L. Portable amperometric immunosensor for histamine detection using Prussian blue-chitosan-gold nanoparticle nanocomposite films. *Biosens. Bioelectron.* **2017**, *98*, 305–309. [CrossRef]

100. Veseli, A.; Vasjari, M.; Arbneshi, T.; Hajrizi, A.; Švorc, L.; Samphao, A.; Kalcher, K. Electrochemical determination of histamine in fish sauce using heterogeneous carbon electrodes modified with rhenium(IV) oxide. *Sens. Actuators B Chem.* **2016**, *228*, 774–781. [CrossRef]

101. Lee, M.-Y.; Wu, C.-C.; Sari, M.I.; Hsieh, Y. A disposable non-enzymatic histamine sensor based on the nafion-coated copper phosphate electrodes for estimation of fish freshness. *Electrochim. Acta* **2018**, *283*, 772–779. [CrossRef]

102. Torre, R.; Costa-Rama, E.; Lopes, P.; Nouws, H.P.A.; Delerue-Matos, C. Amperometric enzyme sensor for the rapid determination of histamine. *Anal. Methods* **2019**, *11*, 1264–1269. [CrossRef]

103. Torre, R.; Costa-rama, E.; Nouws, H.P.A.; Delerue-Matos, C. Diamine oxidase-modified screen-printed electrode for the redox-mediated determination of histamine. *J. Anal. Sci. Technol.* **2020**, *3*, 4–11. [CrossRef]

104. Pérez, S.; Bartrolí, J.; Fàbregas, E. Amperometric biosensor for the determination of histamine in fish samples. *Food Chem.* **2013**, *141*, 4066–4072. [CrossRef] [PubMed]

105. Apetrei, I.M.; Apetrei, C. Amperometric biosensor based on diamine oxidase/platinum nanoparticles/graphene/chitosan modified screen-printed carbon electrode for histamine detection. *Sensors* **2016**, *16*, 422. [CrossRef]

106. Henao-Escobar, W.; Domínguez-Renedo, O.; Alonso-Lomillo, M.A.; Arcos-Martínez, M.J. A screen-printed disposable biosensor for selective determination of putrescine. *Microchim. Acta* **2013**, *180*, 687–693. [CrossRef]

107. Henao-Escobar, W.; Domínguez-Renedo, O.; Alonso-Lomillo, M.A.; Cascalheira, J.F.; Dias-Cabral, A.C.; Arcos-Martínez, M.J. Characterization of a Disposable Electrochemical Biosensor Based on Putrescine Oxidase from Micrococcus rubens for the Determination of Putrescine. *Electroanalysis* **2015**, *27*, 368–377. [CrossRef]

108. Erden, P.E.; Erdoğan, Z.Ö.; Öztürk, F.; Koçoğlu, İ.O.; Kılıç, E. Amperometric Biosensors for Tyramine Determination Based on Graphene Oxide and Polyvinylferrocene Modified Screen-printed Electrodes. *Electroanalysis* **2019**, *31*, 2368–2378. [CrossRef]

109. Apetrei, I.M.; Apetrei, C. The biocomposite screen-printed biosensor based on immobilization of tyrosinase onto the carboxyl functionalised carbon nanotube for assaying tyramine in fish products. *J. Food Eng.* **2015**, *149*, 1–8. [CrossRef]

110. Li, Y.; Hsieh, C.H.; Lai, C.-W.; Chang, Y.-F.; Chan, H.-Y.; Tsai, C.-F.; Ho, J.A.; Wu, L. Tyramine detection using PEDOT:PSS/AuNPs/1-methyl-4-mercaptopyridine modified screen-printed carbon electrode with molecularly imprinted polymer solid phase extraction. *Biosens. Bioelectron.* **2017**, *87*, 142–149. [CrossRef] [PubMed]

111. Dalkıran, B.; Erden, P.E.; Kaçar, C.; Kılıç, E. Disposable Amperometric Biosensor Based on Poly-L-lysine and Fe$_3$O$_4$ NPs-chitosan Composite for the Detection of Tyramine in Cheese. *Electroanalysis* **2019**, *31*, 1324–1333. [CrossRef]

112. Calvo-Pérez, A.; Domínguez-Renedo, O.; Alonso-Lomillo, M.A.; Arcos-Martínez, M.J. Disposable amperometric biosensor for the determination of tyramine using plasma amino oxidase. *Microchim. Acta* **2013**, *180*, 253–259. [CrossRef]

113. Calvo-Pérez, A.; Domínguez-Renedo, O.; Alonso-Lomillo, M.A.; Arcos-Martínez, M.J. Disposable Horseradish Peroxidase Biosensors for the Selective Determination of Tyramine. *Electroanalysis* **2013**, *25*, 1316–1322. [CrossRef]

114. Kaçar, C.; Erden, P.E.; Dalkiran, B.; İnal, E.K.; Kiliç, E. Amperometric biogenic amine biosensors based on Prussian blue, indium tin oxide nanoparticles and diamine oxidase—Or monoamine oxidase–modified electrodes. *Anal. Bioanal. Chem.* **2020**, *412*, 1933–1946. [CrossRef] [PubMed]

115. Henao-Escobar, W.; Román, L.D.T.-D.; Domínguez-Renedo, O.; Alonso-Lomillo, M.A.; Arcos-Martínez, M.J. Dual enzymatic biosensor for simultaneous amperometric determination of histamine and putrescine. *Food Chem.* **2016**, *190*, 818–823. [CrossRef] [PubMed]

116. Henao-Escobar, W.; Domínguez-Renedo, O.; Asunción Alonso-Lomillo, M.; Julia Arcos-Martínez, M. Simultaneous determination of cadaverine and putrescine using a disposable monoamine oxidase based biosensor. *Talanta* **2013**, *117*, 405–411. [CrossRef]

117. Leonardo, S.; Campàs, M. Electrochemical enzyme sensor arrays for the detection of the biogenic amines histamine, putrescine and cadaverine using magnetic beads as immobilisation supports. *Microchim. Acta* **2016**, *183*, 1881–1890. [CrossRef]

118. Alonso-Lomillo, M.A.; Domínguez-Renedo, O.; Matos, P.; Arcos-Martínez, M.J. Disposable biosensors for determination of biogenic amines. *Anal. Chim. Acta* **2010**, *665*, 26–31. [CrossRef]

119. Di Fusco, M.; Federico, R.; Boffi, A.; MacOne, A.; Favero, G.; Mazzei, F. Characterization and application of a diamine oxidase from Lathyrus sativus as component of an electrochemical biosensor for the determination of biogenic amines in wine and beer. *Anal. Bioanal. Chem.* **2011**, *401*, 707–716. [CrossRef]

120. Telsnig, D.; Kalcher, K.; Leitner, A.; Ortner, A. Design of an Amperometric Biosensor for the Determination of Biogenic Amines Using Screen Printed Carbon Working Electrodes. *Electroanalysis* **2013**, *25*, 47–50. [CrossRef]

121. Lange, J.; Wittmann, C. Enzyme sensor array for the determination of biogenic amines in food samples. *Anal. Bioanal. Chem.* **2002**, *372*, 276–283. [CrossRef]

2

The Occurrence of Biogenic Amines and Determination of Biogenic Amine-Producing Lactic Acid Bacteria in *Kkakdugi* and *Chonggak* Kimchi

Young Hun Jin, Jae Hoan Lee, Young Kyung Park, Jun-Hee Lee and Jae-Hyung Mah *(ID)

Department of Food and Biotechnology, Korea University, 2511 Sejong-ro, Sejong 30019, Korea;
younghoonjin3090@korea.ac.kr (Y.H.J.); jae-lee@korea.ac.kr (J.H.L.); eskimo@korea.ac.kr (Y.K.P.);
bory92@korea.ac.kr (J.-H.L.)
* Correspondence: nextbio@korea.ac.kr

Abstract: In this study, biogenic amine content in two types of fermented radish kimchi (*Kkakdugi* and *Chonggak* kimchi) was determined by high performance liquid chromatography (HPLC). While most samples had low levels of biogenic amines, some samples contained histamine content over the toxicity limit. Additionally, significant amounts of total biogenic amines were detected in certain samples due to high levels of putrefactive amines. As one of the significant factors influencing biogenic amine content in both radish kimchi, *Myeolchi-aekjoet* appeared to be important source of histamine. Besides, tyramine-producing strains of lactic acid bacteria existed in both radish kimchi. Through 16s rRNA sequencing analysis, the dominant species of tyramine-producing strains was identified as *Lactobacillus brevis*, which suggests that the species is responsible for tyramine formation in both radish kimchi. During fermentation, a higher tyramine accumulation was observed in both radish kimchi when *L. brevis* strains were used as inocula. The addition of *Myeolchi-aekjeot* affected the initial concentrations of histamine and cadaverine in both radish kimchi. Therefore, this study suggests that reducing the ratio of *Myeolchi-aekjeot* to other ingredients (and/or using *Myeolchi-aekjeot* with low biogenic amine content) and using starter cultures with ability to degrade and/or inability to produce biogenic amines would be effective in reducing biogenic amine content in *Kkakdugi* and *Chonggak* kimchi.

Keywords: kimchi; *Kkakdugi*; *Chonggak* kimchi; radish kimchi; biogenic amines; tyramine; lactic acid bacteria; *Lactobacillus brevis*

1. Introduction

Biogenic amines (BA) have been considered to be toxic compounds in foods. Several authors have proposed the maximum tolerable limits of some toxicologically important BA in foods as follows: histamine, 100 mg/kg; tyramine, 100–800 mg/kg; β-phenylethylamine, 30 mg/kg; total BA, 1000 mg/kg [1,2]. In addition, polyamines such as putrescine and cadaverine have been known to potentiate the toxicity of BA, especially histamine and tyramine, in foods, although they are less toxic [1]. Consumption of foods containing excessive BA may cause symptoms such as migraines, sweating, nausea, hypotension, and hypertension, unless human intestinal amine oxidases—such as monoamine oxidase (MAO), diamine oxidase (DAO), and polyamine oxidase (PAO)—quickly metabolize and detoxify BA [3]. Thus, it is important to know that, although relatively low levels of BA naturally exist in common foods, microbial decarboxylation of amino acids may sometimes lead to a significant increment of BA in fermented or contaminated foods [2]. In lactic acid fermented foods such as cheese and fermented sausage, some species of lactic acid bacteria (LAB) have been considered as producers of BA, particularly tyramine [4]. On the other hand, several reports have indicated that use of LAB starter cultures unable to produce BA may reduce BA accumulation during fermentation and storage [5,6].

Kimchi is a generic term of Korean traditional lactic fermented vegetables. According to Codex standard [7], for preparation of kimchi, salted Chinese cabbage (as a main ingredient) is mixed with seasoning paste consisting of red pepper powder, radish, garlic, green onion, and ginger, and then fermented properly, however, which, in reality, refers to *Baechu* kimchi. Alongside the Chinese cabbage, various vegetables such as radish, ponytail radish, cucumber, and green onion are also used as main ingredients of kimchi depending on kimchi varieties in Korea [8]. Among numerous kimchi varieties prepared with different vegetables, *Baechu* kimchi, *Kkakdugi* (diced radish kimchi), and *Chonggak* kimchi (ponytail radish kimchi) are the most popular varieties of kimchi in Korea [9]. In the meantime, for improving sensory quality of kimchi, various types of salted and fermented seafood (*Jeotgal*) and sauces thereof (*Aekjeot*) are usually used for kimchi preparation in Korea [10]. Particularly, *Myeolchi-jeotgal* (salted and fermented anchovy), *Saeu-jeotgal* (salted and fermented shrimp), *Myeolchi-aekjeot* (a sauce prepared from *Myeolchi-jeotgal*) are commonly used *Jeotgal* and *Aekjeot* [11]. As *Jeotgal* and *Aekjeot* contain high levels of proteins and amino acids, when kimchi is prepared with them, BA accumulation may occur during kimchi fermentation [12]. Hence, several authors have intensively investigated BA content and BA-producing LAB in *Baechu* kimchi [13–15]. On the other hand, there is a lack of study on BA content and BA-producing LAB in *Kkakdugi* and *Chonggak* kimchi, although the two types of radish kimchi are as popular as *Baechu* kimchi in Korea.

In this study, therefore, BA content in *Kkakdugi* and *Chonggak* kimchi was determined to evaluate BA-related risks. Several possible contributing factors to BA content, including physicochemical properties and microbial BA production, were also investigated in the study. Finally, fermentation of both radish kimchi was carried out to determine the most important bacterial species contributing to BA formation in the radish kimchi, employing LAB strains with distinguishable BA-producing activities as fermenting microorganisms. This is the first study describing that *Lactobacillus brevis* is the species responsible for tyramine formation in kimchi variety throughout fermentation period.

2. Materials and Methods

2.1. Sampling

Two types of radish kimchi (*Kkakdugi* and *Chonggak* kimchi) samples of five popular kimchi manufacturers made within 30 days were obtained from the retail markets. After arrival, samples were stored at 4 °C or immediately analyzed for BA content, physicochemical parameters, and microbial measurement.

2.2. Physicochemical Measurements

pH, acidity, salinity, and water activity of *Kkakdugi* and *Chonggak* kimchi samples were determined. The pH of the samples was determined by Orion 3-star Benchtop pH meter (Thermo Scientific, Waltham, MA, USA). Acidity and salinity were measured according to the AOAC method [16]. The water activity was determined by water activity meter (AquaLab Pre; Meter Group, Inc., Pullman, WA, USA).

2.3. Microbial Measurement, Isolation, and Identification of Strains

Lactic acid bacterial counts and total aerobic bacterial counts were determined on de Man, Rogosa, and Sharpe (MRS, Laboratorios Conda Co., Madrid, Spain) agar and Plate Count Agar (PCA, Difco, Becton Dickinson, Sparks, MD, USA). According to manufacturer's instructions, MRS agar was incubated at 37 °C for 48–72 h, and PCA at 37 °C for 24 h. After incubation, enumeration was carried out on plates with 30–300 colonies.

LAB strains were isolated on MRS agar. Individual colonies on MRS agar were randomly selected and streaked on the same media. The single colonies were transferred to MRS broth at 37 °C for 48–72 h.

Then, the cultured broth was stored in the presence of 20% glycerol (v/v) at −80 °C. In *Kkakdugi* and *Chonggak* kimchi samples, 130 and 120 LAB strains were isolated, respectively. The strains were identified by 16s rRNA gene sequence analysis with the universal bacterial primer pair (518F and 805R, Solgent Co., Daejeon, Korea).

2.4. BA Extraction from Samples and Bacterial Cultures for HPLC Analysis

BA extraction from *Kkakdugi* and *Chonggak* kimchi samples was conducted by the methods developed by Eerola et al. [17], with minor modification. The sample broth (5 g) was mixed with 20 mL of perchloric acid (0.4 M). The mixture was incubated at 4 °C for 2 h and centrifuged at $3000×$ g at 4 °C for 10 min. After collecting the supernatant, the pellet was extracted again with equal volumes of perchloric acid under the same conditions. The total volume of supernatant was adjusted to 50 mL with perchloric acid. The extract was filtered using Whatman paper no. 1 and stored before analysis.

BA extraction from bacterial cultures was carried out based on the procedures described by Ben-Gigirey et al. [18,19], with minor modification. A loopful of a strain was inoculated in 5 mL of BA production assay medium. The compositions of BA production assay medium are as follows: MRS broth with 0.5% of L-ornithine monohydrochloride, L-lysine monohydrochloride, L-histidine monohydrochloride monohydrate, and L-tyrosine disodium salt hydrate (all Sigma-Aldrich Chemical Co., St. Louis, MO, USA); 0.0005% of pyridoxal-HCl (Sigma-Aldrich); pH of the broth was adjusted to 5.8 by adding hydrochloride solution (2 M). After incubating the strain at 37 °C for 48 h, 100 μL of the culture was inoculated into the same broth and incubated under the same conditions. Subsequently, after being mixed with 0.4 M perchloric acid at a volume ratio of 1:9, the mixture was incubated at 4 °C for 2 h and stored before analysis.

2.5. Preparation of Standard Solutions for HPLC Analysis

Tryptamine, β-phenylethylamine hydrochloride, putrescine dihydrochloride, cadaverine dihydrochloride, histamine dihydrochloride, tyramine hydrochloride, spermidine trihydrochloride, and spermine tetrahydrochloride (all Sigma-Aldrich) were used for standard solutions, and 1,7-diaminoheptane (Sigma-Aldrich) was applied for an internal standard. The concentrations of all standard solutions were adjusted to 0, 10, 50, 100, and 1000 ppm.

2.6. Derivatization of Extracts and Standards

The procedures of derivatization of BA in the extract were carried out by the method developed by Eerola et al. [17]. Briefly, 200 μL of 2 M sodium hydroxide and 300 μL of saturated sodium bicarbonate were added to 1 mL of the extract/standard solutions. Then, 2 mL of 1% dansyl chloride solution (dissolved in acetone) was mixed with the solution and then incubated for 45 min at 40 °C in dark room. The incubated solution was mixed with 100 μL of 25% ammonium hydroxide and reacted for 30 min at room temperature. The volume of the sample solution was adjusted to 5 mL by adding acetonitrile. The sample solution was centrifuged at $3000×$ g for 5 min, and the supernatant was filtered by using a 0.2 μm-pore-size filter (Millipore Co., Bedford, MA, USA).

2.7. HPLC Analysis

HPLC analysis was carried out according to the procedure developed by Eerola et al. [17] and modified by Ben-Gigirey et al. [18]. YL9100 HPLC system equipped with YL9120 UV–vis detector (all Younglin, Anyang, Korea) was employed and the data were analyzed with Autochro-3000 data system (Younglin). For the gradient HPLC method, 0.1 M ammonium acetate (solvent A; Sigma-Aldrich) and HPLC-grade acetonitrile (solvent B; SK chemicals, Ulsan, Korea) were used as the mobile phases. The chromatographic separation was carried out using Nova-Pak C18 column (4 μm, 4.6 × 150 mm; Waters, Milford, MA, USA) held in 40 °C at a flow rate of 1 mL/min. The gradient elution mode was as follows; 50:50 (A:B) to 10:90 for 19 min, 50:50 at 20 min, isocratic with 50:50 before next analysis. The analysis was conducted at 254 nm, and 10 μL of the sample solution was injected.

The detection limits were within the range of 0.01 to 0.10 mg/kg for food matrices [20]. The validation parameters, including detection limits, of the analytical procedure used in the study were reported in our earlier study [20]. Figure S1 illustrates the procedure, from extraction to HPLC analysis, for BA analysis.

2.8. Fermentation of Two Types of Radish Kimchi: Kkakdugi and Chonggak Kimchi

For preparation of *Kkakdugi* and *Chonggak* kimchi, diced white radish ($2 \times 2 \times 2$ cm^3) or halved ponytail radish were soaked in 10% w/v salt brine for 30 min, respectively. Then, each salted radish was rinsed with tap water three times and drained for 3 h. *Kkakdugi* and *Chonggak* kimchi samples were prepared in triplicate, as shown in Table 1, according to the standard recipes developed by the National Institute of Agricultural Sciences [21]. The salinity of all samples was adjusted to 2.5%. The *Kkakdugi* and *Chonggak* kimchi samples were divided into five experimental groups, respectively, based on the presence or absence of *Myeolchi-aekjeot* and *Saeu-jeotgal* and LAB inoculum. The experimental groups designed for the present study were B group ("Blank" samples prepared with neither *Myeolchi-aekjeot* and *Saeu-jeotgal* nor inoculum), C group ("Control" samples prepared with *Myeolchi-aekjeot* and *Saeu-jeotgal*, but without inoculum), PC group ("Positive Control" samples prepared with *Myeolchi-aekjeot* and *Saeu-jeotgal*, and inoculated with *L. brevis* JCM 1170 as a reference strain), LB group ("L. brevis" samples prepared with *Myeolchi-aekjeot* and *Saeu-jeotgal*, and inoculated with tyramine-producing *L. brevis* strains, i.e., KD3M5 strain for *Kkakdugi* and CG2M15 strain for *Chonggak* kimchi, respectively), and LP group ("L. plantarum" samples prepared with *Myeolchi-aekjeot* and *Saeu-jeotgal*, and inoculated with *L. plantarum* strains, i.e., KD3M15 strain for *Kkakdugi* and CG3M21 strain for *Chonggak* kimchi, respectively). The samples belonging to respective experimental groups were fermented at 25 °C for three days. Changes on the physicochemical and microbial properties, and BA content were measured in triplicate during fermentation.

Table 1. Ingredients used for preparation of *Kkakdugi* and *Chonggak* kimchi.

Ingredients (g)	Salted Radish	Red Pepper Powder	Garlic	Ginger	Sesame Seed	Sugar	Glutinous Rice Paste	Myeolchi-aekjeot	Saeu-jeotgal
Kkakdugi	100	3	3	1.5	1	2	5	2	2
Chonggak kimchi	100	3.5	3	1.5	0.5	1.5	4	2	2

2.9. Statistical Analyses

Statistical analyses were performed with Minitab statistical software version 12.11 (Minitab Inc. State College, PA, USA). The data were presented as means ± standard deviations of the three independent replicates. The mean values were compared by one-way analysis of variance (ANOVA) with Tukey's honest significant difference (HSD) test and a probability (p) values of less than 0.05 were considered statistically significant.

3. Results and Discussion

3.1. Determination of BA Content in Radish Kimchi: Kkakdugi and Chonggak Kimchi

As shown in Table 2, BA content in *Kkakdugi* and *Chonggak* kimchi samples produced by popular manufacturers in Korea was determined, and human health risk of BA in both radish kimchi was estimated based on the suggestions of both Ten Brink et al. [1] and Silla Santos [2]. In all the samples of *Kkakdugi* and *Chonggak* kimchi, low levels of tyramine (<100 mg/kg), tryptamine, β-phenylethylamine, spermidine, and spermine (<30 mg/kg) were detected, which are within safe levels for human consumption. However, one *Kkakdugi* sample (KD2) had 127.78 ± 26.78 mg/kg of histamine, which is over the toxicity limit (100 mg/kg) suggested by Ten Brink et al. [1]. Another *Kkakdugi* sample (KD5) contained putrescine and cadaverine at concentrations of 982.32 ± 19.42 mg/kg and 124.60 ± 108.78 mg/kg, respectively, consequently exceeding the 1000 mg/kg limit for total BA which is considered to provoke toxicity [2]. In *Chonggak* kimchi samples, 131.20 ± 7.90 mg/kg of histamine was detected in one sample (CG5), which also contained 853.7 ± 36.80 mg/kg of putrescine and 112.10 ± 3.60 mg/kg of cadaverine. The amounts of histamine and total BA in the sample were found to exceed toxicity limits. Meanwhile, the BA content detected in both types of radish kimchi samples varied widely in the present study, which is similar to respective BA levels in *Baechu* kimchi reported previously [13,22]. On the other hand, Mah et al. [12] reported lower concentrations of putrescine, cadaverine, histamine, tyramine, spermidine, and spermine in both *Kkakdugi* and *Chonggak* kimchi than those detected in the same kinds of kimchi used in this study. This may be due to the differences in manufacturing methods, main ingredients, and storage conditions between kimchi samples used in the present and previous studies [9]. In the meantime, Mah et al. [12] also reported that the amounts of tyramine and other BA increased during the ripening of *Baechu* kimchi. Therefore, although tyramine was detected at low levels in all the samples of *Kkakdugi* and *Chonggak* kimchi in the present study, the significance and risk of tyramine formation in both types of radish kimchi should not be overlooked.

Table 2. BA content in two types of radish kimchi samples: *Kkakdugi* and *Chonggak* kimchi.

Samples [2]	BA Content (mg/kg) [1]							
	Trp	Phe	Put	Cad	His	Tyr	Spd	Spm
KD1	ND [3]	ND	10.85 ± 1.17 [4]	2.57 ± 0.62	18.75 ± 1.16	2.97 ± 0.33	12.27 ± 0.98	0.56 ± 0.96
KD2	ND	1.93 ± 1.69	563.59 ± 45.64	ND	127.78 ± 26.78	14.73 ± 1.96	12.66 ± 2.75	ND
KD3	ND	ND	19.00 ± 2.00	6.10 ± 0.40	24.50 ± 4.00	10.80 ± 0.40	ND	ND
KD4	ND	0.86 ± 1.49	97.45 ± 77.05	3.15 ± 5.46	40.82 ± 29.05	21.67 ± 17.81	5.30 ± 4.85	3.10 ± 2.82
KD5	ND	15.24 ± 1.87	982.32 ± 19.42	124.60 ± 108.78	67.84 ± 17.46	76.95 ± 4.25	16.76 ± 0.87	1.48 ± 0.08
Average	ND	3.61 ± 6.55	334.64 ± 427.97	27.28 ± 54.44	55.94 ± 44.45	25.42 ± 29.59	9.40 ± 6.68	1.03 ± 1.31
CG1	ND	ND	8.97 ± 2.02	2.38 ± 2.12	38.61 ± 6.03	4.85 ± 4.60	9.22 ± 2.16	20.74 ± 3.47
CG2	ND	ND	3.89 ± 1.68	2.00 ± 0.77	8.24 ± 2.09	0.79 ± 0.69	8.27 ± 2.90	2.12 ± 0.53
CG3	12.30 ± 6.30	ND	175.10 ± 7.30	55.40 ± 2.80	46.30 ± 6.70	18.70 ± 2.40	7.70 ± 5.50	ND
CG4	9.10 ± 7.10	1.10 ± 1.00	303.70 ± 20.20	148.50 ± 9.00	69.30 ± 20.90	11.10 ± 2.20	6.10 ± 3.70	8.30 ± 5.60
CG5	23.70 ± 6.10	2.80 ± 1.20	853.70 ± 36.80	112.10 ± 3.60	131.20 ± 7.90	7.00 ± 2.20	14.00 ± 5.30	ND
Average	9.02 ± 9.86	0.78 ± 1.23	269.07 ± 349.93	64.08 ± 65.51	58.73 ± 46.02	8.49 ± 6.80	9.06 ± 2.99	6.23 ± 8.79

[1] Trp: tryptamine, Phe: β-phenylethylamine, Put: putrescine, Cad: cadaverine, His: histamine, Tyr: tyramine, Spd: spermidine, Spm: spermine; [2] KD: *Kkakdugi* (diced radish kimchi), CG: *Chonggak* kimchi (ponytail radish kimchi); [3] ND: not detected (<0.1 mg/kg); [4] mean ± standard deviation.

According to Tsai et al. [13], a high level of histamine in kimchi may result from the addition of salted and fermented fish products. *Myeolchi-aekjeot* is the most widely used salted and fermented fish product for the preparation of kimchi variety, and approximately 2–4% of *Kkakdugi* (on the basis of weight percent) and 2–5% of *Chonggak* kimchi, respectively, are commonly added to main ingredients during kimchi preparation [21,23–26]. *Saeu-jeotgal* is also added, alone or together with *Myeolchi-aekjeot*, to main ingredients of kimchi, but Mah et al. [12] reported that *Myeolchi-aekjeot* contains a significantly higher level of histamine (up to 1154.7 mg/kg) than *Saeu-jeotgal*. In this study, all radish kimchi samples were prepared with both *Myeolchi-aekjeot* and *Saeu-jeotgal* as ingredients. Altogether, the excessive level of histamine in several radish kimchi samples could be due to the amount of added *Myeolchi-aekjeot* with high histamine content. Unfortunately, the food labels of the samples used in this study just provided the list of ingredients.

An overdose of histamine may provoke undesirable symptoms such as a migraine, sweating, and hypotension [3]. In addition, high levels of putrescine and cadaverine can potentiate histamine toxicity by inhibiting intestinal diamine oxidase and histamine-N-methyltransferase [27] and potentially react with nitrites to form carcinogenic N-nitrosamines [28]. Taking this into account, although most *Kkakdugi* and *Chonggak* kimchi samples seem to be safe for consumption, the fact that several samples contained relatively high levels of putrescine and cadaverine in the present study indicates that it is necessary to monitor and reduce BA content, particularly histamine, putrescine, and cadaverine.

3.2. Physicochemical and Microbial Properties of Radish Kimchi: Kkakdugi and Chonggak Kimchi

To predict possible reasons as to why some samples of two types of radish kimchi contained higher levels of BA, pH, acidity, salinity, water activity (a_w), and lactic acid bacterial and total aerobic bacterial counts of *Kkakdugi* and *Chonggak* kimchi samples were determined. In *Kkakdugi* samples, the values of the parameters were as follows: pH, 4.16 ± 0.17 (minimum to maximum range of 3.94–4.41); acidity (%), 0.86 ± 0.31 (0.51–1.27); salinity (%), 3.36 ± 1.21 (1.40–4.50); a_w, 0.983 ± 0.003 (0.977–0.988); lactic acid bacterial counts, 8.52 ± 0.61 Log CFU/mL (7.88–9.38 Log CFU/mL); total aerobic bacterial counts, 8.37 ± 0.96 Log CFU/mL (6.83–9.32 Log CFU/mL). In case of *Chonggak* kimchi samples, the measured values were as follows: pH, 4.96 ± 1.17 (3.98–6.36); acidity (%), 0.71 ± 0.43 (0.19–1.10); salinity (%), 3.83 ± 1.67 (2.15–6.48); a_w, 0.984 ± 0.004 (0.979–0.991); lactic acid bacterial counts, 7.83 ± 0.48 Log CFU/mL (7.42–8.60 Log CFU/mL); total aerobic bacterial counts, 8.18 ± 1.07 Log CFU/mL (6.88–9.48 Log CFU/mL). The values are in accordance with those of previous reports [13,29]. Linear regression analysis was performed to determine the contributors influencing BA content. Results revealed weak correlations between physiochemical parameters, as well as microbial properties, and BA content (data not shown). Nonetheless, several reports have shown that physicochemical and microbial properties may affect BA content in fermented foods [2,30,31]. Altogether, the results indicate that, besides physicochemical and microbial properties, there are complex factors affecting BA content in both radish kimchi, for instance, kinds of salted and fermented fish products used for kimchi preparation as described above.

3.3. BA Production by LAB Strains Isolated from Radish Kimchi: Kkakdugi and Chonggak Kimchi

BA production by LAB strains isolated from *Kkakdugi* and *Chonggak* kimchi samples was examined to determine BA-producing LAB species in two types of radish kimchi. All the strains showed low production (below the detection limit) of tryptamine, β-phenylethylamine, putrescine, cadaverine, histamine, spermidine, and spermine. However, 39 strains (30%) of 130 LAB isolated from *Kkakdugi* samples produced higher levels of tyramine (287.23–386.17 µg/mL) than other strains (below the detection limit). Among the 120 LAB strains isolated from *Chonggak* kimchi, 16 strains (13%) also showed a stronger tyramine production capability (260.93–339.56 µg/mL), while other strains revealed lower capability (below the detection limit). In addition, the tyramine-producing LAB strains, which were isolated from either *Kkakdugi* or *Chonggak* kimchi samples, revealed a similar ability to produce tyramine, as described right above. Meanwhile, despite the low level of tyramine detected in all the samples of *Kkakdugi* and *Chonggak* kimchi, the fact that parts of LAB strains isolated from both radish kimchi samples were highly capable of producing tyramine supports that tyramine increment may occur during the ripening of the kimchi [12].

To further determine microorganisms responsible for BA formation in radish kimchi at species level, the strains were divided into two groups: (i) 55 tyramine-producing LAB strains (39 strains from *Kkakdugi*; 16 strains from *Chonggak* kimchi) and (ii) 195 LAB strains unable to produce BA. In the two groups, several strains were randomly selected and subsequently identified based on 16s rRNA sequencing analysis. Then, the selected strains able to produce tyramine were all identified as *L. brevis*, which indicates that the species is probably responsible for tyramine formation in both types of radish kimchi. On the other hand, the selected strains unable to produce BA were identified as *Leuconostoc* (*Leu.*) *mesenteroides*, *Weissella cibaria*, *W. paramesenteroides*, *L. pentosus*, and *L. plantarum*. The results are in agreement with previous reports in which *Leuconostoc*, *Weissella*, and *Lactobacillus* spp. were suggested to be responsible for kimchi fermentation [8,32]. Meanwhile, tyramine production by *L. brevis* in various fermented foods, including wine and fermented sausage, as well as *Baechu* kimchi, has been previously reported [14,33,34]. In the reports, tyramine production by *L. brevis* isolated from wine ranged from 441.6 to 1070.0 μg/mL, which is higher than that of the present study. On the contrary, *L. brevis* isolated from fermented sausage and *Baechu* kimchi produced tyramine at the range from 138.51 to 169.47 μg/mL and from 282 to 388 μg/mL, respectively, which are similar or lower than that of this study. In addition, several authors also isolated tyramine-producing *Leu. mesenteroides*, *W. cibaria*, and *W. paramesenteroides* from *Baechu* kimchi [14,15] and *L. plantarum* from wine [35]. Interestingly, as described right above, there are somewhat disparate results between the present and previous studies, which indicates that the strains belonging to the same species may possess different ability to produce tyramine especially depending upon the kinds of foods. Thus, microbial BA production in radish kimchi is likely determined at strain level, probably adapting to the respective food ecosystems, as suggested by previous reports [36,37]. Another implication is that the strains unable to produce BA isolated in the current study have potential as starter cultures for kimchi fermentation. Further investigations are needed to use them as starter cultures, which may involve tests to examine if the strains fulfill the criteria of starter culture, including the technical properties of strains, food safety requirements, and quality expectations [38].

3.4. Changes in Tyramine and Other BA Content during Fermentation of Radish Kimchi: Kkakdugi and Chonggak Kimchi

Fermentation of *Kkakdugi* and *Chonggak* kimchi was performed to investigate the influences of *Myeolchi-aekjeot* (together with *Saeu-jeotgal*) and LAB strains (particularly *L. brevis*) on BA content (especially tyramine) of both radish kimchi. Five groups of *Kkakdugi* and *Chonggak* kimchi samples were prepared based on the presence or absence of *Myeolchi-aekjeot* and types of LAB inocula. *L. brevis* strains of KD3M5 and CG2M15 with the highest tyramine production activity among the identified tyramine-producing LAB strains were used to see if the species is practically responsible for tyramine formation during fermentation of *Kkakdugi* and *Chonggak* kimchi. On the other hand, *L. plantarum* strains of KD3M15 and CG3M21 unable to produce BA were used for two reasons. (i) *L. plantarum*, like *L. brevis*, is predominant species in kimchi [39]. (ii) Differently from *L. brevis*, *L. plantarum* has been found to be negative for tyramine production in the present and previous studies [33,40,41].

As shown in Figures 1 and 2, changes in physicochemical and microbial properties of *Kkakdugi* and *Chonggak* kimchi during the fermentation for 3 days were similar with those of several previous reports [25,29,42]. In detail, the pH of all radish kimchi groups decreased during day 1 of fermentation, and stayed constantly thereafter. On the contrary, the counts of total aerobic bacteria and lactic acid bacteria, and the acidity of all radish kimchi groups increased during day 1 and day 2, respectively, and remained constantly thereafter, which indicates that an appropriate fermentation process of *Kkakdugi* and *Chonggak* kimchi took place. It is mention worthy that the initial pH of C, PC, LB, and LP groups of both radish kimchi was slightly higher than that of B group, which might be because the neutral pH of *Saeu-jeotgal* affected the pH values of the former groups [43]. Nonetheless, the initial acidity of all groups, belonging to either *Kkakdugi* or *Chonggak* kimchi, was similar to each other. The salinity of all radish kimchi groups decreased slightly during fermentation. According to Shin, Ann, and Kim [44], osmosis between radish and broth (containing seasoning paste) occurs during fermentation, which results in a steady reduction of salinity. Regardless of the drop in salinity, water activity of all radish kimchi groups was constant during fermentation. In addition, the initial counts of total aerobic bacteria and lactic acid bacteria of PC, LB, and LP groups inoculated with any of LAB strains were higher than those of B and C groups to be fermented naturally without any inocula, as expected.

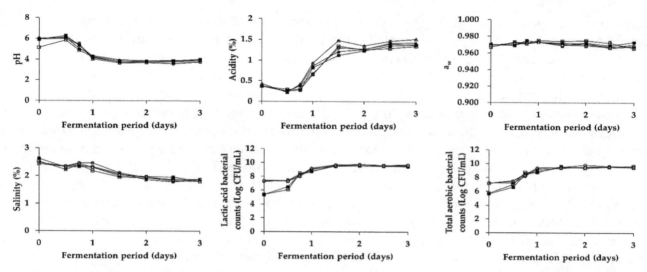

Figure 1. Changes in physicochemical and microbial properties of *Kkakdugi* during fermentation. □: B (no addition of *Myeolchi-aekjeot* and *Saeu-jeotgal*, no inoculum), ■: C (addition of *Myeolchi-aekjeot* and *Saeu-jeotgal*, no inoculum), ▲: PC (addition of *Myeolchi-aekjeot* and *Saeu-jeotgal*, *L. brevis* JCM 1170), △: LB (addition of *Myeolchi-aekjeot* and *Saeu-jeotgal*, *L. brevis* KD3M5), ○: LP (addition of *Myeolchi-aekjeot* and *Saeu-jeotgal*, *L. plantarum* KD3M15).

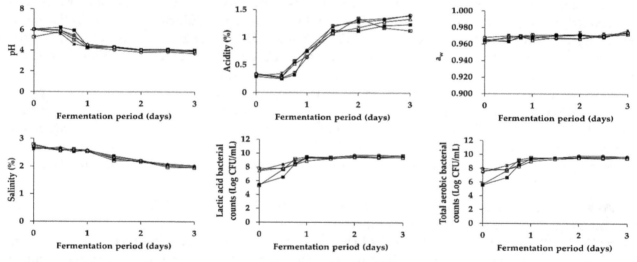

Figure 2. Changes in physicochemical and microbial properties of *Chonggak* kimchi during fermentation. □: B (no addition of *Myeolchi-aekjeot* and *Saeu-jeotgal*, no inoculum), ■: C (addition of *Myeolchi-aekjeot* and *Saeu-jeotgal*, no inoculum), ▲: PC (addition of *Myeolchi-aekjeot* and *Saeu-jeotgal*, *L. brevis* JCM 1170), △: LB (addition of *Myeolchi-aekjeot* and *Saeu-jeotgal*, *L. brevis* CG2M15), ○: LP (addition of *Myeolchi-aekjeot* and *Saeu-jeotgal*, *L. plantarum* CG3M21).

Changes in BA content (except for tryptamine and β-phenylethylamine not detected) during fermentation of *Kkakdugi* and *Chonggak* kimchi were shown in Figures 3 and 4, respectively. There appeared an increment of tyramine content in most groups (except for LP group) of both radish kimchi over the fermentation period, probably resulting from tyramine production by either inoculated or indigenous *L. brevis* strains (refer to Section 3.3). Also, the increment of tyramine content in PC and LB groups was higher than that in B and C groups of both radish kimchi (except for day 3 of *Chonggak* kimchi fermentation). This might be due to higher lactic acid bacterial counts of PC and LB groups, resulting from the inoculation of tyramine-producing *L. brevis* strains, than those of B and C groups of both radish kimchi. In the meantime, tyramine content in B and C groups of *Chonggak* kimchi steadily increased during fermentation, while that in the same groups of *Kkakdugi* increased slightly (but at a low level compared to *Chonggak* kimchi), both of which are likely associated with tyramine production by indigenous LAB strains (probably *L. brevis*). The observations are consistent with previous reports described right below. In short, Choi et al. [45] reported a dramatic increase of tyramine during natural fermentation of *Baechu* kimchi, whereas Kim et al. [46] reported that *Baechu* kimchi had a constantly low level of tyramine during natural fermentation. It is also noteworthy that, in the case of *Chonggak* kimchi, tyramine content in PC and LB groups dramatically increased during day 1 of fermentation, which was higher (and also showed a faster increment) than that in the same groups of *Kkakdugi*. The results, together with the comparison of tyramine content in B and C groups between two types of radish kimchi described above, can be explained by two speculations. The first is the difference in the ability of *L. brevis* strains to produce tyramine. The second is the distinguishable adaptation of the strains to different food ecosystems, i.e., differences in the main ingredients and/or ratio of ingredients in seasoning paste between two types of radish kimchi. Since KD3M5 strain served as an inoculum for *Kkakdugi* revealed a stronger ability to produce tyramine (377.35 ± 4.36 µg/mL) than CG2M15 strain for *Chonggak* kimchi (328.48 ± 2.61 µg/mL) when compared in vitro (refer to Section 3.3), the second speculation seems to be more probable than the first one. In addition, it is well known that bacteria produce BA to neutralize acidic environments as part of homeostatic regulation [47]. In this study, however, both radish kimchi samples of PC and LB groups showed similar patterns of acidity changes, so that the homeostatic regulation was excluded from possible reasons. Either way, there seem to be much complicated cross effects by the combinations of factors influencing the intensity of BA production by LAB during fermentation of kimchi variety. Interestingly, LP group of both radish kimchi had significantly lower levels of tyramine than the other groups. Thus, it seems that *L. plantarum* strains unable to produce BA in vitro not only have incapability of producing BA during fermentation, but also may inhibit tyramine production by indigenous LAB strains. This indicates the applicability of this species as a starter culture for reducing BA in kimchi variety.

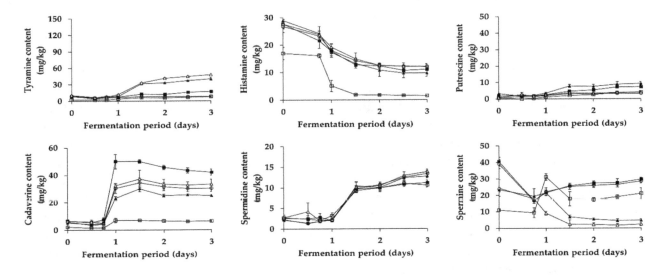

Figure 3. Changes in BA content in *Kkakdugi* during fermentation. □: B (no addition of *Myeolchi-aekjeot* and *Saeu-jeotgal*, no inoculum), ■: C (addition of *Myeolchi-aekjeot* and *Saeu-jeotgal*, no inoculum), ▲: PC (addition of *Myeolchi-aekjeot* and *Saeu-jeotgal*, *L. brevis* JCM 1170), △: LB (addition of *Myeolchi-aekjeot* and *Saeu-jeotgal*, *L. brevis* KD3M5), ○: LP (addition of *Myeolchi-aekjeot* and *Saeu-jeotgal*, *L. plantarum* KD3M15).

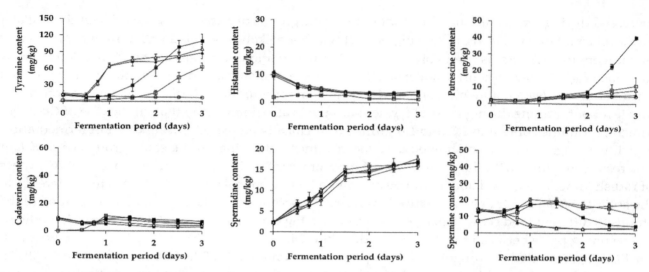

Figure 4. Changes in BA content in *Chonggak* kimchi during fermentation. □: B (no addition of *Myeolchi-aekjeot* and *Saeu-jeotgal*, no inoculum), ■: C (addition of *Myeolchi-aekjeot* and *Saeu-jeotgal*, no inoculum), ▲: PC (addition of *Myeolchi-aekjeot* and *Saeu-jeotgal*, *L. brevis* JCM 1170), △: LB (addition of *Myeolchi-aekjeot* and *Saeu-jeotgal*, *L. brevis* CG2M15), ○: LP (addition of *Myeolchi-aekjeot* and *Saeu-jeotgal*, *L. plantarum* CG3M21).

Differently from tyramine, histamine content in all groups of both radish kimchi gradually decreased during fermentation. This result might be because there were some indigenous LAB strains with histamine-degrading activity. Similarly, Kim et al. [48] reported a significant reduction of histamine content in *Baechu* kimchi inoculated with type strains of different LAB species including *L. sakei*, *L. plantarum*, *Leu. carnosum*, and *Leu. mesenteroides*, when compared with non-inoculated kimchi, suggesting that some LAB stains in kimchi are capable of degrading histamine. Meanwhile, the experimental groups of *Kkakdugi* and *Chonggak* kimchi prepared with *Myeolchi-aekjeot* (C, PC, LB, and LP groups) contained a significantly higher level of histamine than B group, which is in accordance with the suggestion of previous studies [12,22]. In the studies, the authors assumed that histamine level in *Baechu* kimchi could be affected by histamine in *Myeolchi-aekjeot*. Taking this into account, histamine content of *Kkakdugi* and *Chonggak* kimchi in the present study seems to come from *Myeolchi-aekjeot* rather than microbial histamine production during fermentation.

Putrescine and spermidine content steadily increased in all groups of *Kkakdugi* and *Chonggak* kimchi during fermentation, which is in agreement with previous reports [12,46]. There was a small and insignificant difference in putrescine and spermidine content among the groups of both radish kimchi during fermentation, which indicates that LAB strains—including *L. brevis* and *L. plantarum*—produced the polyamines during fermentation. Meanwhile, the initial concentrations putrescine and spermidine in *Kkakdugi* and *Chonggak* kimchi might be come from main ingredients, i.e., white radish and ponytail radish, respectively. In addition, a sharp increment of putrescine was observed during day 3 of fermentation, in the case of C group of *Chonggak* kimchi. To ignore the possibility of outliers, the fermentation experiment was repeatedly performed; however, the same results were observed, and the reason for such observation was not clear.

Somewhat differently from above, cadaverine content in all groups of *Kkakdugi* and *Chonggak* kimchi showed an increment during day 1 of fermentation and slight decline thereafter, although the increased cadaverine amount was mostly higher in *Kkakdugi* than in *Chonggak* kimchi. The difference in the intensity of cadaverine formation between two types of radish kimchi seems to be attributed to the complex combinations of factors described above to explain difference in the kinetics of tyramine formation between two radish kimchi. Interestingly, the initial cadaverine content in C, PC, LB, and LP groups of both radish kimchi was higher than that in B group, which might be come from *Myeolchi-aekjeot* rather than *Saeu-jeotgal*. The speculation is supported by a study by Cho et al. [22]

who reported a significantly higher level of cadaverine in *Myeolchi-aekjeot* (up to 263.6 mg/kg) than that in *Saeu-jeotgal* (up to 7.0 mg/kg). For both radish kimchi, C group contained the highest level of cadaverine, as compared to the other groups, over the fermentation period. This may be explained by a presumption that while cadaverine-producing bacteria derived from *Myeolchi-aekjeot* are probably responsible for cadaverine formation during fermentation of both radish kimchi, LAB strains (*L. brevis* and *L. plantarum*) served as inocula are probably capable of degrading cadaverine. Supporting this presumption, Mah et al. [49] reported that *Bacillus* strains isolated from *Myeolchi-jeotgal* were highly capable of producing cadaverine. Capozzi et al. [50] also reported that *L. plantarum* strains isolated from wine were capable of degrading cadaverine. At present, however, investigations on cadaverine-degrading activity of *L. brevis* are rarely found in literature.

As for change in spermine content, there appeared difference among groups of *Kkakdugi* and *Chonggak* kimchi. In PC and LB groups of both radish kimchi, a gradual decrease of spermine content was observed over the fermentation period, and the content was relatively lower than that in the other groups of both types of radish kimchi. This implies that *L. brevis* could be able to degrade spermine, although relevant reports are scarce to date. It is worth nothing that in B, C, and LP groups, spermine content decreased for day 1 of fermentation and slightly increased thereafter in *Kkakdugi*, whereas that in *Chonggak* kimchi increased for day 1 and slightly decreased thereafter. The different kinetics of spermine formation seems to result from the complex combinations of factors mentioned above. Therefore, it would be interesting in a future study to identify the factors (and combinations thereof) associated with BA formation or degradation by LAB strains during fermentation of *Kkakdugi* and *Chonggak* kimchi. The factors may involve time-related successional changes and/or interactions of microorganisms during fermentation as well as ingredients of foods and metabolic activities of strains [51]. In addition, recent studies suggested that results of in vitro BA production by food fermenting microorganisms were in disagreement with those of BA formation during fermentation of the corresponding foods [52,53]. In the present study, however, *L. brevis* was considered to be responsible for tyramine formation not only in vitro but also during practical fermentation of *Kkakdugi* and *Chonggak* kimchi.

4. Conclusions

The present study indicated that the amounts of BA in most samples of *Kkakdugi* and *Chonggak* kimchi were considered safe for consumption, but some samples contained histamine and total BA at concentrations over toxicity limits ($\geq$100 mg/kg and $\geq$1000 mg/kg, respectively). It was also found that, while *Myeolchi-aekjeot* seems to be an important source of histamine in both types of radish kimchi, *L. brevis* strains isolated from *Kkakdugi* and *Chonggak* kimchi are highly capable of producing tyramine in assay media. On the other hand, the physicochemical and microbial properties of both radish kimchi revealed weak correlations with BA content in the respective kimchi types in the present study. Through the practical fermentation of *Kkakdugi* and *Chonggak* kimchi, it turned out that *L. brevis* is responsible for tyramine formation, and *Myeolchi-aekjeot* influences histamine and cadaverine content in both radish kimchi. Consequently, this study suggests strategies for reducing BA in radish kimchi: the alteration of the ratio of ingredients used for kimchi preparation, particularly reducing ratio of *Myeolchi-aekjeot* to others, and use of starter cultures other than tyramine-producing *L. brevis* strains, especially BA-degrading LAB starter cultures. Studies on other contributing factors influencing the intensity of BA production by LAB are also required to understand complex kinetics of BA formation in the kimchi.

Author Contributions: Conceptualization: J.-H.M.; Investigation: Y.H.J., J.H.L., Y.K.P., and J.-H.L.; Writing—original draft: Y.H.J. and J.-H.M.; Writing—review and editing: Y.H.J. and J.-H.M.; Supervision: J.-H.M.

Acknowledgments: The authors thank Junsu Lee of Department of Food and Biotechnology at Korea University for technical assistance.

References

1. Ten Brink, B.; Damink, C.; Joosten, H.M.L.J.; Huis In't Veld, J.H.J. Occurrence and formation of biologically active amines in foods. *Int. J. Food Microbiol.* **1990**, *11*, 73–84. [CrossRef]

2. Silla Santos, M.H. Biogenic amines: Their importance in foods. *Int. J. Food Microbiol.* **1996**, *29*, 213–231. [CrossRef]

3. Ladero, V.; Calles-Enríquez, M.; Fernández, M.; Alvarez, M.A. Toxicological effects of dietary biogenic amines. *Curr. Nutr. Food Sci.* **2010**, *6*, 145–156. [CrossRef]

4. Marcobal, A.; De Las Rivas, B.; Landete, J.M.; Tabera, L.; Muñoz, R. Tyramine and phenylethylamine biosynthesis by food bacteria. *Crit. Rev. Food Sci. Nutr.* **2012**, *52*, 448–467. [CrossRef]

5. Bover-Cid, S.; Hugas, M.; Izquierdo-Pulido, M.; Vidal-Carou, M.C. Reduction of biogenic amine formation using a negative amino acid-decarboxylase starter culture for fermentation of *Fuet* sausages. *J. Food Prot.* **2000**, *63*, 237–243. [CrossRef]

6. Bover-Cid, S.; Hugas, M.; Izquierdo-Pulido, M.; Vidal-Carou, M.C. Effect of the interaction between a low tyramine-producing *Lactobacillus* and proteolytic *staphylococci* on biogenic amine production during ripening and storage of dry sausages. *Int. J. Food Microbiol.* **2001**, *65*, 113–123. [CrossRef]

7. Codex Alimentarius Commission. *Codex Standard for kimchi, Codex Stan 223-2001*; Food and Agriculture Organization of the United Nations: Rome, Italy, 2001.

8. Park, E.-J.; Chun, J.; Cha, C.-J.; Park, W.-S.; Jeon, C.O.; Bae, J.-W. Bacterial community analysis during fermentation of ten representative kinds of kimchi with barcoded pyrosequencing. *Food Microbiol.* **2012**, *30*, 197–204. [CrossRef]

9. Cheigh, H.-S.; Park, K.-Y. Biochemical, microbiological, and nutritional aspects of kimchi (Korean fermented vegetable products). *Crit. Rev. Food Sci. Nutr.* **1994**, *34*, 175–203. [CrossRef]

10. Jang, K.-S.; Kim, M.-J.; Oh, Y.-A.; Kim, I.-D.; No, H.-K.; Kim, S.-D. Effects of various sub-ingredients on sensory quality of Korean cabbage kimchi. *J. Korean Soc. Food Nutr.* **1991**, *20*, 233–240.

11. Park, D.-C.; Park, J.-H.; Gu, Y.-S.; Han, J.-H.; Byun, D.-S.; Kim, E.-M.; Kim, Y.-M.; Kim, S.-B. Effects of salted-fermented fish products and their alternatives on angiotensin converting enzyme inhibitory activity of *Kimchi* during fermentation. *Korean J. Food Sci. Technol.* **2000**, *32*, 920–927.

12. Mah, J.-H.; Kim, Y.J.; No, H.-K.; Hwang, H.-J. Determination of biogenic amines in *kimchi*, Korean traditional fermented vegetable products. *Food Sci. Biotechnol.* **2004**, *13*, 826–829.

13. Tsai, Y.-H.; Kung, H.-F.; Lin, Q.-L.; Hwang, J.-H.; Cheng, S.-H.; Wei, C.-I.; Hwang, D.-F. Occurrence of histamine and histamine-forming bacteria in kimchi products in Taiwan. *Food Chem.* **2005**, *90*, 635–641. [CrossRef]

14. Kim, M.-J.; Kim, K.-S. Tyramine production among lactic acid bacteria and other species isolated from kimchi. *LWT-Food Sci. Technol.* **2014**, *56*, 406–413. [CrossRef]

15. Jeong, D.-W.; Lee, J.-H. Antibiotic resistance, hemolysis and biogenic amine production assessments of *Leuconostoc* and *Weissella* isolates for kimchi starter development. *LWT-Food Sci. Technol.* **2015**, *64*, 1078–1084. [CrossRef]

16. AOAC. *Official Methods of Analysis of AOAC International*, 17th ed.; AOAC International: Gaithersburg, MD, USA, 2000.

17. Eerola, S.; Hinkkanen, R.; Lindfors, E.; Hirvi, T. Liquid chromatographic determination of biogenic amines in dry sausages. *J. AOAC Int.* **1993**, *76*, 575–577. [PubMed]

18. Ben-Gigirey, B.; De Sousa, J.M.V.B.; Villa, T.G.; Barros-Velazquez, J. Changes in biogenic amines and microbiological analysis in albacore (*Thunnus alalunga*) muscle during frozen storage. *J. Food Prot.* **1998**, *61*, 608–615. [CrossRef]

19. Ben-Gigirey, B.; De Sousa, J.M.V.B.; Villa, T.G.; Barros-Velazquez, J. Histamine and cadaverine production by bacteria isolated from fresh and frozen albacore (*Thunnus alalunga*). *J. Food Prot.* **1999**, *62*, 933–939. [CrossRef]

20. Yoon, H.; Park, J.H.; Choi, A.; Hwang, H.-J.; Mah, J.-H. Validation of an HPLC analytical method for determination of biogenic amines in agricultural products and monitoring of biogenic amines in Korean fermented agricultural products. *Toxicol. Res.* **2015**, *31*, 299–305. [CrossRef]

21. National Institute of Agricultural Sciences. Available online: http://koreanfood.rda.go.kr/kfi/kimchi/kimchi_01 (accessed on 2 February 2019).

22. Cho, T.-Y.; Han, G.-H.; Bahn, K.-N.; Son, Y.-W.; Jang, M.-R.; Lee, C.-H.; Kim, S.-H.; Kim, D.-B.; Kim, S.-B. Evaluation of biogenic amines in Korean commercial fermented foods. *Korean J. Food Sci. Technol.* **2006**, *38*, 730–737.

23. Kim, M.R.; Oh, Y.; Oh, S. Physicochemical and sensory properties of Kagdugi prepared with fermentation northern sand sauce during fermentation. *Korean J. Soc. Food Sci.* **2000**, *16*, 602–608.

24. Park, S.-O.; Kim, W.-K.; Park, D.-J.; Lee, S.-J. Effect of blanching time on the quality characteristics of elderly-friendly *kkakdugi*. *Food Sci. Biotechnol.* **2017**, *26*, 419–425. [CrossRef] [PubMed]

25. Kang, J.-H.; Kang, S.-H.; Ahn, E.-S.; Chung, H.-J. Quality properties of *Chonggak* kimchi fermented at different combination of temperature and time. *J. Korean Soc. Food Cult.* **2003**, *18*, 551–561.

26. Kim, Y.-J.; Jin, Y.-Y.; Song, K.-B. Study of quality change in Chonggak-kimchi during storage, for development of a freshness indicator. *Korean J. Food Preserv.* **2008**, *15*, 491–496.

27. Stratton, J.E.; Hutkins, R.W.; Taylor, S.L. Biogenic amines in cheese and other fermented foods: A review. *J. Food Prot.* **1991**, *54*, 460–470. [CrossRef]

28. Warthesen, J.J.; Scanlan, R.A.; Bills, D.D.; Libbey, L.M. Formation of heterocyclic *N*-nitrosamines from the reaction of nitrite and selected primary diamines and amino acids. *J. Agric. Food Chem.* **1975**, *23*, 898–902. [CrossRef] [PubMed]

29. Mheen, T.-I.; Kwon, T.-W. Effect of temperature and salt concentration on *Kimchi* fermentation. *Korean J. Food Sci. Technol.* **1984**, *16*, 443–450.

30. Lu, S.; Xu, X.; Shu, R.; Zhou, G.; Meng, Y.; Sun, Y.; Chen, Y.; Wang, P. Characterization of biogenic amines and factors influencing their formation in traditional Chinese sausages. *J. Food Sci.* **2010**, *75*, M366–M372. [CrossRef]

31. Özdestan, Ö.; Üren, A. Biogenic amine content of tarhana: A traditional fermented food. *Int. J. Food Prop.* **2013**, *16*, 416–428. [CrossRef]

32. Kim, M.; Chun, J. Bacterial community structure in kimchi, a Korean fermented vegetable food, as revealed by 16S rRNA gene analysis. *Int. J. Food Microbiol.* **2005**, *103*, 91–96. [CrossRef]

33. Landete, J.M.; Ferrer, S.; Pardo, I. Biogenic amine production by lactic acid bacteria, acetic bacteria and yeast isolated from wine. *Food Control* **2007**, *18*, 1569–1574. [CrossRef]

34. Latorre-Moratalla, M.L.; Bover-Cid, S.; Talon, R.; Garriga, M.; Zanardi, E.; Ianieri, A.; Fraqueza, M.J.; Elias, M.; Drosinos, E.H. Vidal-Carou, M.C. Strategies to reduce biogenic amine accumulation in traditional sausage manufacturing. *LWT-Food Sci. Technol.* **2010**, *43*, 20–25. [CrossRef]

35. Arena, M.E.; Fiocco, D.; Manca de Nadra, M.C.; Pardo, I.; Spano, G. Characterization of a *Lactobacillus plantarum* strain able to produce tyramine and partial cloning of a putative tyrosine decarboxylase gene. *Curr. Microbiol.* **2007**, *55*, 205–210. [CrossRef] [PubMed]

36. Bover-Cid, S.; Holzapfel, W.H. Improved screening procedure for biogenic amine production by lactic acid bacteria. *Int. J. Food Microbiol.* **1999**, *53*, 33–41. [CrossRef]

37. Landete, J.M.; Ferrer, S.; Polo, L.; Pardo, I. Biogenic amines in wines from three Spanish regions. *J. Agric. Food Chem.* **2005**, *53*, 1119–1124. [CrossRef] [PubMed]

38. Holzapfel, W.H. Appropriate starter culture technologies for small-scale fermentation in developing countries. *Int. J. Food Microbiol.* **2002**, *75*, 197–212. [CrossRef]

39. Lee, C.-H. Lactic acid fermented foods and their benefits in Asia. *Food Control* **1997**, *8*, 259–269. [CrossRef]

40. Moreno-Arribas, M.V.; Polo, M.C.; Jorganes, F.; Muñoz, R. Screening of biogenic amine production by lactic acid bacteria isolated from grape must and wine. *Int. J. Food Microbiol.* **2003**, *84*, 117–123. [CrossRef]

41. Park, S.; Ji, Y.; Park, H.; Lee, K.; Park, H.; Beck, B.R.; Shin, H.; Holzapfel, W.H. Evaluation of functional properties of lactobacilli isolated from Korean white kimchi. *Food Control* **2016**, *69*, 5–12. [CrossRef]

42. Kim, S.-D.; Jang, M.-S. Effects of fermentation temperature on the sensory, physicochemical and microbiological properties of *Kakdugi*. *J. Kor. Soc. Food Sci. Nutr.* **1997**, *26*, 800–806.

43. Um, M.-N.; Lee, C.-H. Isolation and identification of *Staphylococcus* sp. from Korean fermented fish products. *J. Microbiol. Biotechnol.* **1996**, *6*, 340–346.

44. Shin, Y.-H.; Ann, G.-J.; Kim, J.-E. The changes of hardness and microstructure of Dongchimi according to different kinds of water. *Korean J. Food Cookery Sci.* **2004**, *20*, 86–94.

45. Choi, Y.-J.; Jang, M.-S.; Lee, M.-A. Physicochemical changes in kimchi containing skate (*Raja kenojei*) pretreated with organic acids during fermentation. *Food Sci. Biotechnol.* **2016**, *25*, 1369–1377. [CrossRef] [PubMed]

46. Kim, S.-H.; Kang, K.H.; Kim, S.H.; Lee, S.; Lee, S.-H.; Ha, E.-S.; Sung, N.-J.; Kim, J.G.; Chung, M.J. Lactic acid bacteria directly degrade N-nitrosodimethylamine and increase the nitrite-scavenging ability in kimchi. *Food Control* **2017**, *71*, 101–109. [CrossRef]

47. Arena, M.E.; Manca de Nadra, M.C. Biogenic amine production by *Lactobacillus*. *J. Appl. Microbiol.* **2001**, *90*, 158–162. [CrossRef]

48. Kim, S.-H.; Kim, S.H.; Kang, K.H.; Lee, S.; Kim, S.J.; Kim, J.G.; Chung, M.J. Kimchi probiotic bacteria contribute to reduced amounts of N-nitrosodimethylamine in lactic acid bacteria-fortified kimchi. *LWT-Food Sci. Technol.* **2017**, *84*, 196–203. [CrossRef]

49. Mah, J.-H.; Ahn, J.-B.; Park, J.-H.; Sung, H.-C.; Hwang, H.-J. Characterization of biogenic amine-producing microorganisms isolated from Myeolchi-Jeot, Korean salted and fermented anchovy. *J. Microbiol. Biotechnol.* **2003**, *13*, 692–699.

50. Capozzi, V.; Russo, P.; Ladero, V.; Fernández, M.; Fiocco, D.; Alvarez, M.A.; Grieco, F.; Spano, G. Biogenic amines degradation by *Lactobacillus plantarum*: Toward a potential application in wine. *Front. Microbiol.* **2012**, *3*, 122. [CrossRef]

51. Yılmaz, C.; Gökmen, V. Formation of tyramine in yoghurt during fermentation—Interaction between yoghurt starter bacteria and *Lactobacillus plantarum*. *Food Res. Int.* **2017**, *97*, 288–295. [CrossRef]

52. Nie, X.; Zhang, Q.; Lin, S. Biogenic amine accumulation in silver carp sausage inoculated with *Lactobacillus plantarum* plus *Saccharomyces cerevisiae*. *Food Chem.* **2014**, *153*, 432–436. [CrossRef]

53. Jeon, A.R.; Lee, J.H.; Mah, J.-H. Biogenic amine formation and bacterial contribution in *Cheonggukjang*, a Korean traditional fermented soybean food. *LWT-Food Sci. Technol.* **2018**, *92*, 282–289. [CrossRef]

3

Modeling Some Possible Handling Ways with Fish Raw Material in Home-Made Sushi Meal Preparation

Hana Buchtova [1,*], Dani Dordevic [2,3], Iwona Duda [4], Alena Honzlova [5] and Piotr Kulawik [4]

[1] Department of Meat Hygiene and Technology, Faculty of Veterinary Hygiene and Technology, University of Veterinary and Pharmaceutical Sciences Brno, 61242 Brno, Czech Republic

[2] Department of Plant Origin Foodstuffs Hygiene and Technology, Faculty of Veterinary Hygiene and Technology, University of Veterinary and Pharmaceutical Sciences Brno, 61242 Brno, Czech Republic; dordevicd@vfu.cz

[3] Department of Technology and Organization of Public Catering, South Ural State University, Lenin prospect 76, 454080 Chelyabinsk, Russia

[4] Department of Animal Product Technology, Faculty of Food Technology, University of Agriculture, 31-120 Krakow, Poland; iwona.duda@urk.edu.pl (I.D.); kulawik.piotr@gmail.com (P.K.)

[5] Department of Chemistry, State Veterinary Institute Jihlava, 58601, Jihlava, Czech Republic; honzlova@svujihlava.cz

[*] Correspondence: buchtovah@vfu.cz

Abstract: The aim of this work was to simulate selected ways of handling with raw fish after its purchase. The experiment was designed as three partial simulations: (a) trend in the biogenic amines formation in raw fish caused by breakage of cold chain during the transport after purchase, (b) the use of a handheld gastronomic unit as an alternative method of smoking fish with cold smoke in the household with regard to a possible increase in polycyclic aromatic hydrocarbon content, and (c) whether the cold smoked fish affects selected sensory parameters of nigiri sushi meal prepared by consumers. The material used in the research consisted of: yellowfin tuna (*Thunnus albacares*) sashimi fillets and the Atlantic salmon (*Salmo salar*) fillets with skin. The control (fresh/thawed tuna; without interrupting the cold chain) and experimental (fresh/thawed tuna; cold chain was interrupted by incubation at 35 °C/6 h) samples were stored at 2 ± 2 °C for 8 days and analyzed after 1st, 4th and 8th day of the cold storage. Histamine content was very low throughout the experiment, though one exception was found (thawed tuna without interrupting the cold chain: 272.05 ± 217.83 mg·kg^{-1}/8th day). Tuna samples contained more PAH (4.22 µg·kg^{-1}) than salmon samples (1.74 µg·kg^{-1}). Alarming increases of benzo(a)anthracene (1.84 µg·k^{-1}) and chrysene (1.10 µg·kg^{-1}) contents in smoked tuna were detected.

Keywords: nigiri sushi; polycyclic aromatic hydrocarbons; histamine; household smoker unit

1. Introduction

Currently, sushi meals are becoming popular worldwide [1]. Sushi meals have developed from a simple street food to sophisticated cuisine. Many studies have dealt with the health benefits and health hazards associated with the sushi cuisine [2]. In the past, high attention has been devoted to studies of microbiological [3], chemical [4,5] or parasitic [6] hazards in fishery products, like the toxicological risks of diseases after consumption of raw fish or foodstuffs that include raw fish flesh [1]. Recorded cases of acute gastric anisakiasis are a serious warning to consumers [7]. Sushi belongs to ready to eat foods and is predisposed to contamination with food pathogens, such as *Listeria monocytogenes* [8,9].

Consumer concerns about food safety might disrupt a healthy food choice [10]. Risks associated with the consumption of fish might impose barriers to consumption, though fish is considered an important component of the human diet [11].

A new look at research on sushi meal assessment, including simulation of model of real consumer behavior and culinary practices by chefs, should shift research to a higher level of knowledge. In recent years there has been an increase in collaboration between researchers and chefs in the field of gastronomy [12]. Modern trends of molecular gastronomy that works with human senses, is a fast food preparation method using portable, easy-to-use applications, developed specifically for chefs, to create new unusual flavors.

The biggest interruption in the cold chain occurs after product purchase and during its delivery to the household. Consumer behavior and the ambient temperature largely influence the shelf life and food safety [13].

In recent years, also in the Czech Republic, there has been a growing trend of self-preparation of sushi food by consumers. In our research, based on the buying habits of some consumers and their creativity approach to treat raw fish raw using a handheld smoker unit, we wanted to connect partial studies on fish handling and sushi preparation into one model experiment.

The experiment was designed in the form of three partial simulations: (a) trend in biogenic amines formation caused by severe breakage of cold chain during the transport of fish raw material after purchase to household, (b) the use of a handheld gastronomic unit as an alternative method of smoking fish with cold smoke in the household with regard to a possible increase in polycyclic aromatic hydrocarbons (PAH) content, and (c) whether the cold smoked fish affects selected sensory parameters of nigiri sushi meal prepared by consumers in their households.

2. Materials and Methods

Fresh sashimi fillets of the yellowfin tuna (*Thunnus albacares*, caught, FAO 71 area, category of fishing gear: seines) and fresh fillets with skin of the Atlantic salmon (*Salmo salar*, farmed, Norway) were bought from a retail shop (Ocean48, Brno, Czech Republic).

2.1. Trend in the Biogenic Amines Formation Caused by Severe Breakage of Cold Chain during the Transport of Fish Raw Material after Purchase Place to Household

Tuna sashimi fillet was used to a case study focused on simulating conditions sale of fresh and thawed fish and consumer behavior (compliance/interruption with/of the cold chain) and to determine how this behavior affects the formation of biogenic amines total content and its spectrum (tryptamine TRP, 2-phenylethylamine 2-PHE, putrescine PUT, cadaverine CAD, histamine HIS, tyramine TYR, spermidine SPD, spermine SPR). The experiment was carried out in four separate replicates. Tuna fillets purchased for the fifth repetition had to be excluded from the experiment because of the high histamine content at the start of the storage, which significantly exceeded the limit set out in Regulation (EC) No 2073/2005 [see in Appendix A1] at the start of storage (the possible reason for higher histamine content will be commented on in the Results and Discussion section).

The case study was based on the use of four different sashimi fillets (1, 2, 3, 4). Four types of samples A, B, C, D from each tuna fillet were prepared simultaneously. Characteristics of the samples were as follows: A: control sample, fresh tuna was cold stored at +2 ± 2 °C, without interrupting the cold chain after buying the fish in a store; B: experimental sample, fresh tuna, cold chain of fresh sample was interrupted before the cold storage in a laboratory by incubation of the sample (35 °C/6 h) to simulate the possible consumer behavior in the summer after buying the fish in a store, subsequently samples were cold stored at +2 ± 2 °C; C: experimental sample, thawed tuna, after buying the fish in a store the sample was experimentally frozen (−35 °C) and stored for two weeks in a frozen state (−18 °C), then the samples were thawed in the refrigerator (+2 ± 2 °C/12 h) and subsequently cold stored at +2 ± 2 °C; D: experimental sample, thawed tuna, after buying the fish in a store, the sample was experimentally frozen (−35 °C) and stored two weeks in a frozen state (−18 °C), the cold chain of thawed sample was interrupted by incubation the sample (35 °C/6 h) to simulate the possible consumer behavior in the summer after buying the fish in a store, the samples were subsequently cold stored at

+2 ± 2 °C. All samples (A, B, C, D) were stored at +2 ± 2 °C for 8 days, the samples were analyzed after 1st, 4th and 8th days of cold storage.

The biogenic amines analysis was performed according to the method described by [14]. The chromatographic separation was performed using a Dionex Ultimate 3000 HPLC apparatus (Thermo Scientific, Waltham, MA, USA) with a FLD 3400RS four channel fluorescent detector (Thermo Scientific) and a low pressure gradient pump with a four channel mixer. The detector settings were set to 340 nm for excitation and 540 nm for emission. The separation was performed on a Kromasil 100-5-C18 4.6 × 250 mm column (Akzo Nobel, Amsterdam, The Netherlands) and the column temperature of 30 °C. Flow rate was 0.8 mL/min with two mobile phases: (A) acetonitrile (Merck, Darmstadt, Germany) and (B) water (Merck). The detection limit for each biogenic amine was 0.005 mg·kg^{-1}. The samples were analyzed in duplicate and triplicate and injections into HPLC were carried out on each duplicate (N = 4 × 2 × 3).

2.2. The Part of the Research Consisting of Testing the Use of a Handheld Gastronomic Unit as an Alternative Method of Smoking Fish With Cold Smoke in the Household with Regard to a Possible Increase in Polycyclic Aromatic Hydrocarbons (PAH) Content and Whether the Cold Smoked Fish Affects Selected Sensory Parameters of Nigiri Sushi Meal Prepared by Consumers at Their Household

Tuna and salmon fillets were used for preparation of nigiri sushi with not-smoked (raw, control samples) and smoked (experimental samples) samples. Smoked muscle of both fish was prepared with application of a smoker unit (Super Aladin smoker, Manihi s.r.o., Praha, Czech Republic). Cold smoke (20 °C; Aladin oak chips) was applied on meat surface beneath the glass hatch for 5 min. The experiment was carried out in five separate replicates. Sensory attributes (saltiness, bitterness, juiciness, consistency) of sushi meal and a question focused on examining the fact whether the conscious consumption of smoked fish in sushi can affect the consumer's confidence in the health safety of this food were monitored by a group of trained evaluators and evaluated on the basis of questionnaires.

Nigiri sushi samples with tuna and salmon meat (smoked and not smoked) were prepared in the Sensory Laboratory at the Department of Meat Hygiene and Technology (Faculty of Veterinary Hygiene and Ecology, University of Veterinary and Pharmaceutical Sciences, Brno, Czech Republic). Fillets were frozen according to the Commission regulation (EC) No 853/2004 (Annex III, Section VIII, Chapter III, Part D, Point 2a [see in Appendix A2]) at −40 °C using quick freezing unit F.R.C. BF 031AF (Friulinox, Taiedo di Chions, Italy) to muscle core temperature of −20 °C and were stored in a frozen chamber with regulated temperature (−20 ± 2 °C) for 2 weeks. Then the samples were thawed in refrigerator (+2 ± 2 °C/12 h) and subsequently divided into two parts, the one was used as control (raw not smoked) samples for sushi preparation and the second one was used for cold smoking by Super Aladin smoker and for sushi preparation. The rice was cooked in rice cooker (42507 Design Reiskocher, Gastroback, Hollenstedt, Germany). Information about sushi ingredients are following: sushi rice (short grain variety, Yutaka, Italy), 8% vinegar (apple vinegar, Bzenecky Ocet, Bzenec, Czech Republic), 6% sugar (sugar crystal, producer: Korunni, Hrušovany nad Jevišovkou, Czech Republic) and 2% cooking salt with iodine (NaCl 98%, J 20-34 mg·kg^{-1}, K+S Czech Republic a.s., Olomouc, Czech Republic), wasabi paste (Yutaka, China). The experimental design is shown in Figure 1.

Sushi samples were sensory evaluated in the laboratory equipped according to ISO 8589:2008 [see in Appendix A3]. The protocol consisted of unstructured graphical scales of 100 mm length, with one edge of the scale representing the strongest expressed attribute and the second one the weakest expressed attribute. Saltiness (2% salt addition) was evaluated by respondents' comparison with sushi samples prepared without salt (0 point) and with 3% salt content (100 points) added to rice.

Twenty panelists took part in the sensory evaluation, where they assessed selected parameters: (saltiness, bitterness, juiciness, consistency) and consumer confidence in the food safety (expressed in their own words). Ingredients' weights (g) of nigiri sushi are given in Table 1.

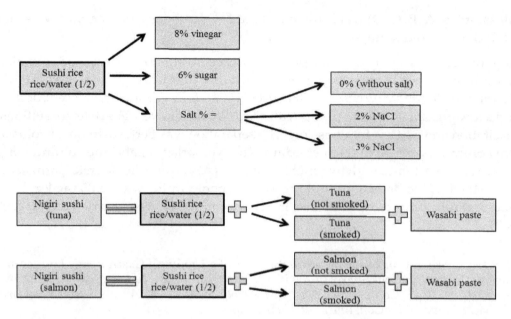

Figure 1. Experimental design of sushi production samples used in the research.

Table 1. Ingredients' portions (%) of nigiri sushi prepared with salmon and tuna.

Fish	Treatment	Rice	Seafood	Wasabi
Tuna	Not smoked	66.84 ± 0.53	37.24 ± 0.94	0.80 ± 0.06
	Smoked	63.02 ± 2.01	36.95 ± 1.04	0.75 ± 0.08
Salmon	Not smoked	66.76 ± 0.36	32.38 ± 0.34	0.86 ± 0.09
	Smoked	65.54 ± 0.25	33.31 ± 0.39	0.84 ± 0.07

The thawed raw tuna and salmon samples were used for chemical analysis (total protein, total fat and dry matter content). The same thawed not smoked and smoked tuna and salmon samples were used for determination of polycyclic aromatic hydrocarbons (PAH).

The total protein content (ISO 937:1978 [see in Appendix A4]) was determined as the amount of organically bound nitrogen (recalculating coefficient f = 6.25) using the analyzer Kjeltec 2300 (FOSS Tecator, Höganäs, Sweden). The total lipid content was determined quantitatively (ISO 1443:1973 [see in Appendix A5]) by extraction in solvents using Soxtec 2055 (FOSS Tecator). The dry matter was determined gravimetrically according to the Czech National Standard (ISO 1442:1997 [see in Appendix A6]) by drying the sample to a constant weight at +103 ± 2 °C (Binder FD 53, Tuttlingen, Germany).

PAH were determined by accredited method (no. 19 Standard operating procedure 8.15A) using HPLC/FLD in the laboratory of State Veterinary Institute Jihlava (Jihlava, Czech Republic). Each sample was analyzed in parallel. A thoroughly homogenized samples, after trituration with anhydrous sodium sulphate p.a. (Lach-Ner, sro, Tovarni 157, 27711, Neratovice, Czech Republic) and after addition of internal standard 2-methylchrysene (Dr. Ehrenstorfer GmbH, Augsburg, Germany) were extracted with diethyl ether. Extracts were filtered through glass fiber filter paper (Cat. No. 516-0867, VWR International bvba, Leuven, Belgien, the solvent was evaporated on a Büchi R-134 rotary evaporator (BÜCHI Labortechnik) AG, Flawil, Switzerland) at a maximum temperature of 30 °C. The residue was carefully blown off with a stream of nitrogen. The extracted fat was dissolved in chloroform (Cat. No. 20034-UT2-M2500-7 (Macron 6754), Lach-Ner, Ltd., Neratovice, Czech Republic). An aliquot of the solution was purified by gel permeation chromatography on a Gilson Aspec XL system (Gilson, Middleton, WI, USA) using a PAH prep column (500 × 8 mm) packed with gel (styrene divinylbenzene copolymer) (Watrex Praha, sro, Carolina Center, Prague, Czech Republic). Purified samples were evaporated to near dryness on a Büchi R-134 rotary evaporator at a maximum temperature of 30 °C or the remaining solvent was blown off with a stream of nitrogen. The residue was dissolved in acetonitrile and used for fluorescence detection by liquid chromatography.

Chromatographic analysis was performed on Waters Alliance e2695 liquid chromatograph with 2475 fluorescence detector (Waters Corporation, Milford, MA, USA) on Waters PAH column (250 mm × 4.6 mm × 5 μm) using gradient elution with mobile phase. Gradient mobile phase consisted out of acetonitrile/redistilled water (75/25 100/0), flow rate 0.7 mL/min, injection 10 μL, column temperature 30 °C. Detection was performed by fluorescence detector with programmable wavelength change (excitation wavelength—265 nm for benzo(a)anthracene and chrysene, 290 nm for benzo(b)fluoranthene and benzo(a)pyrene; wavelength 380 nm for benzo(a)anthracene and chrysene and 430 nm for benzo(b)fluoaranthene and benzo(a)pyrene). PAH Calibration Mix, CRM47940 (Supelco Analytical, Bellefonte, PA, USA) was used for the calibration. The limit of quantification was 0.25–0.29 μg/kg, the repeatability of the method was 10% and yield 65–95%.

2.3. Statistical Analysis

The results of chemical composition were evaluated (mean ± s. d.) in the program Microsoft Office Excel 2007 (Microsoft Corp., Redmond, WA, USA). Statistically significant differences of biogenic amines (BA) spectrum were performed at levels of $\alpha = 0.05$ ($p < 0.05$) using the UNISTAT 6.0 (Unistat® Ltd, London, UK) statistical package (multiple comparison, Tukey's HSD test).

3. Results and Discussion

Fresh fishery products are among the most perishable food commodities Their quality is influenced by a number of factors (the origin-wild/farmed, water temperature and level of environmental pollution, compliance with veterinary and hygienic standards during hunting and after capture, species and health and nutritional status/age/sex/phase of sexual cycle/size/weight of fish, way of treatment-gutting/cutting, initial microbiological contamination, keeping the cold chain) which can vary in time and therefore consequently the quality can vary significantly from batch to batch [15]. Freezing is an excellent way to extend shelf life of fish meat during long-term transport or frozen storage [16]. Offering thawed fishery products is a common way of selling fish in landlocked countries [17]. On the other hand, cold chain breakage can cause potentially serious alimentary intoxication, such as histamine content increment. Therefore, growth and activity of histamine-producing bacteria can become dangerous especially in fish which muscle tissue contain a high concentration (about 10,000 mg·kg^{-1}) of free histidine in muscle tissue as tuna fillets. Though, salmon has a low concentration (about 100 to 200 mg·kg^{-1}) of free histidine [18]. Especially, *Photobacterium phosphoreum* and psychrotolerant bacteria similar to *Morganella morganii* are known to be present in fresh fish tissue and form histamine at low temperatures (under 5 °C) [19]. Appearance freshness is often unlawfully restored by an injection or immersion of fillets in the nitrites solution to change the dark red or brown colour (visually less fresh meat) to red pigmentation. Meat after this application looks fresh but histamine levels can be high. Also, inappropriate practices of freezing with subsequent illegal treatment are often used (tuna fish originally of canning grade can be illegally sold as sushi grade of tuna). According to the regulation (EC) no 853/2004 [see in Appendix A2], fishery products shall be frozen below −18 °C, for canning industry; unprocessed fish initially is tolerated to be frozen in brine at −9 °C. Subsequent illegal treatment of tuna fillets and application of nitrites/nitrates (e.g., salt, additives or vegetable extracts containing high level of nitrites) or using of gas carbon monooxide (CO) is not authorized according to the regulation (EC) no 1333/2008 [see in Appendix A7]. Approximately 25,000 tons of tuna per year undergo this treatment [20].

Besides the fact that consumers may be deceived by the quality of fillets, their health can also be compromised. The high level of histamine can cause allergic syndrome [21], nitrites may lead to formation of nitrosamines that have carcinogenic effects [22]. As we wrote in the Materials and Methods section, fresh tuna fillets purchased for the fifth replicate of the experiment, could be hypothetically treated by any of the above illegal practices. Fresh/thawed samples were stored at $+2 \pm 2$ °C; the samples contained in the 1st day of 781/195, 4th day of 1851/1542 and 8th day of 1717/2914 mg·kg^{-1} of histamine in meat. This fifth fillet was therefore excluded from the further experimentation. However, we have shown that even frozen tuna fillet can become a serious threat to human health due to the increment of BA.

3.1. Trends in Biogenic Amines Formation Caused by Severe Breakage of the Cold Chain during the Transport to Households of Fish Raw Material after Purchase

Significant qualitative and quantitative variability in the observed BA was found between sample groups (A, B, C, D) depending on the sampling day (Table 2) as well as between the individual sampling days within one particular group (Table 3).

Regarding the criteria for foodstuffs (Regulation (EC) No 2073/2005, Chap. I, Point 1.26 [see in Appendix A1]), histamine content was very low throughout the experiment, with one exception (C/8 day) which will be further commented (Table 2). The fresh sample of tuna (A) contained the highest ($p < 0.05$) histamine content in the samples measured after 4 days (Table 3).

The fresh samples (group B) that were exposed to 35 °C/6 h and subsequently cold-stored at $+2 \pm 2$ °C during 8 days had almost the same histamine contents during the storage period (Table 3). Distinctly higher histamine contents were found in C samples (thawed/subsequently cold-stored at $+2 \pm 2$ °C), after 8th day of storage (the mean of four different batches: 272.05 ± 217.83 mg·kg^{-1}). The noticeably high value of the standard deviation (s. d.) draws attention to possible differences in quality among purchased lots of fillets; due to these differences the results of thawed fillets (C) obtained for the 8th day of sampling are presented in Table 4. Based on the partial results for each of the four lots, it can be concluded that the three lots were probably more contaminated with microorganisms capable to decarboxylation activity. Due to their activities, histamine content of three tuna lots (8 days of storage) increased to 354.74 ± 185.67 mg·kg^{-1}; PUT, CAD and TYR contents were significantly ($p < 0.05$) higher and caused five times higher total biogenic amine content (563.32 ± 252.61 mg·kg^{-1}) for these three tuna lots compared to the fourth one (Table 4).

In the fourth lot, the HIS content (23.98 ± 0.93 mg·kg^{-1}) was very low, but compared to the other three lots, the fillet contained significantly more SPD, though the BA total was low overall (108.67 ± 1.67 mg·kg^{-1}) (Table 4). In the thawed samples of group D tuna, that were experimentally exposed to temperatures of 35 °C/6 h and subsequently cold-stored at $+2 \pm 2$ °C during 8 days, the HIS content remained virtually unchanged between 1 and 4 days, and we found only a significant reduction in the HIS content on day 8 (Table 3). Based on higher standard deviation (s. d.) for PUT, CAD, TYR, SPD (Table 2), differences in quality were observed for group D tuna batches (similar to the batches of group C tuna). Three batches were found to have significantly ($p < 0.05$) higher amounts of CAD. Consequently, the total BA content in three batches of fillets was approximately two times higher ($p < 0.05$) in comparison to the fourth fillet (Table 4).

The total BA contents of all sample groups were less than 100 mg·kg^{-1}, except for samples from group C and D/8 days (Table 2, BA sum). Each of studied sample groups (A, B, C, D) showed a different trend depending on the storage time (Table 3). BA contents in fresh (A) and thawed samples (C) without cold chain break taken on the day 1 were very low ($p > 0.05$) (Table 2).

Table 2. Qualitative and quantitative biogenic amine spectrum (in mg·kg^{-1}) for groups (A, B, C, D) depending on the sampling day.

Day	Sample	Tryptamine	2-Phenyl-Ethylamine	Putrescine	Cadaverine	Histamine	Tyramine	Spermidine	Spermine	Sum of Bas
1st	A	0.03 ± 0.05 a	1.05 ± 0.68 a	3.21 ± 1.77 ab	2.99 ± 1.35 ab	2.00 ± 1.40 a	10.21 ± 5.31 a	11.08 ± 8.08 a	0.75 ± 0.44 a	31.31 ± 13.74 a
	B	0.32 ± 0.57 a	2.15 ± 0.99 bc	5.06 ± 6.27 b	4.25 ± 2.24 ab	3.31 ± 3.04 ab	19.31 ± 20.84 a	16.82 ± 3.82 ab	32.45 ± 20.54 b	83.66 ± 38.47 b
	C	0.27 ± 0.23 a	-.27 ± 1.23 ab	0.99 ± 0.41 a	1.48 ± 0.31 a	3.53 ± 1.15 ab	6.71 ± 6.87 a	19.49 ± 6.15 bc	3.16 ± 1.66 a	36.90 ± 11.68 a
	D	0.07 ± 0.05 a	2.61 ± 0.64 c	2.74 ± 1.29 ab	7.23 ± 7.40 b	6.06 ± 3.62 b	10.77 ± 7.34 a	24.10 ± 1.62 c	2.20 ± 0.38 a	55.78 ± 14.66 a
Statistical significance						$p < 0.05$				
4th	A	1.67 ± 1.27 b	3.03 ± 1.45 b	4.29 ± 3.81 ab	6.56 ± 4.87 a	18.55 ± 16.46 b	23.31 ± 10.57 b	22.58 ± 5.69 bc	1.06 ± 1.92 a	81.05 ± 30.95 bc
	B	0.97 ± 1.55 ab	0.98 ± 1.12 a	1.39 ± 0.89 a	5.85 ± 4.71 a	3.73 ± 2.57 a	7.87 ± 4.46 a	23.67 ± 10.86 c	4.65 ± 4.58 b	49.12 ± 23.46 ab
	C	0.09 ± 0.05 a	0.58 ± 0.38 a	3.79 ± 0.62 ab	4.28 ± 3.19 a	3.45 ± 0.67 a	11.77 ± 4.58 a	7.32 ± 3.07 a	1.30 ± 0.22 a	32.59 ± 5.33 a
	D	0.34 ± 0.44 a	2.38 ± 1.26 b	8.10 ± 7.37 b	28.11 ± 28.35 b	13.59 ± 12.59 ab	28.36 ± 15.46 b	15.55 ± 3.85 b	3.39 ± 0.99 ab	99.82 ± 61.65 c
Statistical significance						$p < 0.05$				
8th	A	2.31 ± 1.68 b	1.35 ± 0.46 b	2.87 ± 1.29 a	4.60 ± 2.95 a	6.11 ± 3.95 a	4.95 ± 1.62 a	7.84 ± 2.35 a	21.34 ± 3.31 b	51.37 ± 9.29 a
	B	0.40 ± 0.06 a	0.31 ± 0.09 a	5.23 ± 4.10 a	5.05 ± 3.58 a	3.84 ± 0.38 a	6.66 ± 3.60 a	15.39 ± 11.04 a	4.29 ± 5.44 a	41.16 ± 20.47 a
	C	2.77 ± 2.87 b	C.74 ± 0.46 ab	22.93 ± 15.04 b	75.91 ± 52.06 b	272.05 ± 217.83 b	61.69 ± 20.53 b	12.08 ± 6.69 a	1.50 ± 0.57 a	449.66 ± 297.81 b
	D	0.14 ± 0.16 a	1.41 ± 1.61 b	17.66 ± 12.51 b	45.96 ± 32.37 b	3.11 ± 0.94 a	46.12 ± 19.72 b	16.56 ± 11.60 a	1.79 ± 0.79 a	132.74 ± 56.44 a
Statistical significance						$p < 0.05$				

Statistically significant differences ($p < 0.05$) between values for each day of storage (lower cases "a", "b", "c" in indexes in columns).

Table 3. Statistically significant differences ($p < 0.05$) for a particular biogenic amine (lower case "a", "b" in columns) depending on the sampling day and between biogenic amines (capital letters "A" to "D" in italic in rows) for the sampling day are given.

Day	Sample	Tryptamine	2-Phenylethylamine	Putrescine	Cadaverine	Histamine	Tyramine	Spermidine	Spermine	Sum of Bas
1st	A	a *A*	a *AB*	a *BCD*	a *BCD*	a *BC*	a *D*	a *CD*	a *AB*	a
4th		b *A*	b *AB*	a *AB*	b *ABC*	b *BC*	b *C*	b *C*	a *A*	b
8th		b *AB*	a *A*	a *AB*	ab *ABC*	a *BCD*	a *BC*	a *CD*	b *D*	a
1st	B	a *A*	b *AB*	a *ABC*	a *BCD*	a *ABC*	b *CD*	a *D*	b *D*	b
4th		a *A*	a *A*	a *AB*	a *BCD*	a *ABC*	ab *CD*	a *D*	b *D*	a
8th		a *AB*	a *A*	a *CD*	a *CD*	a *CD*	a *CD*	a *D*	a *BC*	a
1st	C	a *A*	a *AB*	a *ABC*	a *ABC*	a *CD*	a *BCD*	b *D*	b *BCD*	a
4th		a *A*	a *A*	a *BC*	a *BC*	a *BC*	a *C*	a *C*	a *AB*	a
8th		b *A*	a *A*	b *BCD*	b *CD*	b *D*	b *CD*	a *ABC*	a *AB*	b
1st	D	a *A*	a *B*	a *AB*	a *B*	ab *B*	a *BC*	b *C*	a *AB*	a
4th		a *A*	a *AB*	a *BC*	ab *CD*	b *BCD*	b *D*	a *CD*	b *ABC*	ab
8th		a *A*	a *A*	b *CD*	b *D*	a *ABC*	c *D*	a *BCD*	a *AB*	b

Table 4. Qualitative and quantitative BA spectrum (in mg·kg^{-1}) for three batches and for one batch for groups C and D/8th day of sampling in a detailed view.

Day/Sample	Tryptamine	2-Phenyl-Ethylamine	Putrescine	Cadaverine	Histamine	Tyramine	Spermidine	Spermine	Sum of Bas
3/C	3.67 ± 2.77 ^{aA}	0.83 ± 0.51 ^{aA}	27.96 ± 14.04 ^{bA}	97.22 ± 41.03 ^{bA}	354.74 ± 185.67 ^{bB}	68.68 ± 18.96 ^{bA}	8.64 ± 2.91 ^{aA}	1.58 ± 0.64 ^{aA}	563.32 ± 252.61 ^{b}
1/C	0.05 ± 0.00 ^{aA}	0.46 ± 0.01 ^{aAB}	7.83 ± 0.15 ^{aC}	11.97 ± 0.03 ^{aD}	23.98 ± 0.93 ^{aF}	40.71 ± 0.34 ^{aG}	22.39 ± 0.10 ^{bE}	1.28 ± 0.15 ^{aB}	108.67 ± 1.67 ^{a}
S. s.	*	*	$p < 0.05$	$p < 0.05$	$p < 0.05$	$p < 0.05$	$p < 0.05$	*	$p < 0.05$
3/D	0.16 ± 0.18 ^{aA}	1.67 ± 1.80 ^{aAB}	20.30 ± 13.56 ^{aB}	58.93 ± 26.15 ^{bC}	3.36 ± 0.97 ^{aAB}	52.99 ± 17.96 ^{aC}	12.89 ± 11.15 ^{aAB}	1.73 ± 0.91 ^{aAB}	152.04 ± 52.01 ^{b}
1/D	0.06 ± 0.01 ^{aA}	0.62 ± 0.02 ^{aA}	9.75 ± 0.35 ^{aD}	7.04 ± 0.19 ^{aC}	2.36 ± 0.10 ^{aB}	25.51 ± 0.59 ^{aE}	27.59 ± 0.25 ^{aF}	1.94 ± 0.05 ^{aB}	74.87 ± 1.12 ^{a}
S. s.	*	*	$p < 0.05$	$p < 0.05$	$p < 0.05$	$p < 0.05$	*	*	$p < 0.05$

Statistically significant differences ($p < 0.05$) for a particular biogenic amine (lower case "a", "b" in columns) and between biogenic amines (capital letters "A" to "G" in italic in rows) depending of number of batches are given. * values are without statistically significant differences (S. s.).

The BA content in fresh samples (A) from day 4 was significantly higher compared to day 1, followed by a significant ($p < 0.05$) decrease in BA content on the day 8 (Table 3). Thawed samples (C) contained very low levels of BA on day 1 and day 4, while BA samples were significantly ($p < 0.05$) 13 times higher on day 8. In the group of fresh (B) and thawed samples (D), where the cold chain was experimentally broken, the BA formation dynamics was more uniform, but the trend was the opposite. Overall, the BA content of Group B samples gradually decreased (significant differences were observed between days 1 and 4, 8); in contrast, in samples of group D, BA content increased gradually (significant differences were found between days 1 and 8) (Table 3).

The spectrum of BA was formed in larger quantities (above 10 mg·kg^{-1}) mainly TYR and SPD and in isolated cases PUT (C and D/8th day), CAD (D/4th and 8th day). HIS contents were already discussed (A and D/4th day, C/8th day), same as SPR (B and D/1st day, D/4th and 8th day). TRP and 2-PHE levels were very low and oscillated between 0 and 3.0 mg·kg^{-1} in individual samples. Statistically significant ($p < 0.05$) differences between the biogenic amine values are given for each group and each day of the storage in Table 3 (capital letters "A" to "D" in rows).

3.2. Testing of Use of a Handheld Gastronomic Unit as an Alternative Method of Smoking Fish with Cold Smoke in the Household with Regard to a Possible Increase in Polycyclic Aromatic Hydrocarbons (PAH) Content and Whether the Cold Smoked Fish Affects Selected Sensory Parameters of Nigiri Sushi Meal Prepared by Consumers at Their Household

The factors affecting food or meal acceptance among consumers have changed rapidly. Sushi is admired by many consumers worldwide due to its appearance and taste [23–25].

A new non-traditional or unusual treatment of meal can increase consumer interest in its taste and, maybe, in its safety. We have found no previous reports of sushi meal containing tuna or salmon smoked with cold smoke for a very short period. It is possible to predict that Super Aladin smoker will be used more frequently in the practice of molecular gastronomy or home-made sushi preparation to imitate its sensory qualities and bring it closer to that of Philadelphia rolls, which is sometimes prepared with salmon smoked with cold or hot smoke.

Appearance, touch, odor, texture and taste represent the sensory properties of foods/meals. Sensory properties are one of the main factors influencing consumers' acceptance and purchase of meals. Health consciousness (including lowering sodium content, the presence of biogenic amines or smoked products) is another factor that can have significant influence on meal acceptance. Certain sushi ingredients (such as vinegar, wasabi and sugar) provide specific sensory properties to this meal.

The sensory properties of prepared sushi samples, smoked and not smoked, are shown in Table 5. Higher values estimated by panelists for juiciness and consistency indicate worse evaluation of these sensory properties. Bigger values for bitterness are emphasizing savoury intensity. Juiciness of sushi prepared without salt was evaluated with higher values then the rest of sushi samples, though statistical significance was not observed ($p > 0.05$).

Health aspects of certain meal are getting priority over shelf life and nutritional profile [26]. Dealing with issues concerning salt consumption is also important due to the fact that salty foods belong to the group of foods toward which consumers can develop addictive tendencies [27].

Sodium intake according to World Health Organization (WHO) should not exceed 2 g per day or 5 g of natural salt (NaCl) [28]. Worldwide salt intake exceeds this limit and ranges from 9 g/day to 12 g/day, which is equivalent to 3.6 g/day to 4.8 g/day [29].

Published information about cold-smoked tuna is sparse, though cold-smoked tuna processing is similar to that of cold-smoked salmon, in which both processing and product characteristics have been extensively studied [4,18].

Warm or hot smoked seafood is accepted and consumed by consumers due to its unique taste, texture and color. Additionally, due to dehydrating, bactericidal and antioxidant properties, smoking processes increase food shelf life. The problem with smoking of foods is that during the process of smoking considerable amount of polycyclic aromatic hydrocarbons (PAH) can be formed due to incomplete wood combustion. Phenols present in wood smoke belong to desirable molecules, since they positively affect food sensory properties and shelf life, but PAH compounds are undesirable molecules [30].

Table 5. Sensory attributes evaluation for nigiri sushi meal with not smoked and smoked samples of tuna and salmon fillets.

Sushi	Treatment	Salt %	Saltiness	Bitterness	Juiciness	Consistency
tuna	not smoked	0	55.55 ± 20.45	16.25 ± 23.13	32.8 ± 36.15	34.15 ± 36.88
		2		29.6 ± 38.57	16.7 ± 21.72	9.6 ± 8.45
		3		20.00 ± 30.60	27.45 ± 25.43	9.1 ± 10.93
	smoked	0	59.68 ± 17.84	20.09 ± 30.71	26.18 ± 29.02	21.95 ± 26.67
		2		23.45 ± 30.08	18.36 ± 20.27	9.64 ± 6.47
		3		17.91 ± 24.48	25.00 ± 23.34	14.45 ± 15.05
salmon	not smoked	0	53.9 ± 23.68	14.6 ± 10.87	18.25 ± 6.88	18.1 ± 26.23
		2		11.25 ± 9.58	13.55 ± 7.12	9.2 ± 5.71
		3		18.1 ± 14.59	20.25 ± 7.38	11.4 ± 5.21
	smoked	0	48.22 ± 20.33	20.80 ± 17.32	21.25 ± 18.92	19.10 ± 18.58
		2		18.35 ± 19.50	12.10 ± 6.48	7.85 ± 4.54
		3		17.60 ± 12.32	22.00 ± 13.19	16.60 ± 13.51

Saltiness (2% salt addition) was evaluated by respondents' comparison with sushi samples prepared without salt (0 point) and with 3% salt content (100 points) added to rice.

Table 6 shows contents of polycyclic aromatic hydrocarbons (PAH) in the samples of smoked and not smoked tuna and salmon. Tuna samples contained more PAH (4.22 µg·kg^{-1}) than salmon samples (1.74 µg·kg^{-1}).

Table 6. The content of polycyclic aromatic hydrocarbons (PAH) in µg·kg^{-1}.

Fish	Treatment	B(a)a	Chr	B(b)f	B(a)p	Sum of Pah
tuna	not smoked	<0.24	<0.28	<0.28	<0.27	*
	smoked	1.84	1.10	0.52	0.76	4.22
salmon	not smoked	<0.24	<0.28	<0.28	<0.27	*
	smoked	<0.24	0.77	0.37	0.60	1.74

B(a)A—benzo[a]anthracene; CHr—chrysene; B(b)F—benzo[b]fluoranthene; B(a)P—benzo[a]pyrene; * values under limit of detection (LOD).

Despite alarming content findings of B(a)A (1.84 µg·kg^{-1}) and CHr (1.10 µg·kg^{-1}) in smoked tuna, the content of B(a)P and sum of PAH in our samples were lower than the maximum levels written in the regulation (EC) no. 1881/2006, Annex, Section 6 (B(a)P [see in Appendix A8]: 2 µg·kg^{-1}, sum of PAH: 12 µg·kg^{-1}).

The existence of several factors affecting PAH content in smoked fish has been scientifically confirmed. PAHs are produced during combustion processes and smoke formation. The type of used matrix (wood), combustion temperature, smoke generation technique, filtration, temperature and smoke composition. Following factors are also influencing PAH formation: size, treatment and chemical composition of smoked fish. Regarding the effect on food (fish), the diffusion intensity of PAH below its surface into the muscle is relatively low. This fact that PAHs are mainly concentrated in the surface layers means that their content in the food is determined by food surface, same as total weight of the food. Surface/weight ratio is also probably responsible for higher levels of PAH in smoked tuna. Certainly, shape of smoked food (thickness and weight) are influencing PAH levels too. The content of PAHs is also associated with the fat content of food. Reference [31] also found that B(a)A content can differ significantly in dependence of seafood species.

Higher fat content in salmon samples influences higher PAH contents than in tuna samples [4,32,33]. The findings of these authors are not in agreement with the results of our chemical composition analysis

(Table 7). However, the fat content, which we determined in our samples, did correspond to published values [34]. Fluctuations in filtering capacity of the smoker could also influence PAH amounts, though the processes of sample smoking took place in the laboratory under the same conditions (time of smoke/cover of samples with glass lid).

Table 7. Chemical composition of sashimi tuna fillets and salmon fillets with skin (in %).

Samples	Protein	Fat	Dry Matter
tuna	29.02 ± 0.06	0.83 ± 0.01	30.09 ± 0.70
salmon	20.46 ± 0.98	23.46 ± 0.96	43.63 ± 0.00

Beside the antibacterial and antioxidant properties of smoking that are connected with phenolic compounds present in wood smoke, from our results a negative impact of seafood smoking represented as an increase of PAH compounds can be also seen. The importance of PAH level control in food is important, since these compounds are carcinogenic, mutagenic and endocrine disrupting. PAH compounds (there are more than 660 identified PAH compounds) are produced in wood smoke during pyrolysis (depolymerisation) of lignin and then condensation of the lignin components in lignocelluloses at temperature above 350 °C [35,36]. Aside from PAH increment, smoking changes color of food due to Maillard reaction (change coloration occurs due to the reaction of carbonyl groups in smoke with amino groups present on the surface of smoked food [36]. Phenols from wood smoke enter seafood by diffusion and capillary action, changing its flavor, color and prolonging shelf life [30]. PAH compounds in canned smoked tuna were 17.67 $\mu g \cdot kg^{-1}$ [37].

4. Conclusions

The main finding of the research is highlighting a food safety issue that was found by the experiments with a gastronomic smoking unit due to increased amounts of polycyclic aromatic hydrocarbons. The Super Aladin smoker unit is a patented product, but the manufacturer does not comment on the PAH hazards associated with food safety in the user documentation. The producers of this smoker units should at least include in the manuals the maximum smoking time depending on the type of food and its fatness. In this way they would alert users to the possible danger, which is the adhesion of harmful PAH on smoked food. The minimum or maximum exposure time of food to smoke is not specified or restricted for specific food types. For prolonged smoking (up to 24 h!), the manufacturer recommends intermittent repeated batches of smoke under the hatch at multiple time intervals as required (optical smoke density control under the hatch). In addition to commercial hardwood chips (oak, beech, Jack Daniels), users can use other alternative matrices including aromatic oils for the development of smoke. Due to the lipophilic nature of PAH, these substances could hypothetically be added to the smoke and subsequently increase PAH level to even more harmful concentrations. Certainly, that further experiments with household gastronomic smokers, such as, Super Aladin smoker, will probably give broader and more precise picture about all possibilities and issues concerning these types of devices.

In the case of the experiment aimed at monitoring the content of biogenic amines in tuna samples, we found much more favorable results than expected. Serious interruption of thawed tuna samples cold chain after purchase did not result in an increase of biogenic amines to levels that could represent a health risk for consumers. Manufacturers inform consumers that they should store purchased fish at 0–2 °C and consume it within 2 days. According to the results obtained in our experiment, tuna samples could be considered suitable for consumption after 4 days in terms of BA content and even up to 8 days (except for thawed samples without cold chain interruption) after purchase. However, we cannot recommend this practice due to the possibility that purchased tuna could have higher histamine content developed before purchase. The risk of intoxication with histamine becomes more realistic with each new day of storage. Laboratory examinations of fish species associated with high histamine content in muscle should therefore be a normal part of quality controls by sellers so that they do not have to passively rely on written statements from fish suppliers regarding histamine content in commercial or veterinary evidences of their origin.

Author Contributions: H.B. conceived and designed the experiments. H.B. and D.D. performed the basic chemical and sensory analyses and processed all laboratory data including statistical analysis, A.H. carried out the determination of the polycyclic aromatic hydrocarbons include software service and validation of this methodology. I.D. and P.K. carried out the determination of the biogenic amine spectrum include software service and validation of this methodology. The first draft of the manuscript was prepared by H.B. and D.D. and it was revised and substantially improved by D.D. and H.B.

Appendix A

Appendix A.1 Legislation and Procedures

1. Commission Regulation (EC) No 2073/2005 of 15 November 2005 on microbiological criteria for foodstuffs. Available online. https://eur-lex.europa.eu/legal-content/EN/TXT/?qid=1565694175620&uri=CELEX:02005R2073-20190228 (accessed on 28 February 2019).

2. Regulation (EC) No 853/2004 of the European Parliament and of the Council of 29 April 2004 laying down specific hygiene rules for food of animal origin. Available online: https://eur-lex.europa.eu/legal-content/EN/TXT/?qid=1565693935604&uri=CELEX:02004R0853-20190101 (accessed on 1 January 2019).

3. International Organization for Standardization ISO 8589:2008 Sensory analysis— General guidance for the design of test rooms

4. International Organization for Standardization ISO 937:1978 Meat and meat products— Determination of nitrogen content (Reference method)

5. International Organization for Standardization ISO 1443:1973 Meat and meat products— Determination of total fat content. (Reference method)

6. International Organization for Standardization ISO 1442:1997 Meat and meat products— Determination of moisture content (Reference method)

7. Regulation (EC) No 1333/2008 of the European Parliament and of the Council of 16 December 2008 on food additives. Available online: https://eur-lex.europa.eu/legal-content/EN/TXT/?qid=1565693786967&uri=CELEX:02008R1333-20190618 (accessed on 18 July 2019).

8. Commission Regulation (EC) No 1881/2006 of 19 December 2006 setting maximum levels for certain contaminants in foodstuffs. https://eur-lex.europa.eu/legal-content/EN/TXT/?qid=1565694263939&uri=CELEX:02006R1881-20180319 (accessed on 19 March 2018).

References

1. Fusco, V.; den Besten, H.M.W.; Logrieco, A.F.; Rodriguez, F.; Skandamis, P.N.; Stessl, B.; Teixeira, P. Food safety aspects on ethnic foods: Toxicological and microbial risks. *Curr. Opin. Food Sci.* **2015**, *6*, 24–32. [CrossRef]
2. Feng, C.H. Tale of sushi: History and regulations. *Compr. Rev. Food Sci. Food Saf.* **2012**, *11*, 205–220. [CrossRef]
3. Puah, S.M.; Tan, J.A.M.A.; Chew, C.H.; Chua, K.H. Diverse Profiles of Biofilm and Adhesion Genes in Staphylococcus Aureus Food Strains Isolated from Sushi and Sashimi. *J. Food Sci.* **2018**, *83*, 2337–2342. [CrossRef] [PubMed]
4. Storelli, M.M.; Stuffler, R.G.; Marcotrigiano, G.O. Polycyclic aromatic hydrocarbons, polychlorinated biphenyls, chlorinated pesticides (DDTs), hexachlorocyclohexane, and hexachlorobenzene residues in smoked seafood. *J. Food Prot.* **2003**, *66*, 1095–1099. [CrossRef] [PubMed]
5. Rahmani, J.; Miri, A.; Mohseni-Bandpei, A.; Fakhri, Y.; Bjørklund, G.; Keramati, H.; Moradi, B.; Amanidaz, N.; Shariatifar, N.; Khaneghah, A.M. Contamination and Prevalence of Histamine in Canned Tuna from Iran: A Systematic Review, Meta-Analysis, and Health Risk Assessment. *J. Food Prot.* **2018**, *81*, 2019–2027. [CrossRef]

6. EFSA (European Food Safety Authority) Panel on Biological Hazards (BIOHAZ). Scientific Opinion on risk assessment of parasites in fishery products. *EFSA J.* **2010**, *8*, 1543. [CrossRef]

7. Fukita, Y.; Asaki, T.; Katakura, Y. Some like It Raw: An Unwanted Result of a Sushi Meal. *Gastroenterology* **2014**, *146*, E8–E9. [CrossRef] [PubMed]

8. Škaljac, S.; Jokanović, M.; Tomović, V.; Ivić, M.; Tasić, T.; Ikonić, P.; Sojic, B.; Dzinic, N.; Petrović, L. Influence of smoking in traditional and industrial conditions on colour and content of polycyclic aromatic hydrocarbons in dry fermented sausage "Petrovská klobása". *LWT-Food Sci. Technol.* **2018**, *87*, 158–162. [CrossRef]

9. Josewin, S.W.; Ghate, V.; Kim, M.J.; Yuk, H.G. Antibacterial effect of 460 nm light-emitting diode in combination with riboflavin against *Listeria monocytogenes* on smoked salmon. *Food Control.* **2018**, *84*, 354–361. [CrossRef]

10. De Jonge, J.; van Trijp, H.; Renes, R.J.; Frewer, L. Understanding consumer confidence in the safety of food: Its two-dimensional structure and determinants. *Risk Anal.* **2007**, *27*, 729–740. [CrossRef] [PubMed]

11. Verbeke, W.; Sioen, I.; Pieniak, Z.; Van Camp, J.; De Henauw, S. Consumer perception versus scientific evidence about health benefits and safety risks from fish consumption. *Public Health Nutr.* **2005**, *8*, 422–429. [CrossRef] [PubMed]

12. Fooladi, E.; Hopia, A.; Lasa, D.; Arboleya, J.C. Chefs and researchers: Culinary practitioners' views on interaction between gastronomy and sciences. *Int. J. Gastron. Food Sci.* **2019**, *15*, 6–14. [CrossRef]

13. Geczi, G.; Korzenszky, P.; Szakmar, K. Cold chain interruption by consumers significantly reduces shelf life of vacuum-packed pork ham slices. *Acta Aliment. Hung.* **2017**, *46*, 508–516. [CrossRef]

14. Kulawik, P.; Dordevic, D.; Gambuś, F.; Szczurowska, K.; Zając, M. Heavy metal contamination, microbiological spoilage and biogenic amine content in sushi available on the Polish market. *J. Sci. Food Agric.* **2018**, *98*, 2809–2815. [CrossRef] [PubMed]

15. Sharifian, S.; Alizadeh, E.; Mortazavi, M.S.; Moghadam, M.S. Effects of refrigerated storage on the microstructure and quality of Grouper (*Epinephelus coioides*) fillets. *J. Food Sci. Technol. Mys.* **2014**, *51*, 929–935. [CrossRef] [PubMed]

16. Duflos, G.; Le Fur, B.; Mulak, V.; Becel, P.; Malle, P. Comparison of methods of differentiating between fresh and frozen-thawed fish or fillets. *J. Sci. Food Agric.* **2002**, *82*, 1341–1345. [CrossRef]

17. Reis, M.M.; Martinez, E.; Saitua, E.; Rodriguez, R.; Perez, I.; Olabarrieta, I. Non-invasive differentiation between fresh and frozen/thawed tuna fillets using near infrared spectroscopy (Vis-NIRS). *LWT-Food Sci. Technol.* **2017**, *78*, 129–137. [CrossRef]

18. Emborg, J.; Laursen, B.G.; Dalgaard, P. Significant histamine formation in tuna (*Thunnus albacares*) at 2 degrees C - effect of vacuum- and modified atmosphere-packaging on psychrotolerant bacteria. *Int. J. Food Microbiol.* **2005**, *101*, 263–279. [CrossRef]

19. Holland, J. Brussels warns against the illegal treatment of tuna with vegetable extracts. Sea Food Source News 9. 2016. Available online: https://www.seafoodsource.com/news/food-safety-health/brussels-warns-against-the-illegal-treatment-of-tuna-with-vegetable-extracts (accessed on 7 February 2019).

20. European Commission. Food fraud network EU-ccoordinated case: Illegal treatment of Tuna: From canning grade to Sushi grade. 2017. Available online: https://ec.europa.eu/food/sites/food/files/safety/docs/food-fraud_succ-coop_tuna.pdf (accessed on 7 February 2019).

21. Bhangare, R.C.; Sahu, S.K.; Pandit, G.G. Nitrosamines in seafood and study on the effects of storage in refrigerator. *J. Food Sci. Tech. Mys.* **2015**, *52*, 507–513. [CrossRef]

22. De Gennaro, L.; Brunetti, N.D.; Locuratolo, N.; Ruggiero, M.; Resta, M.; Diaferia, G.; Rana, M.; Caldarola, P. Kounis syndrome following canned tuna fish ingestion. *Acta Clin. Belg.* **2017**, *72*, 142–145. [CrossRef]

23. Drewnowski, A.; Moskowitz, H.R. Sensory characteristics of foods: New evaluation techniques. *Am. J. Clin. Nutr.* **1985**, *42*, 924–931. [CrossRef] [PubMed]

24. Desmet, P.M.A.; Schifferstein, H.N.J. Sources of positive and negative emotions in food experience. *Appetite* **2008**, *50*, 290–301. [CrossRef] [PubMed]

25. Dordevic, D.; Buchtova, H. Factors influencing sushi meal as representative of non-traditional meal: Consumption among Czech consumers. *Acta Alim. Hung.* **2017**, *46*, 76–83. [CrossRef]

26. Fellendorf, S.; O'Sullivan, M.G.; Kerry, J.P. Effect of different salt and fat levels on the physicochemical properties and sensory quality of black pudding. *Food Sci. Nutr.* **2017**, *5*, 273–284. [CrossRef] [PubMed]

27. Mies, G.W.; Treur, J.L.; Larsen, J.K.; Halberstadt, J.; Pasman, J.A.; Vink, J.M. Prevalence of food addiction in a large sample of adolescents and its association with addictive substances. *Appetite* **2017**, *118*, 97–105. [CrossRef] [PubMed]

28. World Health Organisation (WHO). *Guideline: Sodium Intake for Adults and Children*; WHO: Geneva, Switzerland, 2012. Available online: https://www.ncbi.nlm.nih.gov/books/NBK133309/ (accessed on 22 June 2018).

29. Rust, P.; Ekmekcioglu, C. Impact of salt intake on the pathogenesis and treatment of hypertension. In *Advances in Experimental Medicine and Biology*, 1st ed.; Islam, M.S., Ed.; Springer: Cham, Switzerland, 2017; Volume 956, pp. 61–84.

30. Remy, C.C.; Fleury, M.; Beauchêne, J.; Rivier, M.; Goli, T. Analysis of PAH residues and amounts of phenols in fish smoked with woods traditionally used in French Guiana. *J. Ethnobiol.* **2016**, *36*, 312–325. [CrossRef]

31. Swastawati, F.; Surti, T.; Agustini, T.W.; Riyadi, P.H. Benzo(α)pyrene potential analysis on smoked fish (Case study: Traditional method and smoking kiln). *KnE Life Sci.* **2015**, *1*, 156–161. [CrossRef]

32. Afolabi, O.A.; Adesula, E.A.; Oke, O.L. Polynuclear aromatic hydrocarbons in some Nigerian preserved freshwater fish species. *J. Agric. Food Chem.* **1983**, *31*, 1083–1090. [CrossRef]

33. Lawrence, J.F.; Weber, D.F. Determination of polycyclic aromatic hydrocarbons in some Canadian commercial fish, shellfish and meat products by liquid chromatography with confirmation by capillary gas chromatography-mass spectrometry. *J. Agric. Food Chem.* **1984**, *32*, 789–794. [CrossRef]

34. Anonymous Salmon vs Tuna. Diffen.com. Diffen LLC, n.d. 2019. Available online: https://www.diffen.com/difference/Salmon_vs_Tuna (accessed on 19 February 2019).

35. Essumang, D.K.; Dodoo, D.K.; Adjei, J.K. Effect of smoke generation sources and smoke curing duration on the levels of polycyclic aromatic hydrocarbon (PAH) in different suites of fish. *Food Chem. Toxicol.* **2013**, *58*, 86–94. [CrossRef]

36. Nithin, C.T.; Ananthanarayanan, T.R.; Yathavamoorthi, R.; Bindu, J.; Joshy, C.G.; Gopal, T.S. Physico-chemical Changes in Liquid Smoke Flavoured Yellowfin Tuna (*Thunnus albacares*) Sausage during Chilled Storage. *Agric. Res.* **2015**, *4*, 420–427. [CrossRef]

37. Novakov, N.J.; Mihaljev, Ž.A.; Kartalović, B.D.; Blagojević, B.J.; Petrović, J.M.; Ćirković, M.A.; Rogan, D.R. Heavy metals and PAHs in canned fish supplies on the Serbian market. *Food Addit. Contam. Part B Surveill.* **2017**, *10*, 208–215. [CrossRef] [PubMed]

4

High Hydrostatic Pressure as a Tool to Reduce Formation of Biogenic Amines in Artisanal Spanish Cheeses

Diana Espinosa-Pesqueira, Maria Manuela Hernández-Herrero and Artur X. Roig-Sagués *

CIRTTA-Departament de Ciència Animal i dels Aliments, Universitat Autònoma de Barcelona, Travessera dels Turons S/N, 08193 Barcelona, Spain; diespe@gmail.com (D.E.-P.); manuela.hernandez@uab.cat (M.M.H.-H.)
* Correspondence: arturxavier.roig@uab.cat

Abstract: Two artisanal varieties of cheese made in Spain, one made of ewes' raw milk and the other of goats' raw milk were selected to evaluate the effect of a high hydrostatic pressure (HHP) treatment at 400 MPa during 10 min at 2 °C on the formation of biogenic amines (BA). These conditions were applied at the beginning of the ripening (before the 5th day; HHP1) and in the case of ewes' milk cheeses also after 15th days (HHP15). BA formation was greatly influenced by HHP treatments in both types of cheese. HHP1 treatments significantly reduced the amounts of BA after ripening, being tyramine and putrescine the most affected BA in goats' milk cheeses and tyramine and cadaverine in ewes' milk cheeses. The BA reduction in the HHP1 samples could be explained by the significant decrease in microbiological counts, especially in the LAB, enteroccocci and enterobacteria groups at the beginning of ripening. The proteolysis in these samples was also affected reducing the amount of free amino acids. Although proteolysis in ewes' milk cheeses HHP15 was similar than in control samples a reduction of BA was observed probably because the decrease caused on microbial counts.

Keywords: biogenic amines; cheese; high hydrostatic pressure

---

1. Introduction

Biogenic amines (BA) are basic nitrogenous compounds formed in different foodstuffs due to the microbial decarboxylation of amino acids. The kind and the amount of BA formed depends on the decarboxylase capability of the bacterial strains present, the availability of substrate amino acids and the physicochemical properties of the matrix [1]. Some aromatic BA, such as histamine (HIS), tyramine (TY), β-phenylethylamine (PHE) and tryptamine (TR), have psychoactive and vasoactive properties that may cause food poisoning when present in foodstuffs, while the diamines putrescine (PU) and cadaverine (CA) can boost the toxic action of aromatic BA [2,3]. Outbreaks caused by HIS are frequent, although not always correctly diagnosed or declared. It causes intoxication—called "histamine food poisoning" or "scombroid food poisoning"—since it is frequently associated with consumption of scombroid fish, especially tuna fish and mackerel [4]. After fish, cheese is the most commonly implicated food product associated with histamine poisoning and outbreaks have been associated with cheeses made from both raw and pasteurized milk [5]. The HIS concentrations in cheeses that were implicated in outbreaks ranged between 850 to 1870 mg/kg [6]. Tyramine is another BA associated with food-borne poisoning, being one of the causative agents of the so called "cheese reaction" [7], causing migraine, headache and, in extreme cases, hypertensive breakdown [5,8,9]. The seriousness of any BA food-borne poisoning depends on the ingested dose and the sensitiveness of the consumer (genetic or acquired) [3,10].

During cheese manufacture several factors may contribute to the accumulation of toxic amounts of BA. Good manufacturing practices to minimize the occurrence of BA-producing microorganisms in raw materials, pasteurization of milk or addition of BA-non-producing starter cultures have been suggested as BA risk mitigation options. Although it has been described that milk pasteurization reduces the level of decarboxylase-positive bacteria, later contamination of milk and curd during cheese manufacturing by decarboxylating bacteria and their subsequent growth and metabolic activity during cheese ripening usually results in BA build up [10–13].

High hydrostatic pressure (HHP) is non-thermal processing method used to extend shelf-life of foods. This induces morphological changes and inhibition of enzymes and genetic mechanisms of microorganism [14]. HHP offers the advantage that can be applied after cheese manipulation is over and no further contamination of curd is expected. The effect of HHP treatments has been evaluated in different kind of cheeses made from cows' milk [15], ewes' milk [16] and goats' milk [17,18], being able to eliminate most of the pathogenic bacteria associated with this product but it may also be useful to eliminate decarboxylating bacteria and avoid BA formation during cheese ripening. Calzada et al. [19] evaluated HHP treatments up to 600 MPa at different ripening stages to control the excessive proteolysis and BA formation on blue veined cheese, observing a lower concentration of TY. In another study [20], it was observed that HHP treatments significantly reduced the BA build up when applied on days 21 and 35 of ripening in "Torta del Casar" type cheese. This kind of cheese is made of raw milk and vegetable rennet, which causes a strong proteolytic activity and leads to extensive caseins breakdown in cheese matrix. Nevertheless, much less information was found about the HHP effect on the BA formation in other type of cheeses made of pressed paste.

The main objective of this survey was to evaluate HHP processing to reduce BA formation in two varieties of artisanal Spanish cheese made of raw milk from ewe and goat. Treatments were applied at different stages of ripening and the consequences of these treatments on the BA formation and proteolytic activity were evaluated during the ripening.

2. Materials and Methods

2.1. Cheese Manufacturing

Two types of artisanal ripened cheeses elaborated in Spain were studied in this work, both made with starter culture, enzymatic curd and pressed paste. The first one was produced from goats' raw milk in the region of Catalonia, northeast of Spain, and the second was made from ewes' raw milk in Castilla y León, central Spain. Three independent batches of each type of cheese were produced following the usual manufacturing procedures used by the manufacturers.

2.2. High-Hydrostatic Pressure (HHP) Treatments

HHP treatments were performed at 400 MPa for 10 min at a temperature of 2 °C using an Alstom HHP equipment (Alstom, Nantes, France) with a 2 L pressure chamber. A mixture of alcohol and water (1:9) was used into the chamber. The pressurization rate and depressurization time were 268 MPa/min and 55 s, respectively. Before processing, goats' milk cheese samples were separated in two batches: samples not HHP treated (Control samples) and samples HHP treated before the 5th day of ripening (HHP1). In the case of ewes' milk cheese samples a third batch for samples that were treated after 15 days of ripening (HHP15) was included. In all cases a portion of about 8.0 cm diameter was obtained and shaped to adequate it to the diameter of the cylinder of the HHP equipment and then vacuum packaged. After the HHP treatments, samples were deprived of the plastic bag and kept into a ripening chamber at 14 °C and 88% of relative humidity to continue with the ripening process for 60 days. Analyses of cheeses were performed at the 5th, 15th, 30th, 45th and 60th day of ripening.

2.3. Microbiological Analysis of Cheeses

Ten grams of each sample were homogenized in 90 mL of sterile Buffered Peptone Water (Oxoid, Basingstoke, Hampshire, UK) with a paddle blender (BagMixer, Interscience, France). Counts of

Lactococcus spp. were made on M-17 agar (Oxoid) supplemented with a bacteriological grade lactose solution (5 g/L, Oxoid) and incubated at 30 °C for 48 h; Lactobacilli were determined on Man Rogose Sharpe agar (MRS, Oxoid) and incubated at 30 °C for 48 h; Enterococci were enumerated using KF *Streptococcus* Agar (Oxoid) supplemented with 2,3,5-trifeniltetrazolium chloride solution 1% (Oxoid) and incubated at 37 °C for 48 h; enumeration of *Enterobactericeae* was performed on Violet Red Bile Glucose Agar (VRBG, Oxoid) and counts of *Escherichia coli* was made on the chromogenic selective media Coli ID (BioMérieux, Marcy l'Etoile, France) and incubated at 30 °C for 24 h; *Staphylococcus aureus* counts were determined on Bair Parker Agar supplemented with rabbit plasma fibrinogen (BP-RPF agar, BioMérieux, Marcy L'Etoile, France) and incubated at 37 °C for 24–48 h.

2.4. Assessment of Proteolysis Activity on Cheeses

Water Soluble Extracts (WSE) of cheese were prepared according to the method described by Kuchroo and Fox [21]. From the WSE adjusted at pH 4.6, Water-Soluble Nitrogen (*WSN*) fraction was obtained and determined by the Dumas combustion method [22]. The nitrogen content of the *WSN* fraction was expressed as a percentage of Total Nitrogen (*TN*), described as the Ripening Index (*RI*), according to the formula:

$$RI = (WSN/TN) \times 100 \tag{1}$$

The measurement of the amino group content was determined on WSE by the Trinitrobenzensulphonic Acid method (TNBS) according to the procedure described by Hernández-Herrero et al. [23]. Results were expressed as mg of L-Leucine per g of cheese. The total Free Amino Acids (FAA) content was determined on WSE by the cadmium-ninhydrin method described by Folkertsmaa and Fox [24] and the results expressed as mg of L-Leucine per g of cheese.

2.5. Determination of Biogenic Amines in Cheese

The RP-HPLC method described by Eerola et al. [25] and modified by Roig Sagués et al. [12] was used to determine the BA content in cheese samples. The BA determined were β-phenylethylamine (PHE), tryptamine (TR), putrescine (PU), cadaverine (CA), histamine (HIS), tyrosine (TY), spermidine (SD) and spermine (SM).

2.6. Statistical Analysis

Analysis of variance (ANOVA) was performed on all data from each batch and treatment of goats' and ewes' milk cheeses at different ripening stages. Comparisons of mean values of physicochemical, microbiological and BA were followed by Duncan test with significance level set on $p < 0.05$. Comparisons of mean values of proteolysis indexes were performed using the Student-Newman-Keuls test with the significance level set at $p < 0.05$. All tests were performed with the SPSS for windows (v.15.01) program (SPSS Inc., Chicago, IL, USA).

3. Results and Discussion

3.1. Effect of HHP Treatment on Microbial Counts

Tables 1 and 2 shows the growth of the microbial groups counts along the ripening process in the goats' and ewes' cheeses, respectively, in both control and HHP processed samples.

Table 1. Changes in microbial population (mean values ± standard deviation as $\log_{10}$ CFU/g) of the goats' raw milk cheese samples with and without high hydrostatic pressure (HHP) treatment during ripening.

Microbial Group	Day of Ripening	Control	HHP1
Lactococci	5	9.55 ± 0.27 [a,A]	7.30 ± 0.22 [B]
	15	8.75 ± 0.34 [b,A]	6.76 ±0.49 [B]
	30	8.47 ± 0.26 [b,A]	7.03 ± 0.64 [B]
	45	8.26 ± 0.20 [b,A]	7.19 ± 0.25 [B]
	60	7.54 ± 0.51 [c,A]	7.05 ± 0.17 [A]
Lactobacilli	5	6.54 ± 0.31 [a,A]	3.13 ± 0.26 [a,B]
	15	8.36 ± 0.42 [b,A]	5.23 ± 1.54 [b,B]
	30	8.28 ± 0.15 [b,A]	6.31 ± 0.33 [c,B]
	45	8.18 ± 0.19 [b,A]	6.80 ± 0.50 [c,B]
	60	7.89 ± 0.45 [b,A]	7.03 ± 0.22 [c,A]
Enterococci	5	6.23 ± 0.24 [a,A]	4.44 ± 0.44 [ab,B]
	15	6.10 ± 0.70 [a,A]	3.99 ± 0.24 [ab,B]
	30	6.17 ± 0.47 [a,A]	4.60 ± 0.96 [b,B]
	45	6.17 ± 0.48 [a,A]	3.45 ± 0.34 [a,B]
	60	5.32 ± 0.64 [a,A]	3.85 ± 0.59 [ab,B]
Enterobacteria	5	6.33 ± 0.26 [a,A]	1.92 ± 0.54 [a,B]
	15	3.15 ± 2.75 [b,A]	2.49 ± 0.36 [a,A]
	30	4.54 ± 0.15 [b,A]	0.86 ± 1.48 [a,B]
	45	3.58 ± 0.40 [b,A]	ND [b,B]
	60	2.94 ± 0.54 [b,A]	ND [b,B]
E. coli	5	2.56 ± 0.62 [b,A]	ND [B]
	15	0.90 ± 0.78 [ab,A]	ND [A]
	30	1.60 ± 1.96 [ab,A]	0.83 ± 1.44 [A]
	45	0.77 ± 0.68 [ab,A]	ND [A]
	60	ND [a,A]	ND [A]
S. aureus	5	2.64 ± 0.15 [a,A]	0.43 ± 0.75 [a,B]
	15	1.06 ± 0.96 [b,A]	ND [A]
	30	1.18 ± 0.31 [b,A]	ND [B]
	45	0.77 ± 1.33 [b,A]	0.33 ± 0.58 [B]
	60	ND [b,A]	ND [A]

ND: not detected (<10 CFU/g); means with different superscript small letter differ significant ($p < 0.05$) in the same column for the same parameter and HHP treatment; means with different superscript capital letter differ significant ($p < 0.05$) in the same row for the same parameter and day of ripening; HHP1: HHP treated the 5th day of the ripening.

Table 2. Changes in microbial population (mean values ± standard deviation as $\log_{10}$ CFU/g) of the ewes' raw milk cheese samples with and without HHP treatment during ripening.

Microbial Group	Day of Ripening	Control	HHP1	HHP15
Lactococci	5	9.34 ± 0.33 [a,A]	7.80 ± 0.31 [a,B]	
	15	9.16 ± 0.20 [a,A]	6.87 ± 0.39 [b,B]	7.25 ± 0.22 [a,B]
	30	9.21 ± 0.18 [a,A]	8.04 ± 0.74 [b,B]	7.67 ± 0.14 [ab,B]
	45	8.74 ± 0.20 [ab,A]	8.07 ± 0.61 [b,B]	7.92 ± 0.39 [ab,B]
	60	8.38 ± 0.16 [b,A]	8.32 ± 0.76 [b,A]	7.57 ± 0.32 [b,B]
Lactobacilli	5	5.57 ± 0.35 [a,A]	4.54 ± 0.64 [a,B]	
	15	7.52 ± 1.07 [b,A]	5.60 ± 0.37 [b,B]	6.19 ± 0.50 [a,B]
	30	8.64 ± 0.41 [c,A]	6.61 ± 0.82 [c,B]	6.78 ± 1.00 [ab,B]
	45	8.68 ± 0.12 [c,A]	8.09 ± 0.55 [d,A]	7.82 ± 0.27 [bc,A]
	60	8.15 ± 0.39 [c,A]	8.12 ± 0.51 [d,A]	7.53 ± 0.15 [c,B]

Table 2. *Cont.*

Microbial Group	Day of Ripening	Control	HHP1	HHP15
Enterococci	5	5.59 ± 0.74 [A]	3.16 ± 1.05 [A]	
	15	6.18 ± 2.14 [AB]	3.50 ± 1.13 [C]	3.90 ± 1.11 [B]
	30	6.16 ± 1.18 [A]	1.76 ± 1.53 [B]	3.96 ± 1.32 [A]
	45	4.63 ± 1.04 [A]	2.76 ± 2.26 [B]	3.27 ± 0.48 [A]
	60	4.23 ± 1.40 [A]	2.58 ± 1.93 [B]	3.88 ± 0.93 [A]
Enterobacteria	5	4.65 ± 0.23 [a,A]	ND [B]	
	15	4.46 ± 0.30 [a,A]	0.29 ± 0.58 [B]	ND $\pm$ ND [B]
	30	4.42 ± 1.92 [a,A]	ND [B]	1.22 ± 1.11 [B]
	45	2.87 ± 1.21 [b,A]	ND [B]	0.57 ± 0.98 [B]
	60	2.35 ± 0.96 [b,A]	ND [B]	ND [B]
E. coli	5	1.16 ± 1.01	ND	
	15	1.15 ± 1.37	ND	ND
	30	0.65 ± 0.79	ND	ND
	45	0.64 ± 0.73	ND	0.78 ± 1.35
	60	0.73 ± 0.91	ND	ND
S. aureus	5	2.35 ± 0.58 [a]	1.07 ± 1.35	
	15	1.74 ± 1.19 [ab]	0.50 ± 0.58	0.33 ± 0.58
	30	1.17 ± 0.35 [ab]	0.29 ± 0.58	0.33 ± 0.58
	45	0.61 ± 0.71 [bc]	ND	ND
	60	0.25 ± 0.50 [c]	ND	ND

ND: Not detected (<10 CFU/g); means with different superscript small letter differ significant ($p < 0.05$) in the same column for the same parameter and HHP treatment; means with different superscript capital letter differ significant ($p < 0.05$) in the same row for the same parameter and day of ripening; HHP1: HHP treated the 5th day of the ripening; HHP15: HHP treated after 15 days of ripening.

In control samples (non HHP-treated) of both types of cheese, Lactoccocci was the main microbial group at the beginning of the ripening, showing counts above 8 $\log_{10}$ CFU/g, probably because strains of *Lactococcus lactis* were added as the starter culture, while lactobacilli counts were significantly lower at the beginning of the ripening, although their counts rose as the ripening progressed, achieving similar counts to lactococci. Enterococci counts remained steady at a level about 6 $\log_{10}$ CFU/g at the beginning of the process, showing a slight decrease around the 15th and the 30th day in the goats' and ewes' cheeses, respectively. *S. aureus*, enterobacteria and *E. coli* counts decreased during the ripening until becoming undetectable in most of the samples. These results are in agreement with those reported by other authors in other varieties of cured cheeses [26–30].

HHP treatments reduced significantly the Lactic Acid Bacteria (LAB) counts. In HHP1 treated samples, lactoccocci counts decreased 2.2 $\log_{10}$ CFU/g and 1.5 $\log_{10}$ CFU/g in the goats' and ewes' milk cheeses, respectively. In the case of lactobacilli mean reductions were 3.41 $\log_{10}$ CFU/g and 1.03 $\log_{10}$ CFU/g. However, these counts recovered along the ripening process and no statistical differences were observed at the end of the work with respect to control samples. Ewes' milk cheeses HHP15 showed a reduction of about 1.91 $\log_{10}$ CFU/g and 1.33 $\log_{10}$ CFU/g for lactoccocci and lactobacilli counts, respectively. In that case the subsequent recovery of these counts was not so clear and they did not achieve the same counts than control samples. Novella-Rodríguez et al. [31] also observed that starter counts were reduced about 2 $\log_{10}$ CFU/g in goats' milk cheeses as a consequence of HHP, although a subsequent increase was found during ripening. Rynne et al. [32] reported significant reductions of about 1.5 $\log_{10}$ CFU/g in starter and non-starter LAB in cheddar cheese treated at 400 MPa on the first day of ripening. In ovine milk ripened cheeses treated on day 1 and 15 at 300 MPa and 400 MPa proved to cause similar reductions although recovery of LAB counts was observed only in samples treated 1st day. HHP treatments also affected significantly ($p < 0.05$) the counts of enteroccocci in both type of cheeses in either HHP1 and HHP15 samples, being unable to recover the initial counts in any case. Arqués et al. [27] reported a reduction of 2.68 $\log_{10}$ CFU/g in enteroccocci counts when a treatment of 400 MPa for 10 min at 10 °C was applied on the 2nd day of ripening to "La Serena" cheeses, remaining constant during the rest of the ripening. Although high counts of enteroccoci have been associated with the unhygienic processing of cheese, their presence is also considered important for the development of the typical aroma and flavour of traditional Mediterranean cheeses. Their counts may range from 10^4 to 10^6 CFU/g in curds and 10^5 to 10^7 CFU/g in ripened cheeses [33]. In the case of *S. aureus* HHP1 treatments also caused significant reductions in both kind of cheeses and HHP15 in ewes' milk cheeses, becoming undetectable in most cases at the last day of the work. Nevertheless, the initial counts were already quite low. Similar results were reported in "La Serena" cheeses treated at 400 MPa on the 2nd day of ripening [27] and in Cheddar cheeses and their slurries treated at 400 MPa for 20 min at 20 °C [34]. *S. aureus* has been described as one of the most HHP resistant non-sporulated bacteria. López-Pedemonte et al. [35] described that at least 500 MPa HHP treatments would be necessary to achieve reductions of about 6 $\log_{10}$ CFU/g.

HHP treatments showed to be very effective reducing Gram-negative bacteria except for HHP1 in goats' milk cheese samples, where a slight growth was noted on the 15th day probably due to a possible recovery of the sub lethal injured cells after the HHP treatment. However, cheese ripening conditions, with low pH, increasing salt concentration and presence of LAB, made difficult this recovery to consolidate and no positive counts were detected later at the 30th and 45th days of ripening. Juan et al. [28] also described reductions above 3 $\log_{10}$ CFU/g of enterobacteria in ewes' milk cheeses after a 400 MPa treatment applied the 1st and the 15th day of ripening. Initial counts of *E. coli* were significantly lower than enterobacteria. O'Reilly et al. [34], Capellas et al. [36] and De Lamo-Castellvi et al. [37] have previously pointed out the greatest sensitivity of *E. coli* to HHP in cheeses with reductions above 6 $\log_{10}$ CFU/g after HHP treatments at 400 MPa.

3.2. Effect of HHP on the Proteolysis of Cheeses

The proteolytic activity parameters measured in the goats' and ewes' cheeses are presented in Tables 3 and 4, respectively.

Table 3. Changes on proteolysis index (mean values ± standard deviation) during the ripening of goats' raw milk cheese samples with and without HHP treatment (TNBS: amino group content determined by the Trinitrobenzensulphonic Acid method, RI: Ripening Index, FFA: free Amino Acid content; TNBS and FAA expressed in mg L-Leucine/g).

Proteolysis Index	Day of Ripening	Control	HHP1
TNBS	5	10.40 ± 2.24 [a]	9.41 ± 1.80 [a]
	15	14.90 ± 1.21 [ab]	15.33 ± 2.15 [ab]
	30	22.54 ± 2.50 [b]	21.14 ± 4.59 [b]
	45	44.89 ± 1.19 [c,B]	32.65 ± 5.81 [c,A]
	60	53.66 ± 2.14 [d,B]	39.96 ± 4.52 [c,A]
RI	5	10.99 ± 3.05 [a]	11.09 ± 2.60 [a]
	15	22.97 ± 1.23 [b]	22.97 ± 1.23 [b]
	30	21.88 ± 0.60 [b]	24.39 ± 3.15 [bc]
	45	22.14 ± 1.78 [b]	25.71 ± 2.15 [bc]
	60	25.07 ± 0.66 [b]	28.03 ± 0.35 [c]
FAA	5	2.08 ± 0.57 [a]	1.48 ± 0.72 [a]
	15	4.85 ± 1.42 [a]	3.71 ± 1.06 [ab]
	30	10.05 ± 2.06 [b,B]	5.90 ± 1.59 [bc,A]
	45	15.07 ± 3.52 [c,B]	8.40 ± 1.43 [cd,A]
	60	15.71 ± 3.06 [c,B]	10.40 ± 1.29 [d,A]

Means with different superscript small letter differ significant ($p < 0.05$) in the same column for the same parameter and HHP treatment; means with different superscript capital letter differ significant ($p < 0.05$) in the same row for the same parameter and day of ripening.

Significant differences between control and HHP1 samples in TNBS were observed from the 45th to the 60th days of ripening in goats' milk cheeses, while the differences in FAA content were mainly noted from day 30 to 60. In both parameters control samples showed the highest values. The ratio of RI displayed a different trend where an increment of around two times was observed during the first 15 days period in control and HHP-treated samples but after this point the proteolysis rate became slower. In the ewes' milk cheeses, an increment on the three-proteolysis index evaluated was observed during the ripening period (Table 4). In the ewes' milk cheeses HHP15 samples presented slightly higher values than control samples, especially in TNBS, although no significant differences were observed at the end of the ripening. Although the HHP1 samples showed an intense proteolysis in the first 30 days of ripening, after this time a decrease on the rate was noticed, obtaining a considerable reduction on the three proteolysis index values at the 60th day, reflecting that pressure treatment caused a deceleration in the rate of proteolysis (Table 4).

Table 4. Evolution of proteolysis index (mean values $\pm$ standard deviation) during the ripening of ewes' raw milk cheese samples with and without HHP treatment.

Proteolysis Index	Day of Ripening	Control	HHP1	HHP15
TNBS	5	7.66 $\pm$ 1.65 [a]	7.92 $\pm$ 0.44 [a]	
	15	13.87 $\pm$ 1.37 [ab]	13.36 $\pm$ 0.82 [a]	15.03 $\pm$ 0.40 [a]
	30	22.13 $\pm$ 8.22 [b,AB]	26.33 $\pm$ 17.48 [b,BC]	34.31 $\pm$ 6.11 [b,C]
	45	34.93 $\pm$ 2.60 [c,BC]	28.00 $\pm$ 2.84 [b,AB]	44.79 $\pm$ 4.93 [bc,C]
	60	47.40 $\pm$ 2.74 [d,BC]	37.59 $\pm$ 6.86 [b,A]	52.50 $\pm$ 2.21 [c,C]
RI	5	9.70 $\pm$ 3.55 [a]	10.44 $\pm$ 4.39 [a]	
	15	19.38 $\pm$ 4.24 [b]	18.51 $\pm$ 4.23 [b]	19.15 $\pm$ 2.57 [a]
	30	24.96 $\pm$ 3.64 [bc]	23.35 $\pm$ 2.38 [b]	25.39 $\pm$ 3.57 [ab]
	45	29.59 $\pm$ 2.97 [c,B]	25.09 $\pm$ 2.95 [b,A]	30.54 $\pm$ 2.46 [b,B]
	60	31.92 $\pm$ 6.63 [c,B]	26.38 $\pm$ 3.77 [b,A]	31.62 $\pm$ 2.21 [b,B]
FAA	5	2.36 $\pm$ 0.04 [a]	2.18 $\pm$ 0.22 [a]	
	15	5.01 $\pm$ 1.27 [a]	3.66 $\pm$ 1.79 [ab]	5.59 $\pm$ 2.94 [a]
	30	11.66 $\pm$ 2.76 [b,B]	6.85 $\pm$ 3.07 [ab,AB]	11.22 $\pm$ 1.11 [b,B]
	45	11.39 $\pm$ 2.69 [b,B]	6.55 $\pm$ 1.10 [ab,AB]	14.86 $\pm$ 4.89 [bc,C]
	60	16.56 $\pm$ 3.51 [c,B]	8.63 $\pm$ 1.71 [b,A]	16.03 $\pm$ 5.14 [c,B]

Means with different superscript small letter differ significant ($p < 0.05$) in the same column for the same parameter and HHP treatment; means with different superscript capital letter differ significant ($p < 0.05$) in the same row for the same parameter and day of ripening; TNBS: amino group content determined by the Trinitrobenzensulphonic Acid method, RI: Ripening Index, FFA: free Amino Acid content; TNBS and FAA expressed in mg L-Leucine/g.

RI values indicated that proteolysis was more intense during the first 15 days of ripening in the goats' and ewes' cheeses in control and HHP treated samples. This proteolytic activity was probably caused by milk and rennet proteinases, being not so clear the role of microbial proteinases. On fact, WSN is produced mainly by the rennet and to a lesser extent by plasmin or cellular proteinases, whereas starter peptidases are primarily responsible for the formation of small peptides and free amino acids [38]. Messens et al. [39] observed that chymosin and plasmin activity in "Gouda" cheese was not influenced by pressure (from 50 to 400 MPa for 20–100 min). Similarly, Malone et al. [40] in a study about HHP effects (100–800 MPa, 5 min, 25 °C) on the activity of proteolytic and glycolytic enzymes, observed that plasmin was insensible to pressure treatments and the chymosin activity was unaffected by treatments up to 400 MPa, decreasing by 50% after an 800 MPa treatment. On the other hand, Juan et al. [28] found in ewes' milk cheese that the chymosin activity decreases depending on the age of the cheese and the pressure applied (above 400 MPa, 1-day old cheeses), whereas plasmin activity was not significantly affected by HHP treatments (200–500 MPa, for 10 min) applied on the 1st and 15th day of ripening.

On ewes' milk cheeses HHP15 samples showed slightly higher values on the three-proteolysis index evaluated than control samples during the ripening period, although no significant differences were observed at the end of the ripening, reflecting that this treatment did not significantly affect the proteolysis process. However, the HHP application during the initial stages of the ripening in ovine and caprine milk cheeses led to a decrease of the proteolysis rate. Similar results were obtained by Juan et al. [16] who in pressurized ovine milk cheeses on the 15th day of ripening obtained similar *WSN/TN* values than control samples, at the end of the ripening but higher than those obtained in samples with HHP-treatment on the 1st day. In contrast, some works reported after application of HHP treatments not differences on proteolysis indexes during ripening in Gouda [39] and cheddar cheeses [32] or an increase of proteolysis in goat milk cheese [41,42]. Starter bacteria are one of the primary sources of ripening intracellular enzymes (proteinases and peptidases). Cellular lysis is required for their release in the cheese matrix [43], being one of the main factors that influence the rate of secondary proteolysis [38]. HHP-induced cell lysis is pressure and strain-dependent [44] and possibly ripening time-HHP dependent. While Messens et al. [39] indicated that, in Gouda cheese, the possible lysis of the starter bacterial cells resulting from the damage suffered at 400 MPa did not increase proteolysis because endocellular enzymes were inactivated by the pressure, O'Reilly et al. [45] pointed out that ripening enzymes in cheddar cheese would probably begin to denature after the application of pressure treatments between 350–400 MPa and Juan et al. [16] and Calzada et al. [30] noticed that HHP-treatments above 400 MPa delay the proteolysis in ewes' and cows' milk, respectively. On the other hand, Saldo et al. [17,42] suggested that in goats' milk cheeses, the release of starter enzymes probably caused an increase in proteolysis two weeks after HHP treatment at 400 MPa for 5 min was applied.

3.3. Effect of HHP on the Formation of BA on Cheeses

Tables 5 and 6 show the values of the total BA content of the goats' and ewes' milk cheeses after HHP treatments. In general, the sum of all BA, including polyamines, showed a constant and significant increase in control samples during ripening from the 5th (59.36 mg/kg) to the 60th day (1156 mg/kg) of ripening in the goats' milk cheese and from the 6th (85.08 mg/kg) to the 60th day (622.39 mg/kg) in ewes' milk cheese samples. This is in agreement with the results reported by other authors [46–51]. The application of pressure during the first stages of ripening (HHP1) caused an initial reduction of BA formation from the 15th day, observing that at the end of the ripening these samples displayed concentrations around 75% lower than control samples. HHP15 ewes' milk cheeses also resulted in a decrease of the BA content with respect to control samples, although this reduction was less pronounced (38% lower than control ones). The BA reduction in the HHP1 ewes' and goats' milk cheeses could be explained because of the significant decrease on microbiological counts observed one day after the treatment (specially of LAB, enteroccoci and enterobacteria) and the lower proteolysis presented in these samples with respect to the control ones, showing a reduction of about 34% and 49% of FAA content, respectively, at the end of the ripening. HHP15 samples also showed lower microbial counts when compared with the untreated cheeses but this did not affect the proteolysis and consequently the release of amino acids. Novella-Rodriguez et al. [52] found that the total BA content found in goats' milk cheeses HHP treated at 400 MPa during 5 min was similar than the untreated cheeses, although TY content was significantly reduced in HHP samples. Ruiz-Capillas et al. [53] observed that HHP treatments at 350 MPa for 15 min used to treat "Chorizo" slices caused a significant decrease in BA content (TY, PU, CA and SM), being the reduction of these amines coincidental with the decrease in microbial counts, especially of LAB.

Table 5. Monoamine and diamines content (mean values ± standard deviation expressed as mg/kg in dry basis) formed during the ripening of goats' raw milk cheese samples with and without HHP treatment.

BA	Day of Ripening	Control	HHP1
TR	5	8.99 ± 7.94 [a]	23.12 ± 5.93 [a]
	15	27.76 ± 15.00 [ab]	10.81 ± 9.37 [a]
	30	95.96 ± 34.38 [c]	82.10 ± 26.33 [b]
	45	68.39 ± 20.42 [bc]	70.11 ± 40.01 [b]
	60	63.69 ± 27.90 [bc]	89.08 ± 49.98 [b]
PHE	5	1.40 ± 0.65 [a]	0.26 ± 0.45 [a]
	15	14.62 ± 1.42 [b]	17.96 ± 4.07 [b]
	30	17.27 ± 5.94 [b]	17.16 ± 4.29 [b]
	45	20.75 ± 4.20 [ab]	14.80 ± 0.45 [ab]
	60	31.13 ± 14.19 [b]	25.56 ± 7.47 [b]
PU	5	4.07 ± 0.51 [a]	3.29 ± 0.42
	15	136.56 ± 21.57 [ab]	59.09 ± 7.34
	30	225.16 ± 22.31 [b]	67.09 ± 8.50
	45	463.88 ± 60.73 [c,B]	42.69 ± 33.89 [A]
	60	476.41 ± 126.21 [c,B]	79.80 ± 19.51 [A]
CA	5	30.20 ± 16.19 [a]	24.07 ± 9.04
	15	29.63 ± 4.35 [a]	26.20 ± 3.13
	30	50.14 ± 10.97 [ab]	35.29 ± 12.70
	45	69.53 ± 16.34 [b]	36.22 ± 17.00
	60	70.45 ± 27.21 [b]	44.22 ± 15.61
HIS	5	1.27 ± 0.56 [a]	1.00 ± 0.87
	15	3.02 ± 0.43 [a]	2.44 ± 0.34
	30	6.38 ± 3.31 [a]	6.27 ± 3.16
	45	18.04 ± 9.36 [b,B]	6.51 ± 1.75 [A]
	60	15.41 ± 7.05 [b,B]	4.85 ± 2.20 [A]
TY	5	10.04 ± 6.80 [a]	6.11 ± 6.95
	15	130.51 ± 42.98 [ab]	18.96 ± 1.07
	30	234.74 ± 69.16 [b,B]	15.59 ± 3.56 [A]
	45	443.87 ± 105.10 [c,B]	16.17 ± 0.68 [A]
	60	491.89 ± 67.45 [c,B]	28.93 ± 5.91 [A]

Means with different superscript small letter differ significant ($p < 0.05$) in the same column for the same parameter and HHP treatment; means with different superscript capital letter differ significant ($p < 0.05$) in the same row for the same parameter and day of ripening. TR: tryptamine, PHE: β-phenylethylamine, PU: putrescine, CA: cadaverine, HIS: histamine, TY: tyramine.

Table 6. Monoamine and diamines content (mean values ± standard deviation expressed as mg/kg in dry basis) formed during the ripening of ewes' raw milk cheese samples with and without HHP treatment.

BA	Day of Ripening	Control	HHP1	HHP15
TR	5	1.66 ± 3.31 [a]	4.92 ± 5.86	-
	15	4.51 ± 3.32 [a]	10.03 ± 7.29	8.76 ± 6.98
	30	9.87 ± 4.43 [ab]	9.83 ± 4.66	6.25 ± 0.80
	45	9.51 ± 3.85 [ab]	12.49 ± 6.76	12.32 ± 8.43
	60	15.73 ± 2.15 [b]	11.06 ± 6.78	11.66 ± 6.50
PHE	5	1.02 ± 1.23 [a]	0.40 ± 0.48	-
	15	2.78 ± 1.44 [a]	2.39 ± 0.56	4.39 ± 1.84
	30	5.67 ± 5.25 [a]	3.04 ± 1.01	3.71 ± 1.45
	45	12.69 ± 5.08 [ab]	4.37 ± 1.52	5.78 ± 1.53
	60	12.74 ± 2.62 [b]	4.43 ± 2.37	13.31 ± 15.37

Table 6. *Cont.*

BA	Day of Ripening	Control	HHP1	HHP15
PU	5	3.62 ± 0.40 [a]	5.20 ± 2.41	-
	15	42.42 ± 23.00 [ab]	11.25 ± 9.80	15.14 ± 11.99
	30	65.89 ± 30.90 [bc,B]	5.45 ± 1.38 [A]	22.33 ± 14.49 [AB]
	45	87.89 ± 76.10 [c,B]	7.81 ± 3.03 [A]	22.92 ± 17.51 [A]
	60	74.89 ± 54.74 [bc,B]	5.24 ± 0.76 [A]	26.08 ± 6.26 [A]
CA	5	62.13 ± 33.00 [a]	55.55 ± 28.54	-
	15	159.07 ± 37.23 [b,B]	48.75 ± 44.16 [A]	87.50 ± 70.43 [AB]
	30	141.63 ± 79.34 [ab,B]	48.40 ± 16.94 [A]	80.65 ± 76.20 [AB]
	45	129.82 ± 67.03 [ab,B]	41.91 ± 22.89 [A]	72.47 ± 29.78 [AB]
	60	105.90 ± 21.87 [ab,B]	28.51 ± 17.32 [A]	80.28 ± 71.97 [AB]
HIS	5	4.81 ± 4.54 [a]	5.33 ± 3.79	-
	15	39.98 ± 11.42 [b]	12.71 ± 10.97	29.14 ± 22.84 [a]
	30	74.34 ± 29.42 [c,B]	6.38 ± 2.77 [A]	47.52 ± 31.34 [ab,B]
	45	81.66 ± 46.30 [c,B]	5.94 ± 1.25 [A]	58.99 ± 3.14 [b,B]
	60	91.02 ± 5.73 [c,C]	7.07 ± 4.30 [A]	57.65 ± 13.16 [ab,B]
TY	5	4.37 ± 0.71 [a]	3.15 ± 1.94 [a]	-
	15	123.72 ± 40.62 [ab,B]	13.79 ± 5.28 [b,A]	115.79 ± 55.44 [B]
	30	222.32 ± 84.26 [bc,B]	22.16 ± 10.34 [b,A]	162.64 ± 36.21 [A]
	45	268.02 ± 130.91 [c,B]	80.49 ± 126.16 [ab,A]	147.73 ± 30.57 [AB]
	60	277.30 ± 114.08 [c,B]	32.69 ± 16.54 [b,A]	147.62 ± 26.64 [AB]

Means with different superscript small letter differ significant ($p < 0.05$) in the same column for the same parameter and HHP treatment; means with different superscript capital letter differ significant ($p < 0.05$) in the same row for the same parameter and day of ripening. TR: tryptamine, PHE: β-phenylethylamine, PU: putrescine, CA: cadaverine, HIS: histamine, TY: tyramine.

TY and PU were the main BA formed in untreated goats' milk cheeses, showing concentrations of about 492 and 476 mg/kg, respectively, at the end of the ripening. Whereas in ewes' milk control samples the predominant BA were TY and CA with 277 and 106 mg/kg, respectively. Several authors reported, in variable ranges, TY (88.6–445 mg/kg), HIS (not detected–697 mg/kg), PU (74.15–446.5 mg/kg) and CA (44–269.77 mg/kg) as the most abundant BA in goats' and ewes' milk ripened cheeses [26,46,48,49,54–56]. These variable contents depended on the type of cheese, length of the ripening period, the manufacturing process and the type of microorganisms present (starters and non-starter bacteria with decarboxylase activity). Post-ripening processing (cutting, slicing and grating) also has an important influence on the presence of decarboxylating bacteria in cheese and the formation of BA, such as HIS, may be greater than in entire cheeses [57]. TY levels in pressurized cheeses were lower than those presented in control cheeses. This aromatic amine increased slowly until the 15th day of ripening on goats' milk HHP1 samples and remained constant throughout the rest of the ripening. At the end of the ripening the concentrations were about 93% lower than in control samples. Likewise, ewes' milk HHP1 samples reached concentrations of 33 mg/kg at the 60th day, being 88% lower than control samples. The application of HHP treatment on the 15th day of ripening resulted in a slight decrease of TY levels in ewes' milk cheeses and the final content was not significantly different than control samples at the 60th day of ripening. The reduction of the TY levels in pressurized goats' and ewes' milk cheese samples coincided mainly with the decrease in LAB and enteroccocci counts and with the amount of FAA. Novella-Rodriguez et al. [52] observed HHP-treated goats' milk cheeses showed a significant lower TY content than untreated samples, attributing this behaviour to the reducing effect of HHP on the microbiological counts, especially on the non- starter LAB, although levels of TY in this kind of cheeses were similar. High "in vitro" capability to form TY has been described in different species of *Enterococcus* spp. and LAB isolated from cheese samples [12,58–61].

PU amounts in control goats' milk cheeses were almost the same than TY but HHP1-treatments caused a decrease on the first 15 days, remaining almost stable throughout the rest of the ripening. PU levels were about 83% lower in the HHP1 samples than in control samples at the 60th day. The pressure application in ewes' milk cheeses also affected the PU amounts formed, showing that HHP1 treatments limited the production of this diamine around 93% compared with control cheeses at the end of the ripening. HHP15 showed to be less efficient reducing the formation of PU. With respect to CA, untreated and HHP-treated goats' milk cheeses displayed amounts below 100 mg/kg without appreciate significant differences between them. In control ewes' milk cheeses this diamine increased mainly during the first 15 days, while in HHPI and HHP15 cheeses remained without significant changes throughout the ripening. The amounts of TR and PHE increased during the ripening in control goats' milk cheese samples without significant differences in relation to HHP1-treated samples, while low amounts were detected in ewes' cheeses remaining practically constant during ripening and without showing significant differences between treatments. Formation of PU and CA is usually associated with Gram negative bacteria, although some strains of *Enterococcus* spp. also have shown this capability "in vitro" [61]. Some LAB, such as *Lactococcus lactis*, are also able to form PU via the agmatine deiminase [62]. The application of the HHP treatments in the early stages of maturation causes an important reduction of both the Gram-negative microbiota, which practically disappeared and the enterococci that could not recover their initial counts during the rest of the ripening, which would explain that both PU and CA, together with TY, are the AB that present the greatest reduction. When treatments are applied after 15 days of maturation, the decarboxylating capacity of these microorganisms in the early stages of maturation is still present and consequently there would be a lesser effect on the formation of these AB. LAB are also affected by the HHP treatments, although later they are able to recover their initial counts, which is important to develop the cheese's own characteristics, although they may also be responsible for the residual decarboxylating activity [30].

In goats' raw milk control cheeses the concentration of HIS was very low at the beginning of ripening but increased later reaching its maximum after 45 days (18 mg/kg). No significant changes were observed until the 60th day. In HHP1-treated samples HIS showed a similar behaviour but in this case, differences were found after the 45th day displaying levels 68% lower than control samples at the 60th day of ripening. HHP1 ewes' milk cheese samples showed a reduction of 92% of HIS when compared with control samples. This treatment was more efficient than the HHP15. Diverse authors reported low HIS amounts in cheese (below 100 mg/kg), relating the production of this BA with some LAB [12,31,54,56,63,64]. Some works have also described *Enterobacteriaceae* strains able to decarboxylate histidine in diverse foodstuffs [12,58,65–69]. Enterococci have also been related with histamine formation in cheeses [48,59,70].

In cheeses made from raw goat milk, low levels of polyamines were found at the beginning of ripening, increasing their concentration very slightly during ripening until the 60th day (Table 7). However, they showed an increase in their amount when HHP1- treatment was applied. Higher levels of SD were observed in ewes' milk cheeses showing maximum amounts the 30th day of ripening (about 64 mg/kg in control cheeses) followed by a decrease at the end of the ripening. However, HHP1 treatment caused a significant increase of this polyamine at the 30th day, reaching maximum levels of about 122 mg/kg. SM was detected at constant amounts throughout the ripening in both control and HHP treated samples. SD has been reported as the main polyamine in some cheeses [10,46,55]. Novella-Rodríguez et al. [26] found in HHP treated goat cheeses an increase of polyamines, especially SD but no data were reported to elucidate the cause of this phenomenon. Polyamines are described as natural amines of non-microbial origin present generally at a lower concentration than other BA of bacterial origin. No toxic effects have been attributed to them although some authors have mentioned their importance for the intestine cell growth and proliferation in childhood [3,71].

Table 7. Polyamine content (mean values ± standard deviation expressed as mg/kg in dry basis) during the ripening of goats' and ewes' raw milk cheese samples with and without HHP treatment. (SD: spermidine, SM: spermine).

	Day of Ripening	Control	HHP1	HHP15
Goats' milk chesses				
SD	5	1.13 ± 0.51 [a]	0.32 ± 0.50 [a]	-
	15	1.92 ± 0.18 [aA]	4.34 ± 0.21 [b,B]	-
	30	2.09 ± 0.24 [abA]	3.70 ± 0.89 [b,B]	-
	45	4.32 ± 1.75 [c]	3.28 ± 1.30 [b]	-
	60	3.92 ± 1.49 [bcA]	6.53 ± 1.99 [c,B]	-
SM	5	2.25 ± 2.07 [a]	1.47 ± 1.48 [a]	-
	15	4.35 ± 1.25 [ab,A]	7.80 ± 0.60 [bc,B]	-
	30	3.03 ± 0.75 [ab,A]	6.83 ± 0.46 [b,B]	-
	45	5.57 ± 3.91 [b,B]	2.16 ± 0.30 [a,A]	-
	60	3.90 ± 0.75 [ab,A]	10.46 ± 2.18 [c,B]	-
Ewes' milk cheeses				
SD	5	0.73 ± 0.65 [a]	0.51 ± 0.90 [a]	-
	15	55.94 ± 29.00 [ab]	90.23 ± 67.99 [bc]	73.15 ± 38.57 [b]
	30	64.22 ± 8.22 [b,A]	122.25 ± 48.16 [c,B]	65.61 ± 10.97 [b,A]
	45	14.74 ± 14.76 [ab]	52.04 ± 36.91 [ab]	34.49 ± 19.56 [ab]
	60	30.14 ± 23.04 [ab]	19.49 ± 33.57 [a]	17.36 ± 9.35 [a]
SM	5	6.76 ± 5.17	5.94 ± 4.03 [a]	-
	15	14.13 ± 12.84	18.69 ± 14.57 [a]	8.00 ± 6.99 [a]
	30	25.18 ± 18.41	28.39 ± 7.52 [b]	39.50 ± 32.16 [b]
	45	17.57 ± 9.32	13.13 ± 7.26 [ab]	20.08 ± 21.91 [ab]
	60	14.66 ± 17.02	15.27 ± 17.76 [a]	26.92 ± 17.93 [ab]

Means with different superscript small letter differ significant ($p < 0.05$) in the same column for the same parameter and HHP treatment; means with different superscript capital letter differ significant ($p < 0.05$) in the same row for the same parameter and day of ripening.

4. Conclusions

The effectiveness of HHP treatments depends on different factors, such as the type of cheese, the stage of ripening, the HHP processing conditions applied, the kind and number of microorganisms present. In this work, the use of HHP applied at the initial phases of ripening affected significantly the microorganisms responsible of forming BA, and, in consequence, reduced its content, especially of TY and PU in goats' milk cheeses and TY and CA in ewes' milk cheeses, assuring the safety of this product for the most BA sensitive consumers. As previously mentioned, HIS and TY are the BA that most frequently have been related with food-borne outbreaks, being suggested threshold values in cheese between 200–400 mg/kg of HIS [6,72,73] and between 100–800 mg kg of TY [73,74]. TY dose of 6 mg [6] or above 20 mg of HIS [75] have been suggested as toxic to very sensitive individuals. If we consider the consumption of 30 g of cheese as a serving the goats' and ewes' cheeses analysed in this work would provide around 15 and 8.3 mg of TY, respectively, that can be dangerous for the most sensible consumers but HHP1-treated cheeses would provide with less than 1 mg of TY considering the same size of serving.

Author Contributions: Conceived and designed the experiments: D.E.-P., M.M.H.-H. and A.X.R.-S.; performed the experiments: D.E.-P.; analysed the data: D.E.-P. and M.M.H.-H.; wrote the paper: D.E.-P. and A.X.R.-S.

References

1. Naila, A.; Flint, S.; Fletcher, G.; Bremer, P.; Meerdink, G. Control of biogenic amines in food—Existing and emerging approaches. *J. Food Sci.* **2010**, *75*, 139–150. [CrossRef] [PubMed]

2. Fernández, E.J.Q.; Ventura, M.T.M.; Sagués, A.X.R.; Jerez, J.J.R.; Herrero, M.M.H. Aminas biogenas en queso: Riesgo toxicológico y factores que influyen en su formación. *Alimentaria* **1998**, *294*, 59–66.

3. Mariné Font, A. *Les Amines Biògenes en els Aliments: Història i Recerca en el Marc de les Ciències de L'alimentació*; Institut d'Estudis Catalans, Secció de Ciències Biològiques: Barcelona, Spain, 2005; ISBN 8472837882.

4. Sumner, S.S.; Taylor, S.L. Detection Method for Histamine-Producing, Dairy-Related Bacteria using Diamine Oxidase and Leucocrystal Violet. *J. Food Prot.* **1989**, *52*, 105–108. [CrossRef]

5. Stratton, J.E.; Hutkins, R.W.; Taylor, S.L. Biogenic Amines in Cheese and other Fermented Foods: A Review. *J. Food Prot.* **1991**, *54*, 460–470. [CrossRef]

6. Daniel Collins, J.; Noerrung, B.; Budka, H.; Andreoletti, O.; Buncic, S.; Griffin, J.; Hald, T.; Havelaar, A.; Hope, J.; Klein, G.; et al. Scientific Opinion on risk based control of biogenic amine formation in fermented foods. *EFSA J.* **2011**, *9*, 2393. [CrossRef]

7. Shalaby, A.R. Significance of biogenic amines to food safety and human health. *Food Res. Int.* **1996**, *29*, 675–690. [CrossRef]

8. Taylor, S.L.; Eitenmiller, R.R. Histamine food poisoning: Toxicology and clinical aspects. *CRC Crit. Rev. Toxicol.* **1986**, *17*, 91–128. [CrossRef] [PubMed]

9. Nordisk Ministerråd; Nordisk Råd. *Present Status of Biogenic Amines in Foods in Nordic Countries*; Nordisk Ministerråd: Copenhagen, Denmark, 2002; ISBN 9289307730.

10. Novella-Rodriguez, S.; Veciana-Nogues, M.T.; Izquierdo-Pulido, M.; Vidal-Carou, M.C. Distribution of Biogenic Amines and Polyamines in Cheese. *J. Food Sci.* **2003**, *68*, 750–756. [CrossRef]

11. Joosten, H.M.L.J. Conditions allowing the formation of biogenic amines in cheese. *Neth Milk Dairy J.* **1988**, *4*, 329–357.

12. Roig-Sagués, A.; Molina, A.; Hernández-Herrero, M. Histamine and tyramine-forming microorganisms in Spanish traditional cheeses. *Eur. Food Res. Technol.* **2002**, *215*, 96–100. [CrossRef]

13. Kalac, P.; Abreu Gloria, M.B. Biogenic amine in cheeses, wines, beers and sauerkraut. In *Biological Aspects of Biogenic Amines, Poliamines and Conjugates*; Dandrifosse, G., Ed.; Transworld Research Network: Kerala, 2009; pp. 267–285, ISBN 9788178952499.

14. Hoover, D.G.; Metrick, C.; Papineau, A.M.; Farkas, D.F.; Knorr, D. Biological effects of high hydrostatic pressure on food microorganisms. *Food Technol.-Chic.* **1989**, *43*, 99–107.

15. Evert-Arriagada, K.; Hernández-Herrero, M.M.; Juan, B.; Guamis, B.; Trujillo, A.J. Effect of high pressure on fresh cheese shelf-life. *J. Food Eng.* **2012**, *110*, 248–253. [CrossRef]

16. Juan, B.; Ferragut, V.; Guamis, B.; Buffa, M.; Trujillo, A.J. Proteolysis of a high pressure-treated ewe's milk cheese. *Milchwissenschaft* **2004**, *59*, 616–619.

17. Saldo, J.; McSweeney, P.L.H.; Sendra, E.; Kelly, A.L.; Guamis, B. Proteolysis in caprine milk cheese treated by high pressure to accelerate cheese ripening. *Int. Dairy J.* **2002**, *12*, 35–44. [CrossRef]

18. Delgado, F.J.; González-Crespo, J.; Cava, R.; Ramírez, R. Changes in microbiology, proteolysis, texture and sensory characteristics of raw goat milk cheeses treated by high-pressure at different stages of maturation. *LWT—Food Sci. Technol.* **2012**, *48*, 268–275. [CrossRef]

19. Calzada, J.; Del Olmo, A.; Picon, A.; Gaya, P.; Nuñez, M. Proteolysis and biogenic amine buildup in high-pressure treated ovine milk blue-veined cheese. *J. Dairy Sci.* **2013**, *96*, 4816–4829. [CrossRef] [PubMed]

20. Calzada, J.; del Olmo, A.; Picón, A.; Gaya, P.; Nuñez, M. Reducing Biogenic-Amine-Producing Bacteria, Decarboxylase Activity, and Biogenic Amines in Raw Milk Cheese by High-Pressure Treatments. *Appl. Environ. Microbiol.* **2013**, *79*, 1277–1283. [CrossRef] [PubMed]

21. Kuchroo, C.N.; Fox, P.F. Soluble nitrogen in Cheddar cheese: Comparison of extraction procedures. *Milchwissenschaft* **1982**, *37*, 331–335.

22. *IDF (International Dairy Federation) Standard 185: 2002(E), I; 14891:2002(E) Milk and milk products—Determination of nitrogen content—Routine method using combustion according to the Dumas principle*; IDF: Brussels, Belgium, 2002.

23. Hernández-Herrero, M.M.; Roig-Sagués, A.X.; López-Sabater, E.I.; Rodríguez-Jerez, J.J.; Mora-Ventura, M.T. Protein hydrolysis and proteinase activity during the ripening of salted anchovy (*Engraulis encrasicholus* L.). A microassay method for determining the protein hydrolysis. *J. Agric. Food Chem.* **1999**, *47*, 3319–3324. [CrossRef] [PubMed]

24. Folkertsma, B.; Fox, P.F. Use of the Cd-ninhydrin reagent to assess proteolysis in cheese during ripening. *J. Dairy Res.* **1992**, *59*, 217–224. [CrossRef]

25. Eerola, S.; Hinkkanen, R.; Lindfors, E.; Hirvi, T. Liquid chromatographic determination of biogenic amines in dry sausages. *J. AOAC Int.* **1993**, *76*, 575–577. [PubMed]

26. Novella-Rodríguez, S.; Veciana-Nogués, M.T.; Roig-Sagués, A.X.; Trujillo-Mesa, A.J.; Vidal-Carou, M.C. Comparison of biogenic amine profile in cheeses manufactured from fresh and stored (4 degrees C, 48 hours) raw goat's milk. *J. Food Prot.* **2004**, *67*, 110–116. [CrossRef] [PubMed]

27. Arqués, J.L.; Garde, S.; Gaya, P.; Medina, M.; Nuñez, M. Short Communication: Inactivation of Microbial Contaminants in Raw Milk La Serena Cheese by High-Pressure Treatments. *J. Dairy Sci.* **2006**, *89*, 888–891. [CrossRef]

28. Juan, B.; Ferragut, V.; Buffa, M.; Guamis, B.; Trujillo, A.J. Effects of High Pressure on proteolytic enzymes in cheese: Relationship with the proteolysis of ewe milk cheese. *J. Dairy Sci.* **2007**, *90*, 2113–2125. [CrossRef] [PubMed]

29. Cabezas, L.; Sánchez, I.; Poveda, J.M.; Seseña, S.; Palop, M.L. Comparison of microflora, chemical and sensory characteristics of artisanal Manchego cheeses from two dairies. *Food Control* **2007**, *18*, 11–17. [CrossRef]

30. Calzada, J.; del Olmo, A.; Picon, A.; Gaya, P.; Nuñez, M. Effect of High-Pressure Processing on the Microbiology, Proteolysis, Biogenic Amines and Flavour of Cheese Made from Unpasteurized Milk. *Food Bioprocess Technol.* **2015**, *8*, 319–332. [CrossRef]

31. Novella-Rodriguez, S.; Veciana-Nogues, M.T.; Trujillo-Mesa, A.J.; Vidal-Carou, M.C. Profile of Biogenic Amines in Goat Cheese Made from Pasteurized and Pressurized Milks. *J. Food Sci.* **2002**, *67*, 2940–2944. [CrossRef]

32. Rynne, N.M.; Beresford, T.P.; Guinee, T.P.; Sheehan, E.; Delahunty, C.M.; Kelly, A.L. Effect of high-pressure treatment of 1 day-old full-fat Cheddar cheese on subsequent quality and ripening. *Innov. Food Sci. Emerg. Technol.* **2008**, *9*, 429–440. [CrossRef]

33. Foulquié Moreno, M.R.; Sarantinopoulos, P.; Tsakalidou, E.; De Vuyst, L. The role and application of enterococci in food and health. *Int. J. Food Microbiol.* **2006**, *106*, 1–24. [CrossRef] [PubMed]

34. O'Reilly, C.E.; O'Connor, P.M.; Kelly, A.L.; Beresford, T.P.; Murphy, P.M. Use of hydrostatic pressure for inactivation of microbial contaminants in cheese. *Appl. Environ. Microbiol.* **2000**, *66*, 4890–4896. [CrossRef] [PubMed]

35. López-Pedemonte, T.; Roig-Sagués, A.X.; Lamo, S.D.; Gervilla, R.; Guamis, B. High hydrostatic pressure treatment applied to model cheeses made from cow's milk inoculated with Staphylococcus aureus. *Food Control* **2007**, *18*, 441–447. [CrossRef]

36. Capellas, M.; Mor-Mur, M.; Sendra, E.; Pla, R.; Guamis, B. Populations of aerobic mesophils and inoculated E. coli during storage of fresh goat's milk cheese treated with high pressure. *J. Food Prot.* **1996**, *59*, 582–587. [CrossRef]

37. De Lamo-Castellví, S.; Capellas, M.; Roig-Sagués, A.X.; López-Pedemonte, T.; Hernández-Herrero, M.M.; Guamis, B. Fate of Escherichia coli strains inoculated in model cheese elaborated with or without starter and treated by high hydrostatic pressure. *J. Food Prot.* **2006**, *69*, 2856–2864. [CrossRef] [PubMed]

38. Fox, P.F. Proteolysis During Cheese Manufacture and Ripening. *J. Dairy Sci.* **1989**, *72*, 1379–1400. [CrossRef]

39. Messens, W.; Estepar-Garcia, J.; Dewettinck, K.; Huyghebaert, A. Proteolysis of high-pressure-treated Gouda cheese. *Int. Dairy J.* **1999**, *9*, 775–782. [CrossRef]

40. Malone, A.S.; Wick, C.; Shellhammer, T.H.; Courtney, P.D. High Pressure Effects on Proteolytic and Glycolytic Enzymes Involved in Cheese Manufacturing. *J. Dairy Sci.* **2003**, *86*, 1139–1146. [CrossRef]

41. Trujillo, A.; Buffa, M.; Casals, I.; Fernández, P.; Guamis, B. Proteolysis in goat cheese made from raw, pasteurized or pressure-treated milk. *Innov. Food Sci. Emerg. Technol.* **2002**, *3*, 309–319. [CrossRef]

42. Saldo, J.; Sendra, E.; Guamis, B. High Hydrostatic Pressure for Accelerating Ripening of Goat's Milk Cheese: Proteolysis and Texture. *J. Food Sci.* **2000**, *65*, 636–640. [CrossRef]

43. Thomas, T.D.; Pritchard, G.G. Proteolytic enzymes of dairy starter cultures. *FEMS Microbiol. Lett.* **1987**, *46*, 245–268. [CrossRef]

44. Malone, A.S.; Shellhammer, T.H.; Courtney, P.D. Effects of high pressure on the viability, morphology, lysis, and cell wall hydrolase activity of Lactococcus lactis subsp. cremoris. *Appl. Environ. Microbiol.* **2002**, *68*, 4357–4363. [CrossRef] [PubMed]

45. O'Reilly, C.E.; Kelly, A.L.; Oliveira, J.C.; Murphy, P.M.; Auty, M.A.; Beresford, T.P. Effect of varying high-pressure treatment conditions on acceleration of ripening of cheddar cheese. *Innov. Food Sci. Emerg. Technol.* **2003**, *4*, 277–284. [CrossRef]

46. Ordóñez, A.I.; Ibáñez, F.C.; Torre, P.; Barcina, Y. Formation of biogenic amines in Idiazabal ewe's-milk cheese: Effect of ripening, pasteurization, and starter. *J. Food Prot.* **1997**, *60*, 1371–1375. [CrossRef]

47. Gardini, F.; Martuscelli, M.; Caruso, M.C.; Galgano, F.; Crudele, M.A.; Favati, F.; Guerzoni, M.E.; Suzzi, G. Effects of pH, temperature and NaCl concentration on the growth kinetics, proteolytic activity and biogenic amine production of Enterococcus faecalis. *Int. J. Food Microbiol.* **2001**, *64*, 105–117. [CrossRef]

48. Galgano, F.; Suzzi, G.; Favati, F.; Caruso, M.; Martuscelli, M.; Gardini, F.; Salzano, G. Biogenic amines during ripening in 'Semicotto Caprino' cheese: Role of enterococci. *Int. J. Food Sci. Technol.* **2001**, *36*, 153–160. [CrossRef]

49. Martuscelli, M.; Gardini, F.; Torriani, S.; Mastrocola, D.; Serio, A.; Chaves-López, C.; Schirone, M.; Suzzi, G. Production of biogenic amines during the ripening of Pecorino Abruzzese cheese. *Int. Dairy J.* **2005**, *15*, 571–578. [CrossRef]

50. Combarros-Fuertes, P.; Fernández, D.; Arenas, R.; Diezhandino, I.; Tornadijo, M.E.; Fresno, J.M. Biogenic amines in Zamorano cheese: factors involved in their accumulation. *J. Sci. Food Agric.* **2016**, *96*, 295–305. [CrossRef] [PubMed]

51. Poveda, J.M.; Molina, G.M.; Gómez-Alonso, S. Variability of biogenic amine and free amino acid concentrations in regionally produced goat milk cheeses. *J. Food Compost. Anal.* **2016**, *51*, 85–92. [CrossRef]

52. Novella-Rodríguez, S.; Veciana-Nogués, M.T.; Saldo, J.; Vidal-Carou, M.C. Effects of high hydrostatic pressure treatments on biogenic amine contents in goat cheeses during ripening. *J. Agric. Food Chem.* **2002**, *50*, 7288–7292. [CrossRef] [PubMed]

53. Ruiz-Capillas, C.; Jiménez Colmenero, F.; Carrascosa, A.V.; Muñoz, R. Biogenic amine production in Spanish dry-cured "chorizo" sausage treated with high-pressure and kept in chilled storage. *Meat Sci.* **2007**, *77*, 365–371. [CrossRef] [PubMed]

54. Valsamaki, K.; Michaelidou, A.; Polychroniadou, A. Biogenic amine production in Feta cheese. *Food Chem.* **2000**, *71*, 259–266. [CrossRef]

55. Pinho, O.; Ferreira, I.M.P.L.V.O.; Mendes, E.; Oliveira, B.M.; Ferreira, M. Effect of temperature on evolution of free amino acid and biogenic amine contents during storage of Azeitão cheese. *Food Chem.* **2001**, *75*, 287–291. [CrossRef]

56. Pintado, A.I.E.; Pinho, O.; Ferreira, I.M.P.L.V.O.; Pintado, M.M.E.; Gomes, A.M.P.; Malcata, F.X. Microbiological, biochemical and biogenic amine profiles of Terrincho cheese manufactured in several dairy farms. *Int. Dairy J.* **2008**, *18*, 631–640. [CrossRef]

57. Ladero, V.; Fernandez, M.; Alvarez, M.A. Effect of post-ripening processing on the histamine and histamine-producing bacteria contents of different cheeses. *Int. Dairy J.* **2009**, *19*, 759–762. [CrossRef]

58. Pircher, A.; Bauer, F.; Paulsen, P. Formation of cadaverine, histamine, putrescine and tyramine by bacteria isolated from meat, fermented sausages and cheeses. *Eur. Food Res. Technol.* **2007**, *226*, 225–231. [CrossRef]

59. Leuschner, R.G.K.; Kurihara, R.; Hammes, W.P. Formation of biogenic amines by proteolytic enterococci during cheese ripening. *J. Sci. Food Agric.* **1999**, *79*, 1141–1144. [CrossRef]

60. Perin, L.M.; Belviso, S.; Dal Bello, B.; Nero, L.A.; Cocolin, L. Technological properties and biogenic amines production by bacteriocinogenic lactococci and enterococci strains isolated from raw goat's milk. *J. Food Protect.* **2017**, *80*, 151–157. [CrossRef] [PubMed]

61. Torracca, B.; Pedonese, F.; Turchi, B.; Fratini, F.; Nuvoloni, R. Qualitative and quantitative evaluation of biogenic amines in vitro production by bacteria isolated from ewes' milk cheeses. *Eur. Food Res. Technol.* **2018**, *244*, 721–728. [CrossRef]

62. del Rio, B.; Redruello, B.; Ladero, V.; Fernandez, M.; Cruz Martin, M.; Alvarez, M.A. Putrescine production by *Lactococcus lactis* subsp. cremoris CECT 8666 is reduced by NaCl via a decrease in bacterial growth and the repression of the genes involved in putrescine production. *Int. J. Food Microbiol.* **2016**, *232*, 1–6. [CrossRef] [PubMed]

63. Joosten, H.M.L.J.; Northolt, M.D. Detection, growth, and amine-producing capacity of lactobacilli in cheese. *Appl. Environ. Microbiol.* **1989**, *55*, 2356–2359. [PubMed]

64. Burdychova, R.; Komprda, T. Biogenic amine-forming microbial communities in cheese. *FEMS Microbiol. Lett.* **2007**, *276*, 149–155. [CrossRef] [PubMed]

65. Roig-Sagues, A.X.; Hernandez-Herrero, M.; Lopez-Sabater, E.I.; Rodriguez-Jerez, J.J.; Mora-Ventura, M.T. Histidine decarboxylase activity of bacteria isolated from raw and ripened Salchichon, a Spanish cured sausage. *J. Food Prot.* **1996**, *59*, 516–520. [CrossRef]

66. Roig-Sagués, A.X.; Hernàndez-Herrero, M.M.; López-Sabater, E.I.; Rodríguez-Jerez, J.J.; Mora-Ventura, M.T. Evaluation of three decarboxylating agar media to detect histamine and tyramine-producing bacteria in ripened sausages. *Lett. Appl. Microbiol.* **1997**, *25*, 309–312. [CrossRef] [PubMed]

67. Silla Santos, M.H. Amino acid decarboxylase capability of microorganisms isolated in Spanish fermented meat products. *Int. J. Food Microbiol.* **1998**, *39*, 227–230. [CrossRef]

68. Hernández-Herrero, M.M.; Roig-Sagués, A.X.; Rodríguez-Jerez, J.J.; Mora-Ventura, T.M. Halotolerant and Halophilic Histamine-Forming Bacteria Isolated during the Ripening of Salted Anchovies (*Engraulis encrasicholus*). *J. Food Prot.* **1999**, *62*, 509–514. [CrossRef] [PubMed]

69. Özogul, F.; Özogul, Y. The ability of biogenic amines and ammonia production by single bacterial cultures. *Eur. Food Res. Technol.* **2007**, *225*, 385–394. [CrossRef]

70. Tham, W.; Karp, G.; Danielsson-Tham, M.L. Histamine formation by enterococci in goat cheese. *Int. J. Food Microbiol.* **1990**, *11*, 225–229. [CrossRef]

71. Bardócz, S. Polyamines in food and their consequences for food quality and human health. *Trends Food Sci. Technol.* **1995**, *6*, 341–346. [CrossRef]

72. Rauscher-Gabernig, E.; Grossgut, R.; Bauer, F.; Paulsen, P. Assessment of alimentary histamine exposure of consumers in Austria and development of tolerable levels in typical foods. *Food Control* **2009**, *20*, 423–429. [CrossRef]

73. Benkerroum, N. Biogenic Amines in Dairy Products: Origin, Incidence, and Control Means. *Compr. Rev. Food Sci. Food Saf.* **2016**, *15*, 801–826. [CrossRef]

74. Karovicova, J.; Kohajdova, Z. Biogenic Amines in Food. *ChemInform* **2005**, *36*, 70–79. [CrossRef]

75. Vind, S.; Søndergaard, I.; Poulsen, L.K.; Svendsen, U.G.; Weeke, B. Comparison of methods for intestinal histamine application: Histamine in enterosoluble capsules or via a duodeno-jenunal tube. *Allergy* **1991**, *46*, 191–195. [CrossRef] [PubMed]

5

Biogenic Amines, Phenolic, and Aroma-Related Compounds of Unroasted and Roasted Cocoa Beans with Different Origin

Umile Gianfranco Spizzirri [1], Francesca Ieri [2,*], Margherita Campo [2], Donatella Paolino [3], Donatella Restuccia [1] and Annalisa Romani [2]

[1] Department of Pharmacy, Health and Nutritional Sciences, University of Calabria, I-87036 Rende (CS), Italy
[2] Department of Statistic, Informatics and Applications "G. Parenti" (DiSIA)—University of Florence, Phytolab Laboratory, via Ugo Schiff 6, 50019 Sesto Fiorentino (FI), Italy
[3] Department of Experimental and Clinical Medicine, University of Catanzaro "Magna Græcia", 88100 Catanzaro, Italy
* Correspondence: francesca.ieri@unifi.it

Abstract: Biogenic amines (BAs), polyphenols, and aroma compounds were determined by chromatographic techniques in cocoa beans of different geographical origin, also considering the effect of roasting (95, 110, and 125 °C). In all samples, methylxantines (2.22–12.3 mg kg^{-1}) were the most abundant followed by procyanidins (0.69–9.39 mg kg^{-1}) and epicatechin (0.16–3.12 mg kg^{-1}), all reduced by heat treatments. Volatile organic compounds and BAs showed variable levels and distributions. Although showing the highest BAs total content (28.8 mg kg^{-1}), Criollo variety presented a good aroma profile, suggesting a possible processing without roasting. Heat treatments influenced the aroma compounds especially for Nicaragua sample, increasing more than two-fold desirable aldehydes and pyrazines formed during the Maillard cascade and the Strecker degradation. As the temperature increased, the concentration of BAs already present in raw samples increased as well, although never reaching hazardous levels.

Keywords: cocoa nibs; roasting; bioactive amines; polyphenols; volatile organic compounds; geographical areas

1. Introduction

Cocoa beans represent the basic raw material in the production of chocolate and cocoa-based products. During processing, cocoa beans undergo important manipulations, including fermentation and roasting, which drastically influence the quality of the final product. During fermentation, cocoa beans are exposed to the action of various microorganisms and enzymes, while in the roasting process, high temperatures determine important modifications on the cocoa bean's composition [1].

Cocoa is a food rich in polyphenols, mainly flavonoids, procyanidins, and flavan-3-ols. The preservation or enhancement of cocoa procyanidins is of great importance since these compounds, despite their poor bioavailability, have been related to the health beneficial effects of cocoa, particularly in cardiovascular diseases [2]. Polyphenol and xanthine content in cocoa seeds changes during ripening and during the processing phases [3]. Microbial activity during fermentation and the drying process contribute to settling the amounts of theobromine and caffeine and their relative abundances, polyphenol amounts, in particular of catechin and epicatechin, and the amounts of organic acids, sugars, mannitol, ethanol, and alkaloids, thus influencing the quality and the biological properties of the finished product [4]. At the beginning of fermentation, during the first three days, the highest contents of total phenolic compounds and total anthocyanins prevailed in cocoa beans. Finally, at the end of cocoa beans fermentation, the lowest contents of total phenolic compounds, and total anthocyanins, were observed [5].

In addition, roasting temperature has been seen to have effects on the flavanols amounts causing losses and structural modifications, particularly epimerization of both monomers and polymers. At high roasting temperatures, a progressive loss of (−)-epicatechin and (+)-catechin and an increase in (−)-catechin were observed as a result of heat-related epimerization from (−)-epicatechin; additionally, a temperature-related epimerization of procyanidin dimers has been reported [6,7]. These structural modifications could have negative effects on the biological properties of the product, being (−)-epicatechin the most bioavailable isomer and (−)-catechin the one with the lowest bioavailability. Roasting processes may also cause reduction of the content of hydroxycynnamic compounds, clovamide in particular [8], with a possible further reduction of the antioxidant activity.

The secret of the flavor of chocolate, so highly appreciated worldwide, resides mainly in its volatile aromatic fraction. Its complex composition depends on the cocoa bean genotype and is the consequence of several processes [9,10]. To date, descriptive studies have identified >600 volatile compounds in cocoa and chocolate products [11,12], mainly pyrazines, esters, amines and amides, acids, and hydrocarbons. Besides cocoa flavor precursors, also toxic molecules, such as biogenic amines (BAs), can be produced during processing. BAs are a class of organic, basic, nitrogenous compounds with low molecular weight, are usually part of bioactive molecules of cocoa beans and derivatives. They mainly derived by decarboxylation of corresponding amino acids, due to the action of suitable enzymes widely distributed in spoilage bacteria and other microorganisms, as well as in naturally occurring and/or artificially added bacteria involved in food fermentation [5,13]. Although several amines (i.e., natural polyamines) are present in living cells and contribute to promoting many human physiological functions, these compounds can represent a serious health hazard for humans, when present in food in significant amounts or ingested in the presence of potentiating factors, such as amine oxidase-inhibiting drugs, alcohol, and gastrointestinal diseases. Then, their attendance in foodstuffs is often undesirable, because often associated with several of pathological syndromes, such as headaches, respiratory distress, heart palpitations, hypo- or hypertension and several allergenic disorders [14]. In particular, tyramine, β-phenylethylamine, and histamine have been considered as the initiators of hypertension and dietary-induced migraines, while the neurotransmitter serotonin is essential in the regulation of appetite, body temperature, and sleep [15].

Generally speaking, a variety of factors determining cocoa and cocoa derivatives quality are strongly related to the cocoa beans processing, from the opening of the fruit until the end of industrial processes [16]. In addition, qualitative characteristics of the cocoa beans are a consequence of the differences in the farming practices regarding growing, fermenting, and drying the cocoa beans, with significant differences sometimes found in samples from the same country [16,17].

Recently, it has been reported that in cocoa beans the amino acid oxidative decarboxylation can be also obtained during food processing suggesting a new chemical, heat-induced formation of BAs [18]. It follows that, in addition to the amino acid catabolism produced by microorganisms, amino acids can also be degraded chemically as a consequence of thermal treatment of foods [19]. These reactions are responsible for the formation of taste and flavor compounds, reducing, at the same time, the concentration of essential amino acids and contributing to the accumulation of compounds that may be dangerous for consumers, such as BAs [20]. It follows that two main reasons can be underlined accounting for the analysis of BAs in foods: first their potential toxicity; second the possibility of using them as food quality markers as their concentration can be related with the hygienic-sanitary quality of the process and with the freshness of the raw materials and the processed products. As BAs have been widely exploited as important indicators of safety and quality in a variety of foods, many papers appeared in recent years reporting their quantitative determination in many fermented and non-fermented foods, including cocoa and its derivatives [13].

In this work, the quantitative determination of biogenic amines, xanthine, and polyphenol molecules was performed by liquid chromatography (LC) techniques on cocoa beans from different origin. The aroma profile of cocoa nibs was investigated by headspace solid-phase micro-extraction (HS-SPME) combined with gas chromatography-mass spectrometry (GC-MS). Additionally, the same compounds were monitored after roasting process at different temperatures to establish a possible correlation between heating and different biomolecules profiles.

2. Materials and Methods

2.1. Samples

Seven cocoa beans samples from different world areas and years were considered as reported in Table 1. All fruits studied in this work were considered as well fermented. Approximately 500 g portions of Coviriali and O'Payo cocoa beans of uniform size were random selected from 50 kg of each nib variety peeled and roasted in a forced airflow-drying oven ROASTER CENTOVENTI vertiflow® system (Selmi Chocolate Machinery, Cuneo, Italy) at Meraviglie S.r.l. (Verona, Italy). Applied in these studies, heat treatment parameters were chosen to obtain a range of roasted beans with acceptable physico-chemical and sensory properties. The parameters of thermal processing were optimized to avoid over-roasting, varying time of roasting as soon as bean cracking occurs, as follows: temperatures of 95, 110, and 125 °C for 60, 30, and 20 min respectively. Samples from all bean types were prepared using differentiated methods according to the analyses carried out as reported in the subsections of the experimental section.

Table 1. List of samples. ^{nr}not roasted; ^{r1}95 °C; ^{r2}110 °C; ^{r3}125 °C.

ID	Sample	Year	Variety	Country	Region
1^{nr}	CAMINO VERDE	2015	Nacional Forastero	Ecuador	Guayas
2^{nr}	MADAGASCAR	2015	Trinitario	Madagascar	Sambirano
3^{nr}	COVIRIALI	2015	Forastero	Peru	Junín
4^{nr}, 4^{r1}, 4^{r2}, 4^{r3}	COVIRIALI	2016	Forastero	Peru	Junín
5^{nr}	CHENI	2016	Forastero	Peru	Satipo
6^{nr}, 6^{r1}, 6^{r2}	O'PAYO	2016	Trinitario	Nicaragua	North Caribbean Coast Autonomous Region
7^{nr}	CRIOLLO	2016	Criollo	Indonesia	Bali

2.2. Chemicals

BAs spermine (SPM, tetrahydrochloride), spermidine (SPD, trihydrochloride), putrescine (PUT, dihydrochloride), histamine (HIM, dihydrochloride), tyramine (TYR, hydrochloride), β-phenylethylamine (PHE, hydrochloride), cadaverine (CAD, hydrochloride), as well as dansyl chloride, ammonia (30%), perchloric acid, and LC solvents (acetonitrile and methanol LC grade) were purchased from Sigma-Aldrich (Milford, MA, USA). Ultrapure water was obtained from Milli-Q System (Millipore Corp., Milford, MA, USA). Filters (0.45 and 0.20 μm) were purchased from Sigma-Aldrich. SPE C18 cartridges (0.5 g) were obtained from Supelco Inc. (Bellefonte, PA, USA). All GC chemicals were from Sigma-Aldrich (Milford, MA, USA). The HPLC grade standards (±)-catechin hydrate, theobromine, caffeic acid, quercetin-3-glucoside were purchased from Sigma-Aldrich (Milford, MA, USA).

2.3. Samples Preparation for the Analysis of Polyphenols and Xanthines

The peeled cocoa beans were crushed in a mortar, then 1.0 g accurately weighed of crushed material was extracted in 10.0 mL of a solution EtOH:H_2O 70:30 at pH 3.2 by addition of HCOOH, at room temperature, for 24 h under stirring. The solid material was removed by filtration under vacuum and the extracts analyzed by high performance liquid chromatography coupled with diode array detection and electrospray ionization mass spectrometer (HPLC-DAD-ESI-MS) and by high performance liquid chromatography coupled with diode array detection and fluorescence detector (HPLC-DAD-FLD).

2.4. Chromatographic Conditions Xanthine and Polyphenol Determination

The method used for the quali-quantitative analysis, and described below, was optimized according to literature data and previous studies of this research group about polyphenolic compounds and xanthines, and modifying them based on the specific results [8,21–23]. The analyses were performed using a HP-1200 Liquid Chromatograph with a DAD and a fluorescence detector and a HP-1100 MSD API Electrospray (Agilent Technologies, Palo Alto, CA, USA) operating in negative and positive ionization mode. Gas temperature was 350 °C, flow rate 10.0 L/min, nebulizer pressure 30 psi, quadrupole temperature 300 °C, capillary voltage 3500 V, and fragmentor 120 eV.

For the chromatographic separation a Luna C18 250 × 4.60 mm, 5 µm (Phenomenex, Torrance, CA, USA) column was used operating at 26 °C. A multistep linear solvent gradient starting from 95% H_2O at pH 3.2 by addition of HCOOH (A), up to 100% CH_3CN (B) was performed with a flow rate of 0.8 mL min^{-1} over a 63 min period. In detail, the applied gradient started with 95% A; from 95% A to 85% A in 20 min; isocratic elution 85% A until 30 min; from 85% A to 75% A in 9 min; isocratic elution 75% A until 47 min; from 75% A to 15% A in 2 min; isocratic elution 15% A until 53 min; from 15% A to 0% A in 2 min; 0% A until 60 min; from 0% A to the initial conditions in 3 min. Figure 1 reports on the chromatographic profile of raw O'Payo bean extract (2016) by DAD at 280 nm (A) and 330 nm (B) and FLD ex. 280 nm; em. 315 nm (C).

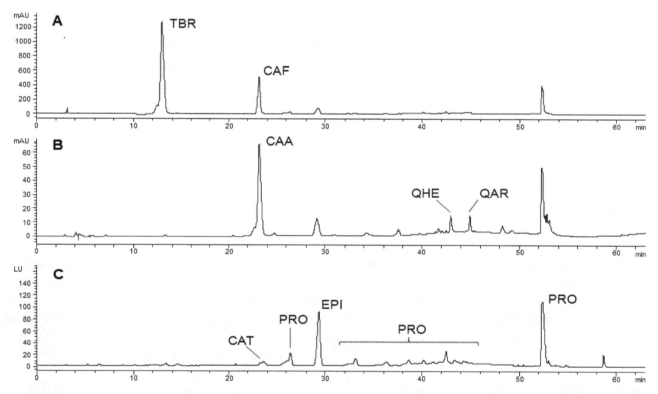

Figure 1. Chromatographic profile of raw O'Payo bean extract (2016): diode array detection (DAD) 280 nm (**A**) and 330 nm (**B**); fluorescence detector (FLD) ex. 280 nm; em. 315 nm (**C**). TBR—Theobromine; CAF—Caffeine; PRO—Procyanidins; CAA—Caffeoyl aspartic acid; QHE—Quercetin hexoside; QAR—Quercetin arabinoside; CAT—Catechin; EPI—Epicatechin.

2.5. Calibration

Quantitation of xanthines, flavonols, hydroxycinnamic derivatives, and procyaninides was performed by HPLC-DAD using five-point regression curves built with the available standards. Curves with an r^2 > 0.9998 were considered. Calibration was performed at the wavelength of the maximum UV-Vis absorbance, applying the correction of molecular weight. In particular, the extinction coefficient of each quantified compound being comparable to that of the specific standard used for its calibration, the weights in mg were calculated by multiplying the weight obtained from the calibration by a correction factor given by the ratio between the molecular weight of the compound and the molecular weight of the standard used for its calibration. In particular, xanthines were calibrated at 280 nm using theobromine as reference; procyanidins were calibrated at 280 nm using catechin hydrate as reference; hydroxycinnamic derivatives were calibrated at 330 nm using caffeic acid as reference; flavanols were calibrated at 350 nm using quercetin as reference. The quantitation of catechin and epicatechin was performed by HPLC-FLD, using a five point calibration curve (r^2 = 0.9999) built with standard solutions of catechin hydrate. The fluorescence detector was set as follows: excitation wavelength 280 nm; emission wavelength 315 nm [24].

2.6. HS-SPME-GC-MS Analyses

Headspace solid-phase micro-extraction combined with gas chromatography–mass spectrometry (HS-SPME-GC-MS) was selected as the most suitable technique to recover and analyze the Volatile Organic Compounds (VOCs) in peeled cocoa beans samples. Samples were ground and homogenous powders were obtained. A total of 1 g of the powdered sample, was placed into a 20-mL screw cap vial fitted with PTFE/silicone septa. An Internal Standard (IS) in suitable amount was added to each sample (IS: ethylacetate-D8; 1-Butanol-D10; ethyl hexanoate-D11; 5-methyl-hexanol; acetic acid-D3; Hexanoic acid-D11; 3,4-Dimethylphenol;). The Internal Standard was used for normalizing the analyte responses over the area of the IS, to minimize the instrumental error during the time of analysis.

After some trials aimed at optimizing amounts of sample, exposure time, and temperature, SPME conditions were set as follows: after 5 min of equilibration at 60 °C, VOCs were absorbed exposing a 1-cm divinilbenzene/carboxen/polydimethylsiloxane SPME fiber (DVB/CAR/PDMS by Supelco) for 15 min into the vial headspace under orbital shaking (500 rpm) and then immediately desorbed at 280 °C in a gas chromatograph injection port operating in split less mode. The chromatographic analysis was performed in a GC system coupled to quadrupole mass spectrometry using an Agilent 7890a GC equipped with a 5975C MSD (Agilent Technologies, Palo Alto, CA, USA). The separation of analytes was achieved by an Agilent DB InnoWAX column (length 50 m, id 0.20 μm, df 0.40 μm). Chromatographic conditions were: initial temperature 40 °C, then 10 °C min^{-1} up to 260 °C, hold for 6.6 min. Compounds were tentatively identified by comparing calculated Kovats retention index and mass spectra of each peak with those reported in mass spectral databases, namely the standard NIST08/Wiley98 libraries. Quadrupole MS operated in full-scan mode from which the specific ions of the analyte were extracted. Only the compounds with higher intensity were identified in order to select major compounds over a complex mixture of VOCs. Each sample was analyzed in triplicate.

2.7. Amine Standard Solutions and Calibration

A calibration curve was built starting from 1.0 mg mL^{-1} standard solution of each amine in purified water and preparing 12 BAs standard mixtures to a final volume of 25 mL employing HClO$_4$ 0.6 mol L^{-1}. Amine final concentrations were 0.1, 0.5, 0.8, 2.0, 4.0, 5.0, 10.0, 16.0, 25.0, 50.0, 75.0, and 100.0 μg mL^{-1}. The comparison between the retention times of peaks of samples and standard solutions allowed the identification of each BA. Standard concentration against peak area allowed to build a calibration plot, and six independent replicates for each concentration level were performed. Moreover, the matrix effect was evaluated by comparison of external calibration plots, depicting concentration of standard solutions versus peak area, with standard addition method plots, depicting peak area versus concentration of standard solutions added to the sample. No significant matrix effect was recorded because of the slopes of the two plots were not significantly different. Quantitative determination was then accomplished by direct interpolation in the external calibration plot of each BA. Chromatogram of a standard mixture of BAs is displayed in Figure 2A.

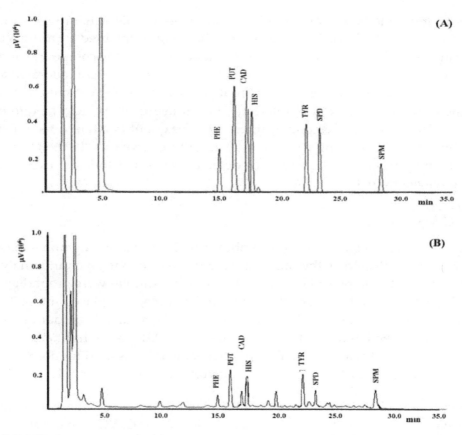

Figure 2. LC-UV chromatogram of biogenic amines (BAs) standard mixture at concentration of 100 µg mL^{-1} (**A**) and sample 1 (**B**). The chromatogram was obtained employing gradient conditions as specified in the Materials and Methods section.

2.8. BAs Extraction and Purification

The extraction of BAs from peeled cocoa beans samples was performed by adding 20 mL of HClO$_4$ 0.6 mol L^{-1} to about 5.0 g of grounded sample, in a 50.0 mL test tube. The mixture was homogenized (vortex at 40 Hz for 40 min), centrifuged (10,000× g for 20 min), filtered (syringe filter 0.20 µm), collected in a plastic vial and purified by SPE on a C$_{18}$ sorbent (conditioning: 2.0 mL of H$_2$O and 2.0 mL (two times) of CH$_3$OH; loading: 5.0 mL of the basified sample; washing: 2.0 mL of NH$_4$OH at pH 11.0; eluting: 2.0 mL (two times) of CH$_3$OH). Nitrogen gas was employed to dry eluting solution providing a solid residue that was re-dissolved in a plastic test tube with 1.3 mL of extraction solvent.

To perform recovery experiments sample 1 was spiked, before the extraction procedure, with an aliquot of standard mixture of BAs. Specifically, 5.0 g of peeled cocoa beans were spiked with 1.0 mL of 25.0 mg L^{-1} BAs standard solution. Method validation was obtained in terms of recovery percentages, linearity, intra- and inter-day repeatability, limits of quantification and limits of detection (LOQs and LODs), to confirm analytical suitability [25].

Dansylation reaction was performed by adding at 1.0 mL of standard solution (or acid sample extract spiked with BAs or acid sample extract) 200 µL of NaOH 2.0 mol L^{-1}, 300 µL of saturated NaHCO$_3$ solution and 2.0 mL of dansyl-chloride solution (10.0 mg mL^{-1} in acetone prepared just before use). After 30 min, dansyl-chloride in excess was removed with 100 µL of NH$_4$OH 25% (v/v) and the suspension filtered by a 0.45 µm syringe filters. Finally, 20 µL was injected for LC-UV analysis. Figure 2B shows a chromatogram of sample 1.

2.9. Chromatographic Conditions for BAs Quantification

Jasco PU-2080 instrument equipped with a Rheodyne 7725 injector with a 20 mL sample loop and a gradient pump (PU-2089 plus, Jasco Inc., Easton, MD, USA) was employed to obtain the chromatograms. The system was interfaced with an UV detector operating at $\lambda = 254$ nm (UV-2075, Jasco Inc., Easton, MD, USA). Data were collected and analyzed with an integrator Jasco-Borwin1. A reverse-phase C18 column (250 mm × 4.6 I.D., 5 mm) (Supelco Inc., Bellefonte, PA, USA) equipped with a C18 guard-pak (10 mm × 4.6 I.D., 5 mm) were used (Supelco Inc., Bellefonte, PA, USA) for separation of BAs. Two solvent reservoirs containing (A) purified water and (B) acetonitrile were used to separate all the BAs with a gradient elution which began with 3 min of isocratic program A-B 50:50 (v/v) reaching after 20 min A-B 10:90 (v/v). Then 3 min of isocratic elution was carried out and 4 min further where necessary to restore again the starting conditions (A-B 50:50, v/v). A constant flow at 1.2 mL min^{-1} was employed.

2.10. Statistical Analyses

All analyses were performed in triplicate and data were expressed as mean ± relative standard deviations (RSD). Studies of the correlation coefficient and linear regression, calculation of average, assessment of repeatability, standard deviation, and RSD were performed using Microsoft Excel 2010 software. Significance was performed using a one-way analysis of variance (ANOVA) test, employing Duncan's multiple range test at significance level $p < 0.05$.

3. Results and Discussion

3.1. Polyphenol Content in Cocoa Beans

To identify the phenolic compounds and xanthines, UV-Vis absorption spectra, mass spectra and literature data were used and combined for tentative identification of the analytes. In the present study, catechin and epicatechin were quantified through calibration by using a FLD detector, whereas oligomeric and polymeric procyanidins were quantified through DAD calibration. FLD calibration was used for catechin and epicatechin because of its higher sensibility and specificity, needed for catechin in particular, that often partially coelutes with caffeine, and is always present in low amounts with respect to this latter (average amount of catechin with respect to caffeine: 5.2% in the analyzed samples). In these conditions, the fact that both catechin and caffeine have also very similar wavelengths of maximum UV-Vis absorption hinders a correct quantification of the flavanol by using a DAD detector. Conversely, unlike caffeine, catechin and epicatechin emit a very intense fluorescence signal in the experimental conditions (excitation wavelength 280 nm; emission wavelength 315 nm), easily detectable and measurable also in presence of methylxanthines [24,26]. On the other hand, for the calibration of the other procyanidins a DAD detector is needed, because fluorescence detection is insensitive to procyanidins containing a gallic acid ester and/or gallocatechins as monomeric units [23]. In this case, the use of a fluorescence detector could lead to an underestimation of the total procyanidin content.

In Table 2 polyphenol distributions and total amounts in fermented, not-roasted cocoa beans of different origin are reported. Xanthines, theobromine in particular, are the most abundant compounds and their amounts decrease with increasing roasting temperature, after an initial slight increase by roasting at 95 °C, probably due to a further loss of water after the drying process.

Among the raw samples analyzed, the highest in polyphenols were 1^{nr} (Camino verde 2015), 3^{nr} (Coviriali 2015) and 6^{nr} (O'Payo 2016), respectively with 10.67, 10.53, and 12.96 mg g^{-1} total polyphenols. In all samples, it is possible to observe a clear predominance of epicatechin with respect to catechin; the highest [epicatechin]:[catechin] ratio (27.9) was found for the raw sample from Madagascar (2015), but its low content in polyphenols (1.44 mg g^{-1}) suggests a lower quality with respect to the other samples under study. The other two samples harvested in 2015 (Camino verde and Coviriali) appear not to contain catechin, so it was impossible to evaluate the ratio even though the epicatechin amount appears to be high in particular for the Camino verde sample (2.84 mg g^{-1}). The lowest [epicatechin]/[catechin] ratio was found for the raw samples Cheni and Criollo (5.0 for both the samples).

Table 2. HPLC/DAD quantitative analysis of methylxantines and polyphenols in cocoa nibs. Results in mg g^{-1} vegetal material. Data obtained from triplicate analysis with relative standard deviation (SD) 2–5%.

	1nr	2nr	3nr	5nr	7nr	4nr	4r1	4r2	4r3	6nr	6r1	6r2
Theobromine	8.65 f	1.95 a	7.81 de	7.69 d	7.61 d	8.49 f	8.52 f	7.68 d	7.14 c	8.08 e	10.08 g	6.68 b
Caffeine	2.01 c	0.27 a	1.89 c	1.70 b	0.96 b	3.81 e	2.81 d	2.18 c	1.60 b	2.60 d	2.04 c	2.03 c
Catechin	nd a	0.01 b	nd a	0.17 g	0.06 c	0.20 h	0.15 f	0.10 d	0.14 f	0.24 i	0.12 e	0.12 e
Epicatechin	2.84 i	0.16 a	0.84 e	0.87 e	0.32 b	1.98 g	1.08 f	0.58 c	0.73 d	3.12 h	2.22 h	0.69 d
Procyanidins	7.58 h	1.00 b	9.39 j	8.15 i	0.62 a	2.47 f	1.29 d	1.17 c	1.01 b	9.17 j	5.17 g	1.92 e
Quercetin hexoside	0.03 c	0.02 b	0.03 c	0.05 d	0.02 b	<LOQ a	0.07 e	0.05 d	0.05 d	0.03 c	0.05 d	0.02 b
Quercetin arabinoside	0.04 e	0.03 d	0.04 e	0.05 f	0.02 c	<LOQ a	0.09 h	0.06 g	0.06 g	0.05 f	0.10 i	0.01 b
Caffeoyl aspartic acid	0.12 b	0.13 b	0.24 c	<LOQ a	0.24 c	0.51 h	0.43 g	0.76 i	0.38 f	0.34 e	0.29 d	0.28 d
Hydroxycinnamic der.	0.06 b	0.09 c	nd a	nd a	nd a	nd a	nd a	nd a	nd a	nd a	nd a	nd a
N-caffeoyl-L-DOPA	<LOQ a	<LOQ a	<LOQ a	<LOQ a	<LOQ a	<LOQ a	<LOQ a	<LOQ a	<LOQ a	<LOQ a	<LOQ a	<LOQ a
Epicatechin:catechin ratio		27.9		5.0	5.0	9.7	7.0	6.0	5.3	12.9	18.7	5.9
Toal xanthines	10.66 d	2.22 a	9.71 c	9.39 c	8.58 b	12.30 f	11.33 e	9.86 c	8.74 b	10.68 d	12.12 f	8.71 b
Total polyphenols	10.67 i	1.44 b	10.55 i	9.29 h	1.28 a	5.16 f	3.11 e	2.72 d	2.37 c	12.96 j	7.95 g	3.04 e

Different letters express significant differences ($p < 0.05$). nrnot roasted; r1 95 °C; r2 110 °C; r3 125 °C. nd means not detected and <LOQ means under limit of quantification.

For Coviriali 2016 and O'Payo 2016 beans it was possible to compare the polyphenols contents after roasting at 110 °C, confirming that the total amounts of polyphenols significantly decrease with respect to the raw samples; Coviriali beans were roasted also at a higher temperature (125 °C) with a further decreasing of total polyphenols. According to previously reported data [27,28], also [epicatechin]/[catechin] ratios follow the same trend depending on the roasting process at high temperature. Procyanidins, quantified as catechin equivalents, are the most abundant polyphenols in all samples and their amounts also appear to be negatively influenced by the roasting process [28]. Two flavonolic compounds were found, quercetin hexoside and quercetin arabinoside, in low amounts and apparently not depending on the roasting temperature. The only hydroxycynnamic derivative present in the extracts was caffeoyl aspartic acid [3].

The characteristics of cocoa beans and derived food products, such as hydrophilic and volatile secondary metabolites profile, organoleptic properties, antioxidant activity etc., depend not only on fermentation and processing methods, but also on several variables related to genetics, geographical regions of cultivation, agronomical practices, and climatic conditions [29,30]. In particular, concerning phenolic compounds, the low amount of total polyphenols (1.28 mg g^{-1}) found for Criollo variety is reliable based on literature data that identify this variety as the one with the lowest content of polyphenols [31,32]. The Forastero variety includes also Nacional Forastero that is the Forastero variety cultivated in northern Ecuador [31]. The analyzed Forastero samples were harvested in 2015 (Nacional Forastero from Ecuador, sample 1^{nr}, and Forastero from Junìn region, Peru, sample 3^{nr}) and 2016 (Forastero from Junìn region, sample 4^{nr}, and Forastero from Satipo region, Peru, sample 5^{nr}). For Forastero beans harvested in 2015, no significant difference was found between their contents of total polyphenols; interestingly, catechin was not detected in either of the two samples. Conversely, statistically significant differences were found between epicatechin and procyanidins contents. For Forastero beans harvested in 2016, total polyphenols are higher in the sample from Satipo region than in the one from Junìn region, but it must be taken into consideration that the total polyphenols content varies also according to climate variations, thus possibly from one year to another, as it can be seen for the two Junìn region samples of 2015 and 2016 (10.55 and 5.16 mg/g total polyphenols respectively). Moreover, a small difference was found between their contents of catechin, whereas the difference between epicatechin amounts is more evident. The differences between polyphenolic compositions are evident for the two samples of Trinitario variety from Sambirano region, Madagascar (sample 2^{nr}) and Bali, Indonesia (sample 6^{nr}). Sample 6^{nr}, not roasted, is the highest in polyphenols among all the samples analyzed in the present study (12.96 mg/g total polyphenols), and its polyphenols content consists mainly of procyanidins (9.17 mg/g). Sample 2^{nr} shows a consistently lower total polyphenols content (1.44 mg/g), of which 1.00 mg/g procyanidins remaining the most represented polyphenolic subclass. Monomeric flavanols are also higher in the sample from Indonesia, but the epicatechin/catechin ratio is better for the other sample (12.9 vs. 27.9). Again, it must be noted that the samples 2^{nr} and 6^{nr} differ not only concerning their geographical origins but also regarding the years of production (2015 and 2016).

3.2. VOCs in Cocoa Beans

After a successful fermentation process, it is necessary to reduce the water content of the cocoa seeds to between 5% and 8% and this is achieved by drying [33]. The drying process is not only important in preserving the cocoa seeds but also plays a very crucial role in the development of cocoa flavor and the overall quality of the raw cocoa seeds. Cocoa is dried to minimize the formation of molds and to reduce the acid level and astringency of the beans. Then the roasting of the cocoa seeds takes place in the consumer countries. Among the cacao beans analyzed, the 7^{nr} sample (Criollo) beans will be used by a chocolate maker that produces raw chocolate, so beans will be not roasted and never reach temperatures of more than 42 °C. Especially for this product, the quality of cocoa beans is a very big determinant of the final taste of the chocolate. Cocoa flavor resides in volatile fraction, which is composed of a complex mixture of up to 600 compounds with new research continuously increasing this number [12].

HS-SPME coupled to GC-MS has proven a valuable tool for analysis of volatile and semi-volatile compounds from cocoa and chocolate products. The technique is very sensitive to experimental conditions and in this study, the DVB/CAR-PDMS fiber was found to afford the most efficient extraction of both volatile and semi-volatile compounds from the analyte's headspace according to literature [34].

The HS-SPME-GC-MS analysis of raw cocoa seeds allowed the extraction of a complex mixture of VOCs and the compounds with higher intensity were selected and reported in Table 3. The key VOCs considered in this work belonged to the class of alcohols, aldehydes, esters, acids, ketones, pyrazines, and terpenes and they have all been previously reported in other works [12,34].

The main VOCs were associated with the aroma of vinegar (acetic acid) and with the aroma of roasted and nutty (tetramethyl-pyrazine). High level of acetic acid could influence in a negative way the final aroma of chocolate, moreover pyrazines were considered important contributors to the desirable chocolate aroma [12] and changed during roasting [35]. Additionally, the aldehydes 2-methylbutanal and 3-methylbutanal are reported to have a strong influence on the chocolate flavor [12]. Phenylethyl alcohol and 3-Methylbutyl acetate were considered key aroma compounds and were associated to the odor of floral, rose and sweet, fruity, banana, respectively. There were significant differences among the types of beans, in particular for the abovementioned VOCs (Table 3). The 2[nr] (Madagascar) sample showed highest level of acetic acid and tetramethyl-pyrazine, instead of 1[nr] (Camino Verde) showing the lowest values. O'Payo beans (sample 6[nr]) showed highest levels of the aldehydes with chocolate and almond aroma, 2-methylbutanal, 3-methylbutanal and benzaldehyde. Cheni beans (sample 5[nr]) showed highest level of banana flavor (3-Methylbutyl acetate) and Coviriali beans (sample 4[nr]) the highest value of rose aroma (Phenylethyl alcohol).

The 7[nr] sample (Criollo) showed a large number of VOCs, not reported in Table 3, as 2-Heptanol for the class of alcohols, as 2-Pentanol acetate for esters and as 2-Heptanone for ketones. The high variety of VOCs, the low level of acetic acid and the good quantity of pyrazines and aldehydes confirmed the high quality of this variety and the possible use without roasting to produce raw chocolate.

Forastero, Criollo, Trinitario, and Nacional, the variety grown in Ecuador, exhibit differences in flavor characteristics that can be attributed to original variety but also growing conditions and geographical origin [36]. The extent to which other factors such as climate and soil chemical compositions influence the formation of flavor precursors and their relationships with final flavor quality remains unclear [36]. Some authors have studied the influence of cocoa's origin on the composition in volatile compounds and profile comparison allowed beans, liquor, and chocolate from various geographical origins to be distinguished [29,37].

Trinitario samples from Madagascar (2[nr]) and Nicaragua (6[nr]) exhibited high differences in many of key VOCs considered and also Forastero samples from different areas of Peru, 4[nr] and 5[nr], showed significant differences in volatile composition.

Cocoa beans underwent the roasting process by means of a dry heat treatment, changing chocolate flavor. Flavor precursors developed during fermentation interact in the roasting process. Aldehydes and pyrazines are among the major compounds formed during roasting. They are formed through the heat induced Maillard reaction and Strecker degradation of amino acids and sugars [35]. The roasting process not only generated and increased the concentration of some flavor compounds through pyrolysis of sugars, but also reduced the amount of minor compounds affecting the final quality of chocolate [38]. The degree of chemical changes depends on the temperature applied during the process [38]. The HS-SPME-GC-MS analysis of roasted beans confirmed changes in VOCs during roasting as the decrease of acetic acid, especially in samples 4 where a higher level was present, and the increase of pyrazines in both samples (Table 3). Additionally, 3-methylbutanal increased with roasting, especially at 95 and 110 °C, while at 125 °C there was a return to initial values, due to the prevalence of the volatilization phenomenon compared to the production one. The loss of minor compounds that influence the chocolate's aroma as phenylethyl alcohol and benzaldehyde was observed by increasing the roasting temperature, in particular at 125 °C.

Table 3. Key aroma volatiles in raw samples and in roasted samples together with their odor descriptors, normalized peak area from Q (quantitation)-ion, and Internal Standard (IS). Data obtained from triplicate analysis with SD < 5%.

Odor Descriptor	Compound Name	1[nr]	2[nr]	4[nr]	5[nr]	6[nr]	7[nr]	4[r1]	4[r2]	4[r3]	6[r1]	6[r2]
	Alcohols											
Fruity, creamy, buttery floral, rose	2,3-Butanediol	0.03 b	0.36 c	0.63 e	1.31 f	3.96 h	nd a	0.44 d	0.45 d	0.37 c	3.22 g	7.76 i
	Phenylethyl alcohol	34.98 e	16.39 b	72.20 i	5.20 a	42.72 f	23.03 c	77.84 j	128.69 k	48.64 h	45.53 g	27.08 d
	Aldehydes											
Aldehydic, chocolate	2-methyl-Butanal	nd a	nd a	nd a	0.003 b	0.01 c	0.003 b	nd a	nd a	nd a	0.01 c	0.01 c
aldehydic, chocolate	3-methyl-Butanal	0.02 b	0.01a	0.02 b	0.01 a	0.05 e	0.03 c	0.03 c	0.04 d	0.02 b	0.13 f	0.13 f
sweet, bitter, almond, cherry	Benzaldehyde	16.24 g	4.30b	12.25 d	1.61 a	46.55 k	12.98 e	15.28 f	26.92 i	10.23 c	33.36 j	22.69 h
	Esters											
Sweet, fruity, banana	3-Methylbutyl acetate	3.46 a	15.96 h	7.61 e	23.13 j	6.54 c	17.85 i	9.63 g	9.38 f	4.53 b	7.64 e	6.96 d
floral, rose, honey, tropical	Acetic acid, 2-phenylethyl ester	3.38 b	5.96 c	22.19 i	2.46 a	10.54 d	13.80 f	17.02 g	41.48 j	18.03 h	11.39 e	10.69 d
	2,3-Butanediol diacetate	0.04 b	1.39 i	0.44 f	1.46 k	0.26 c	nd a	0.38 e	0.42 f	0.74 h	0.32 d	0.62 g
	Acids											
Acidic	2-methyl-Propanoic acid	0.17 e	0.07 b	0.16 e	0.26 f	0.38 i	0.30 g	0.09 c	0.06 a	0.15 d	0.36 h	0.32 g
Cheese, pungent, fruity	3-methyl-Butanoic acid	0.31 f	0.11 b	0.22 d	0.27 e	0.50 i	0.45 h	0.13 c	0.10 a	0.22 d	0.45 h	0.40 g
vinegar	Acetic acid	4.39 a	14.32 j	12.01 i	7.85 f	7.32 e	7.94 fg	6.18 c	5.23 b	8.06 h	6.50 d	8.01 g
fatty, sweat, cheese	Hexanoic acid	nd a	0.36 e	0.73 f	0.09 b	0.26 c	nd a	0.36 e	0.38 e	0.09 b	0.25 c	0.30 d
	4-hydroxy-Butanoic acid	0.04 c	0.03 b	0.06 d	0.02 a	0.02 a	0.03 b	0.02 a	0.02 a	0.03 ab	0.02 a	0.02 a
	Ketones											
Sweet, pungent, caramel buttery, milky, fatty	2,3-Butanedione	0.02 a	0.53 h	0.09 d	0.08 c	0.17 g	0.04 b	0.08 c	0.04 b	0.08 cd	0.15 f	0.13 e
	3-hydroxy-2-Butanone	0.20 a	1.76 g	1.40 e	1.52 f	2.39 i	0.45 b	1.38 e	0.92 c	1.31 d	3.22 k	1.91 h
Sweet, pungent, mimosa, almond	Acetophenone	0.68 c	0.58 b	1.04 e	0.08 a	3.46 h	0.61 b	0.83 d	1.01 e	0.69 c	2.69 g	1.27 f
musty, nutty, coumarin	2-acetylpyrrole	0.04 a	0.40 d	0.10 b	0.09 b	0.17 c	0.53 e	0.57 f	1.10 h	0.62 g	1.58 h	1.45 i
	Pyrazines											
Nutty, cocoa, roasted, peanut	Trimethyl-Pyrazine	0.36 a	4.63 i	1.64 c	1.83 d	1.57 b	2.29 e	2.37 f	2.62 g	3.28 h	7.67 k	7.60 j
Nutty, cocoa, peanut-like	Tetramethyl-Pyrazine	0.78 a	76.59 k	31.99 g	38.88 h	22.90 d	18.78 b	28.25 f	19.54 c	40.39 i	58.60 j	28.08 e
	Terpenes											
Citrus	Limonene	0.16 c	nd a	0.03 c	0.02 b	Nd a	0.05 d	0.02 b	0.02 b	0.02 b	nd a	nd a

Different letters express significant differences ($p < 0.05$). nd means not detected. [nr] not roasted. [r1] 95 °C; [r2] 110 °C; [r3] 125 °C.

3.3. BAs in Cocoa Beans

It can be assumed that, prior to roasting, the bacterial decarboxilation of the amino acids plays the main role in the BAs production in fresh cocoa beans. In fact, during fermentation, cocoa proteins can be hydrolyzed by microorganisms to release free amino acids, although their total amount can considerably vary [39]. Usually, low amounts of total free amino acids, mostly acidic, were detected in the unfermented seeds. It has been shown that after fermentation, acidic free amino acids decreased, while total free amino acids, as well as hydrophobic free amino acids, increased [40]. The latter aspect seems to be related to the characteristics of the aspartic endoprotease and the carboxypeptidase present in cocoa beans, as the first preferentially attacks hydrophobic amino acids of the storage proteins and the second releases single hydrophobic amino acids [40]. Considering the different optimal temperature and pH of these enzymes, proteolysis primarily depends on the fermentation conditions: duration and intensity of acidification, temperature, and aeration [39]. Once free amino acids are released, they can undergo decarboxylase activity by some bacterial enzymes to form amines [41].

Microbiota evolution during cocoa bean fermentation has been studied extensively, also owing to its importance in the formation of the precursor compounds of the cocoa flavor [42]. It was found that yeasts, filamentous fungi, lactic and acetic acid bacteria as well as members of the genus Bacillus, are typically present, all of them being able to produce BAs [43]. As a consequence of the protection mechanism of bacteria against the acid medium, decarboxylase activity is favored by low pH values during fermentation [44]. Moreover, contaminating bacteria can also decarboxylate amino acids to support a further accumulation of BAs.

In Table 4 BA distributions and total amounts in fermented, not-roasted cocoa beans of different origin are reported. Quantities of total BAs ranged from 13 mg kg^{-1} in sample 3a to 28.8 mg kg^{-1} in sample 7^{nr}, never reaching hazardous concentrations. Total BAs concentrations collected in Table 4 are in agreement with a recent study, who recorded the evolution of BAs in fresh cocoa beans over a fermentation period of seven days, reaching a maximum level of 39.6 mg kg^{-1} at the fourth day of fermentation [5]. On the contrary, considering samples of the same geographical origin, Oracz and Nebesny (2014) reported for raw cocoa beans from Ecuador and Indonesia, much lower total BAs content, not exceeding 5.0 and 6.0 mg kg^{-1}, respectively. However, only five BAs were considered in this study, neglecting natural polyamines PUT, SPM, and SPD, as well as, HIS and CAD, representing in our study the most abundant compounds [44].

As can be seen from data in Table 4 some variations are present, depending on the sample. This is not surprising, as it was already underlined that, when analyzing samples of cocoa beans from different countries, several attributes can be very different [45]. In fact, wide variations have been obtained considering cocoa beans coming from big producing countries, much more emphasized in samples obtained from smaller producing countries. Moreover, differences were not only country-dependent, but also farmer-dependent, as significant discrepancies were found in quality attributes of cocoa beans from the same country [39]. To this regard, it is noteworthy that samples 3^{nr} and 4^{nr} were collected from the same farm but harvested respectively in 2015 (sample 3^{nr}) and 2016 (sample 4^{nr}). As it can be seen, BAs profiles and concentrations did not significantly differ, implying a high degree of farming standardization.

Table 4. Biogenic amines (BAs) in fermented and roasted cocoa beans samples. Results in mg kg^{-1} vegetal material.

BAs	1^{nr}	2^{nr}	3^{nr}	4^{nr}	4^{r1}	4^{r2}	4^{r3}	5^{nr}	6^{nr}	6^{r1}	6^{r2}	7^{nr}
PHE	1.1 ± 0.1 [bc]	nd [a]	nd [a]	nd [a]	nd [a]	nd [a]	1.3 ± 0.2 [cd]	1.5 ± 0.1 [d]	1.0 ± 0.1 [b]	1.5 ± 0.1 [d]	2.1 ± 0.1 [e]	nd [a]
PUT	3.4 ± 0.2 [d]	1.5 ± 0.1 [a]	2.5 ± 0.1 [bc]	2.6 ± 0.1 [c]	4.0 ± 0.1 [e]	5.6 ± 0.2 [g]	7.3 ± 0.2 [h]	2.4 ± 0.1 [b]	1.5 ± 0.1 [a]	4.5 ± 0.2 [f]	10.9 ± 0.3 [i]	1.5 ± 0.1 [a]
CAD	1.8 ± 0.1 [c]	1.1 ± 0.1 [a]	1.3 ± 0.1 [b]	1.3 ± 0.1 [b]	1.8 ± 0.1 [c]	2.0 ± 0.1 [d]	2.1 ± 0.2 [d]	1.2 ± 0.1 [ab]	1.1 ± 0.1 [a]	1.9 ± 0.1 [cd]	2.6 ± 0.2 [f]	2.1 ± 0.1 [e]
HIS	4.1 ± 0.1 [c]	5.3 ± 0.2 [d]	3.5 ± 0.1 [b]	3.1 ± 0.1 [a]	8.1 ± 0.2 [f]	10.0 ± 0.3 [g]	10.3 ± 0.2 [g]	3.5 ± 0.1 [b]	5.6 ± 0.2 [de]	10.0 ± 0.2 [g]	12.8 ± 0.3 [h]	5.9 ± 0.2 [e]
TYR	4.9 ± 0.2 [d]	1.8 ± 0.1 [b]	nd [a]	nd [a]	nd [a]	6.7 ± 0.2 [e]	7.3 ± 0.2 [f]	4.9 ± 0.2 [d]	4.3 ± 0.1 [c]	10.8 ± 0.3 [g]	11.1 ± 0.3 [g]	4.8 ± 0.1 [d]
SPD	1.3 ± 0.1 [b]	nd [a]	nd [a]	nd [a]	nd [a]	nd [a]	nd [a]	1.9 ± 0.1 [c]	6.7 ± 0.2 [d]	9.1 ± 0.2 [f]	9.7 ± 0.2 [g]	7.3 ± 0.2 [e]
SPM	5.8 ± 0.2 [c]	4.5 ± 0.1 [b]	5.9 ± 0.2 [c]	6.0 ± 0.1 [c]	6.5 ± 0.2 [d]	6.5 ± 0.2 [d]	7.3 ± 0.2 [e]	3.5 ± 0.1 [a]	6.1 ± 0.2 [cd]	8.8 ± 0.2 [f]	9.1 ± 0.2 [f]	7.2 ± 0.2 [e]
Total	22.4 ± 0.2 [e]	14.2 ± 0.2 [b]	13.2 ± 0.4 [a]	12.9 ± 0.1 [a]	20.5 ± 0.2 [d]	30.9 ± 0.8 [h]	35.6 ± 0.5 [i]	18.9 ± 0. 1 [c]	26.3 ± 0.2 [f]	46.6 ± 0.7 [j]	58.3 ± 0.3 [k]	28.8 ± 0.4 [g]

The values are expressed as means ± SD of three independent experiments. Different letters express significant differences ($p < 0.05$). nd means not detected or below limit of quantitation. [nr]not roasted; [r1]95 °C; [r2]110 °C; [r3]125 °C. PHE—β-phenylethylamine, PUT—putrescine, CAD—cadaverine, HIS—histamine, TYR—tyramine, SPD—spermidine, SPM—spermine.

Considering BAs profiles, the data obtained in this study clearly showed that BAs present in all samples at higher concentrations were SPM (3.5–7.2 mg kg^{-1}), HIS (3.1–5.3 mg kg^{-1}), PUT (1.5–3.4 mg kg^{-1}), and CAD (1.1–2.1 mg kg^{-1}), while TYR (not detected (nd)–4.9 mg kg^{-1}), SPD (nd–7.3 mg kg^{-1}) and PHE (nd–1.5 mg kg^{-1}) were present more rarely and at variable concentrations. The presence of natural polyamines in the cocoa beans is expected since they are ubiquitous in plants and all living organisms. It is also known the ability of the bacteria to produce some amines, e.g., TYR, as a protection against the acidic environment, while low levels of PHE in cocoa and derivatives seem to be associated with their aphrodisiac effects and mood lifting [46]. Guillen-Casla et al. (2012) [20] reported that TYR, PHE, serotonin, and HIS were the main amines in cocoa beans, although also PUT, dopamine, and ethanolamine have also been determined. Comparison among samples of same geographical origin (samples 1 and 7), displayed comparable (Ecuador) or higher (Indonesia) amounts of PHE, while our samples always showed much higher concentration of TYR for both raw cocoa beans [44]. In addition, do Carmo Brito et al. (2017) [5] found different results. Only tryptamine, TYR, SPD, and SPM were present during fermentation of fresh cocoa beans, while CAD and PUT where always undetectable in all the analyzed samples. SPD and SPM concentrations increased from the beginning to the end of fermentation, while TYR reached its maximum level at the fourth day of fermentation, decreasing afterward to initial contents [5].

It can be concluded that the differences already recorded for total BAs concentrations are much more evident when considering BAs profiles. This is a very common situation already underlined for many other foods supporting BAs accumulation. Considering that the aminogenesis takes origin from multiple and complex variables, all of which interact, a direct overlapping of the data arising from different studies (or from different samples of the same study, if they are not produced in the same way) is generally difficult to accomplish. Many parameters concerning either the hygienic conditions of the raw materials or the production process, as well as the preservation techniques, influence BAs levels and distributions [5,16,44].

In Table 4 the evolution of BAs concentration evaluated for sample 4^{nr} and 6^{nr} at different roasting temperatures (95, 110 and 125 °C samples 4^{r1}, 4^{r2}, 4^{r3} and 95 and 110 °C for sample 6^{r1}, 6^{r2}) is reported. As can be seen, the roasting temperature is strictly related to the amine total amount, reaching for sample 6^b the maximum level of 58.3 mg kg^{-1}, in agreement with Oracz and Nebesny (2014) [44]. They underlined that temperature and relative humidity of air during roasting influenced the BAs concentrations and profiles a lot. As can be noted from Table 4, sample 6^{nr} contained all the considered amine before roasting. Each amine concentration raised after processing, although to a different extent. In particular, PUT concentration showed the highest increasing factor (7.3) followed by TYR, CAD, HIS, PHE amounts with increasing factors between 2.1 and 2.6. SPD and SPM contents recorded the lowest enhancing, both with an increasing factor of 1.5. Although, with different profiles and distributions in comparison with Oracz and Nebesny (2014) [44], data obtained in our study confirmed the influence of the thermal processing on the increase of BAs concentrations in cocoa beans. This effect has been related to the transformation of free amino acids caused by the treatment at high temperature. In fact, it is now well established that during the Strecker degradation, the thermal decarboxylation of amino acids can occur in the presence of α-dicarbonyl compounds formed during the Maillard reaction or lipid peroxidation [19]. To this regard, literature data confirmed that asparagine, phenylalanine, and histidine changed in the corresponding amines 3-aminopropionamide, PHE, and HIS either in cocoa or in model systems [47].

As far as samples 4^{nr}-^{r3} are concerned, the impact of temperature on BAs concentrations during roasting showed a different behavior. Once again, all the amines exhibited higher quantities at the end of the thermal treatment mainly HIS (increasing factor 3.3) and PUT (increasing factor 2.8), followed by CAD (increasing factor 1.6) and SPM (increasing factor 1.3). To this regard, Hidalgo et al. (2013) [47] reported that the thermal degradation of histidine was more easily produced in comparison with that of phenylalanine. This effect could explain the higher increasing factors of HIS set against with those of PHE, for both samples series 6^{nr}-^{r2} and 4^{nr}-^{r3}.

As can be seen in Table 4, before roasting, sample 4^{nr} did not contain TYR and PHE, both appearing respectively only after a thermal treatment at 110 and 125 °C, supporting the idea that PHE is generated mainly by thermal decarboxylation of phenylalanine and not by biochemical reactions [44]. Additionally, in the case of TYR, traces of this compound, absent in green coffee (Rio quality), were detected in roasted samples after 16 min at 220 °C [48], thus demonstrating its "thermogenic" formation.

The different situations underlined by data in Table 4 probably depend on the complexity of the heat-induced formation of BAs. In fact, amines and amino acid-derived Strecker aldehydes, are simultaneously produced in food products during roasting, due to parallel pathways through the same key intermediates. Reactive carbonyl compounds started these degradations and the ratio between both aldehydes and amines generated is related to the carbonyl compound involved in the reaction and the experimental conditions, including amount of oxygen, pH, temperature, time, as well as the presence of other compounds such as antioxidants or amino acids [19]. In particular, additional amino acids were shown to play an important role in the preferential formation of either Strecker aldehydes or amino acid-derived amines by amino acid degradation in the presence of reactive carbonyl compounds. In this sense, the formation of PHE and phenylacetaldehyde in mixtures of phenylalanine, a lipid oxidation product, and a second amino acid was studied to determine the role of the second amino acid in the degradation of phenylalanine produced by lipid-derived reactive carbonyls. The presence of the second amino acid usually increased the formation of the amine and reduced the formation of the Strecker aldehyde to a differ extent depending on the considered amino acid [19]. The reasons for this behavior are not fully understood, although the obtained results suggested that they seem to be related to the other functional groups (mainly amino or similar groups) present in the side chains of the amino acid. To this regard, the limited aldehydes concentrations, especially at 125 °C (Table 3), could support this hypothesis.

Finally, the effect of antioxidants on BAs formation during roasting should be also considered. To this regard, the effect of the presence of phenolic compounds [49] on the degradation of phenylalanine, initiated by lipid-derived carbonyls was studied, to determine the structure-activity relationship of phenolics on the protection of amino compounds against modifications produced by carbonyl compounds. The obtained results showed that, among the different phenolic compounds assayed, the most efficient phenolic compounds were flavan-3-ols followed by single m-diphenols. The efficiency of these molecules was dependent on their ability to rapidly trap the carbonyl compounds. In this way the reaction of the carbonyl compound with the amino acid was avoided. This implies that the carbonyl-phenol reactions involving lipid-derived reactive carbonyls can be produced more rapidly than carbonyl-amine reactions, supporting the idea that antioxidants can provide a protection of amino compounds during thermal treatments of cocoa beans. In this sense, the loss of flavan-3-ols as the roasting temperature increased (Table 2) might be responsible of the limited BAs accumulation in the roasted cocoa beans.

4. Conclusions

Many classes of compounds present in cocoa nibs can be evaluated as indicators of quality and safety of raw materials and consequently of the final products.

In particular, along with a high [epicatechin]/[catechin] ratio, indicating a better bioavailability of flavanols, a high content on polyphenols could be considered as a favorable attribute of cocoa beans. This is related either to the health qualities of these compounds or to their capacity of preserving other compounds from chemical oxidation or enzymatic degradation, thus increasing stability and general characteristics of the product. According with the obtained results and taking into consideration that cocoa beans used for producing chocolate are usually roasted, the sample 6^{nr} (O'Payo 2016) appears to be the one with the best quality, showing a good content in polyphenols also after roasting at 110 °C (12.96 mg g^{-1} raw sample vs. 3.04 mg g^{-1} roasted sample. On the contrary, although showing the highest [epicatechin]:[catechin] ratio (27.9), the sample 2^{nr} seems to possess a lower quality among considered samples, in relation to its low polyphenols content (1.44 mg g^{-1}).

The monitoring of the volatile aromatic fraction, as reported for the not volatile one, suggested the same conclusions. Sample 2^{nr}, showed lower quality having high levels of acids that influence, in a negative way, the final aroma of chocolate. Besides, among the analyzed raw samples, low level of acetic acid and the highest levels of the aldehydes with chocolate and almond aroma confirmed the high quality of sample 6^{nr} (O'Payo 2016), as described from polyphenols analysis. The analysis of roasted beans confirmed changes in VOCs during roasting, as the decrease of acetic acid, especially in sample 6, and the increase of pyrazines associated with the nutty, cocoa, peanut-like aroma. The roasting temperature at 125 °C seemed to cause a loss of some minor compounds involved in the aroma of chocolate such as alcohols, aldehydes, and esters, resulting therefore excessive for the tested variety.

Considering BAs as cocoa quality markers as well, their total levels seem to indicate an opposite trend in comparison to that underlined by polyphenols and aroma compounds. In fact, among raw cocoa nibs, sample 6^{nr} showed the second higher BAs total concentration, indicating a medium quality among considered samples. However, it should be underlined that, after roasting at 110 °C, amine total amounts showed an increasing factor of 2.22 (6^{nr} vs. 6^{r2}) and of 2.37 (4^{nr} vs. 4^{r2}) implying, among the analyzed samples, a lower attitude of sample 6^{nr} to form amines during heat treatment. Anyway, from the food safety point of view, not alarming BAs amounts were found in all samples, both raw and roasted. All BAs concentrations increased after roasting, although to a different extent depending on the sample and on the considered amine. The latter aspect supports the idea that heat induced amines formation/accumulation probably during the Strecker degradation where aldehydes and amines compete to be formed, and at the same time BAs accumulation was lowered by the polyphenols intervention.

Author Contributions: Conceptualization, A.R. and D.R.; Methodology, U.G.S., F.I., M.C., D.R. and A.R.; Software, U.G.S., F.I. and M.C.; Validation, U.G.S., F.I. and M.C.; Formal Analysis, U.G.S., F.I., M.C. and D.P.; Investigation, U.G.S., F.I., M.C. and D.P.; Resources, U.G.S., F.I., M.C., D.R. and A.R.; Data Curation, U.G.S., F.I., M.C. and D.P.; Writing—Original Draft Preparation, U.G.S., F.I., M.C., D.R. and A.R.; Writing—Review and Editing, U.G.S., F.I., M.C., D.R. and A.R.; Visualization, U.G.S., F.I. and M.C.; Supervision, A.R., D.R.; Project Administration, A.R.; Funding Acquisition, A.R.

Acknowledgments: We thank Giorgio Sergio of Meraviglie S.r.l., Sommacampagna (VR), Italy for supplying us with cocoa beans and technical support and we thank Sixtus (BANDO A—POR CREO FESR 2014–2020. Linea 1.1.2) and project BIOSINOL—PSR 2014–2200 for financial support.

References

1.   Schwan, R.F.; Wheals, A.E. The microbiology of cocoa fermentation and its role in chocolate quality. *Crit. Rev. Food Sci. Nutr.* **2004**, *44*, 205–221. [CrossRef] [PubMed]

2.   Ding, E.L.; Hutfless, S.M.; Ding, X.; Girotra, S. Chocolate prevention of cardiovascular disease: A systematic review. *Nutr. Metab.* **2006**, *3*, 1–12. [CrossRef] [PubMed]

3.   Pereira-Caro, G.; Borges, G.; Nagai, C.; Jackson, M.C.; Yokota, T.; Crozier, A.; Ashihara, H. Profiles of Phenolic Compounds and Purine Alkaloids during the Development of Seeds of *Theobroma cacao* cv. Trinitario. *J. Agric. Food Chem.* **2013**, *61*, 427–434. [CrossRef] [PubMed]

4.   Camu, N.; De Winter, T.; Addo, S.K.; Takrama, J.S.; Bernaert, H.; De Vuyst, L. Fermentation of cocoa beans: Influence of microbial activities and polyphenol concentrations on the flavour of chocolate. *J. Sci. Food Agric.* **2008**, *88*, 2288–2297. [CrossRef]

5.   do Carmo Brito, B.D.N.; Campos Chisté, R.; da Silva Pena, R.; Abreu Gloria, M.B.; Santos Lopes, A. Bioactive amines and phenolic compounds in cocoa beans are affected by fermentation. *Food Chem.* **2017**, *228*, 484–490. [CrossRef] [PubMed]

6.   Hurst, W.J.; Krake, S.H.; Bergmeier, S.C.; Payne, M.J.; Miller, K.B.; Stuart, D.A. Impact of fermentation, drying, roasting and Dutch processing on flavan-3-ol stereochemistry in cacao beans and cocoa ingredients. *Chem. Cent. J.* **2011**, *5*, 53–62. [CrossRef]

7. Kothe, L.; Zimmermann, B.F.; Galensa, R. Temperature influences epimerization and composition of flavanol monomers, dimers and trimers during cocoa bean roasting. *Food Chem.* **2013**, *141*, 3656–3663. [CrossRef]

8. Arlorio, M.; Locatelli, M.; Travaglia, F.; Coïsson, J.D.; Del Grosso, E.; Minassi, A.; Appendino, G.; Martelli, A. Roasting impact on the contents of clovamide (N-caffeoyl-L-DOPA) and the antioxidant activity of cocoa beans (*Theobroma cacao* L.). *Food Chem.* **2008**, *106*, 967–975. [CrossRef]

9. Bailey, S.; Mitchell, D.; Bazinet, M.; Weurman, C. Studies of the volatile components of different varieties of cocoa beans. *J. Food Sci.* **1962**, *27*, 165–170. [CrossRef]

10. Clapperton, J.; Yow, S.; Chan, J.; Lim, D.; Lockwood, R.; Romanczyk, L.; Hammerstone, J. The contribution of genotype to cocoa (*Theobroma cacao* L.) flavour. *Trop. Agric.* **1994**, *71*, 303–308.

11. Van der Wals, B.; Kettenes, D.; Stoffelsma, J.; Sipma, G.; Semper, A. New volatile components of roasted cocoa. *J. Agric. Food Chem.* **1971**, *19*, 276–280.

12. Counet, C.; Callemien, D.; Ouwerx, C.; Collin, S. Use of gas chromatography-olfactometry to identify key odorant compounds in dark chocolate: Comparison of samples before and after conching. *J. Agric. Food Chem.* **2002**, *50*, 2385–2391. [CrossRef] [PubMed]

13. Restuccia, D.; Spizzirri, U.G.; Puoci, F.; Picci, N. Determination of biogenic amine profiles in conventional and organic cocoa-based products. *Food Addit. Contam. Part A Chem. Anal. Control Expo. Risk Assess.* **2015**, *32*, 1156–1163. [CrossRef] [PubMed]

14. Jairath, G.; Singh, P.K.; Dabur, R.S.; Rani, M.; Chaudhari, M. Biogenic amines in meat and meat products and its public health significance: A review. *J. Food Sci. Technol.* **2015**, *52*, 6835–6846. [CrossRef]

15. Araujo, Q.R.D.; Gattward, J.N.; Almoosawi, S.; Parada Costa Silva, M.D.G.C.; Dantas, P.A.D.S.; Araujo Júnior, Q.R.D. Cocoa and human health: From head to foot—A review. *Crit. Rev. Food Sci. Nutr.* **2016**, *56*, 1–12. [CrossRef] [PubMed]

16. Restuccia, D.; Spizzirri, U.G.; De Luca, M.; Parisi, O.I.; Picci, N. Biogenic amines as quality marker in organic and fair-trade cocoa-based products. *Sustainability* **2016**, *8*, 856. [CrossRef]

17. Bandanaa, J.; Egyir, I.S.; Asante, I. Cocoa farming households in Ghana consider organic practices as climate smart and livelihoods enhancer. *Agric. Food Secur.* **2016**, *5*, 29. [CrossRef]

18. Ormanci, H.B.; Arik Colakoglu, F. Changes in biogenic amines levels of lakerda (*Salted Atlantic Bonito*) during ripening at different temperatures. *J. Food Process. Preserv.* **2017**, *41*, e12736. [CrossRef]

19. Hidalgo, F.J.; León, M.; Zamora, R. Amino acid decarboxylations produced by lipid-derived reactive carbonyls in amino acid mixtures. *Food Chem.* **2016**, *209*, 256–261. [CrossRef]

20. Guillén-Casla, V.; Rosales-Conrado, N.; León-González, M.E.; Pérez-Arribas, L.V.; Polo-Díez, L.M. Determination of serotonin and its precursors in chocolate samples by capillary liquid chromatography with mass spectrometry detection. *J. Chromatogr. A* **2012**, *1232*, 158–165. [CrossRef]

21. Romani, A.; Ieri, F.; Turchetti, B.; Mulinacci, N.; Vincieri, F.F.; Buzzini, P. Analysis of condensed and hydrolysable tannins from commercial plant extracts. *J. Pharm. Biomed. Anal.* **2006**, *41*, 415–420. [CrossRef]

22. Sànchez-Rabaneda, F.; Jàuregui, O.; Casals, I.; Andrès-Lacueva, C.; Izquierdo-Pulido, M.; Lamuela-Raventòs, R.M. Liquid chromatographic/electrospray ionization tandem mass spectrometric study of the phenolic composition of cocoa (*Theobroma cacao*). *J. Mass Spectrom.* **2003**, *38*, 35–42. [CrossRef]

23. Hammerstone, J.F.; Lazarus, S.A.; Schmitz, H.H. Procyanidin Content and Variation in Some Commonly Consumed Foods. *J. Nutr.* **2000**, *130*, 2086S–2092S. [CrossRef] [PubMed]

24. Shumov, L.; Bodor, A. An industry consensus study on an HPLC fluorescence method for the determination of (±)-catechin and (±)-epicatechin in cocoa and chocolate products. *Chem. Cent. J.* **2011**, *5*, 39–45. [CrossRef] [PubMed]

25. Spizzirri, U.G.; Parisi, O.I.; Picci, N.; Restuccia, D. Application of LC with evaporative light scattering detector for biogenic amines determination in fair trade cocoa-based products. *Food Anal. Methods* **2016**, *9*, 2200–2209. [CrossRef]

26. Hümmer, W.; Schreier, P. Analysis of proanthocyanidins. *Mol. Nutr. Food Res.* **2008**, *52*, 1381–1398. [CrossRef] [PubMed]

27. Payne, M.J.; Hurst, W.J.; Miller, K.B.; Rank, C.; Stuart, D.A. Impact of fermentation, drying, roasting, and Dutch processing on epicatechin and catechin content of cacao beans and cocoa ingredients. *J. Agric. Food Chem.* **2010**, *58*, 10518–10527. [CrossRef]

28. Ioannone, F.; Di Mattia, C.D.; De Gregorio, M.; Sergi, M.; Serafini, M.; Sacchetti, G. Flavanols, proanthocyanidins and antioxidant activity changes during cocoa (*Theobroma cacao* L.) roasting as affected by temperature and time of processing. *Food Chem.* **2015**, *174*, 256–262. [CrossRef]

29. Counet, C.; Ouwerx, C.; Rosoux, D.; Collin, S. Relationship between procyanidin and flavor contents of Cocoa liquors from different origins. *J. Agric. Food Chem.* **2004**, *52*, 6243–6249. [CrossRef]

30. Di Mattia, C.D.; Sacchetti, G.; Mastrocola, D.; Serafini, M. From Cocoa to Chocolate: The impact of Processing on In Vitro Antioxidant Activity and the effects of Chocolate on Antioxidant Markers In Vivo. *Front. Immunol.* **2017**, *8*, 1207. [CrossRef]

31. Oracz, J.; Zyzelewicz, D.; Nebesny, E. The content of polyphenolic compounds in cocoa beans (*Theobroma cacao* L.), depending on variety, growing region and processing operations: A review. *Crit. Rev. Food Sci. Nutr.* **2015**, *55*, 1176–1192. [PubMed]

32. Tomas-Barberaän, F.A.; Cienfuegos-Jovellanos, E.; Marìn, A.; Muguerza, B.; Gil-Izquierdo, A.; Cerdà, B.; Zafrilla, P.; Morillas, J.; Mulero, J.; Ibarra, A.; et al. A New Process to Develop a Cocoa Powder with Higher Flavonoid Monomer Content and Enhanced Bioavailability in Healthy Humans. *J. Agric. Food Chem.* **2007**, *55*, 3926–3935. [CrossRef]

33. Afoakwa, E.O.; Quao, J.; Takrama, J.; Simpson Budu, A.; Saalia, F.K. Chemical composition and physical quality characteristics of Ghanaian cocoa beans as affected by pulp pre-conditioning and fermentation. *J. Food Sci. Technol.* **2013**, *50*, 1097–1105. [CrossRef]

34. Ducki, S.; Miralles-Garcia, J.; Zumbé, A.; Tornero, A.; Storey, D.M. Evaluation of solid-phase micro-extraction coupled to gas chromatography–mass spectrometry for the headspace analysis of volatile compounds in cocoa products. *Talanta* **2008**, *74*, 1166–1174. [CrossRef] [PubMed]

35. Heinzler, M.; Eichner, K. The role of amodori compounds during cocoa processing—Formation of aroma compounds under roasting conditions. *Z. Lebensm.-Unters.-Forsch.* **1992**, *21*, 445–450.

36. Kongor, J.E.; Hinneha, M.; Van deWalle, D.; Ohene Afoakwa, E.; Boeckx, P.; Dewettinck, K. Factors influencing quality variation in cocoa (*Theobroma cacao*) bean flavour profile—A review. *Food Res. Int.* **2016**, *82*, 44–52. [CrossRef]

37. Cambrai, A.; Marcic, C.; Morville, S.; Sae Houer, P.; Bindler, F.; Marchioni, E. Differentiation of chocolates according to the Cocoa's geographical origin using chemometrics. *J. Agric. Food Chem.* **2010**, *58*, 1478–1483. [CrossRef] [PubMed]

38. Ramli, N.; Hassan, O.; Said, M.; Samsudin, W.; Idris, N.A. Influence of roasting conditions on volatile flavor of roasted malaysian cocoa beans. *J. Food Process. Preserv.* **2006**, *30*, 280–298. [CrossRef]

39. Rohsius, C.; Matissek, R.; Lieberei, R. Free amino acid amounts in raw cocoas from different origins. *Eur. Food Res. Technol.* **2006**, *222*, 432–438. [CrossRef]

40. Adeyeye, E.I.; Akinyeye, R.O.; Ogunlade, I.; Olaofe, O.; Boluwade, J.O. Effect of farm and industrial processing on the amino acid profile of cocoa beans. *Food Chem.* **2010**, *118*, 357–363. [CrossRef]

41. Granvogl, M.; Bugan, S.; Schieberle, P. Formation of amines and aldehydes from parent amino acids during thermal processing of cocoa and model systems: New insights into pathways of the Strecker reaction. *J. Agric. Food Chem.* **2006**, *54*, 1730–1739. [CrossRef] [PubMed]

42. Lima, L.J.R.; Almeida, M.H.; Nout, M.J.R.; Zwietering, M.H. Theobroma cacao L., "the food of the gods": Quality determinants of commercial cocoa beans, with particular reference to the impact of the fermentation. *Crit. Rev. Food Sci. Nutr.* **2011**, *52*, 731–761.

43. Schwan, R.F.; Pereira, G.V.M.; Fleet, G.H. Microbial activities during cocoa fermentation. In *Cocoa and Coffee Fermentations*; Schwan, R.F., Fleet, G.H., Eds.; Taylor & Francis: London, UK, 2014; pp. 125–135.

44. Oracz, J.; Nebesny, E. Influence of roasting conditions on the biogenic amine content in cocoa beans of different *Theobroma cacao* cultivars. *Food Res. Int.* **2014**, *55*, 1–10. [CrossRef]

45. Caligiani, A.; Cirlini, M.; Palla, G.; Ravaglia, R.; Arlorio, M. GC/MS detection of chiral markers in cocoa beans of different quality and geographic origin. *Chirality* **2007**, *19*, 329–334. [CrossRef] [PubMed]

46. Shukla, S.; Park, H.K.; Kim, J.K.; Kim, M. Determination of biogenic amines in Korean traditional fermented soybean paste (Doenjang). *Food Chem. Toxicol.* **2010**, *48*, 1191–1195. [CrossRef] [PubMed]

47. Hidalgo, F.J.; Navarro, J.L.; Delgado, R.M.; Zamora, R. Histamine formation by lipid oxidation products. *Food Res. Int.* **2013**, *52*, 206–213. [CrossRef]

48. Oliveira, S.D.; Franca, A.S.; Gloria, M.B.A.; Borges, M.L.A. The effect of roasting on the presence of bioactive amines in coffees of different qualities. *Food Chem.* **2005**, *90*, 287–291. [CrossRef]

49. Hidalgo, F.J.; Delgado, R.M.; Zamora, R. Protective effect of phenolic compounds on carbonyl-amine reactions produced by lipid-derived reactive carbonyls. *Food Chem.* **2017**, *229*, 388–395. [CrossRef]

6

Impact of Biogenic Amines on Food Quality and Safety

Claudia Ruiz-Capillas * and Ana M. Herrero

Department of Products, Institute of Food Science, Technology and Nutrition, ICTAN-CSIC,
Ciudad Universitaria, 28040 Madrid, Spain; ana.herrero@ictan.csic.es
* Correspondence: claudia@ictan.csic.es

Abstract: Today, food safety and quality are some of the main concerns of consumer and health agencies around the world. Our current lifestyle and market globalization have led to an increase in the number of people affected by food poisoning. Foodborne illness and food poisoning have different origins (bacteria, virus, parasites, mold, contaminants, etc.), and some cases of food poisoning can be traced back to chemical and natural toxins. One of the toxins targeted by the Food and Drug Administration (FDA) and European Food Safety Authority (EFSA) is the biogenic amine histamine. Biogenic amines (BAs) in food constitute a potential public health concern due to their physiological and toxicological effects. The consumption of foods containing high concentrations of biogenic amines has been associated with health hazards. In recent years there has been an increase in the number of food poisoning cases associated with BAs in food, mainly in relation to histamines in fish. We need to gain a better understanding of the origin of foodborne disease and how to control it if we expect to keep people from getting ill. Biogenic amines are found in varying concentrations in a wide range of foods (fish, cheese, meat, wine, beer, vegetables, etc.), and BA formation is influenced by different factors associated with the raw material making up food products, microorganisms, processing, and conservation conditions. Moreover, BAs are thermostable. Biogenic amines also play an important role as indicators of food quality and/or acceptability. Hence, BAs need to be controlled in order to ensure high levels of food quality and safety. All of these aspects will be addressed in this review.

Keywords: biogenic amines; food products; food quality; food safety; quality control; quality indexes; public health; legislation–regulation; analytical determination

1. Biogenic Amines and Food Safety

Food safety is one of the main concerns of consumer and health agencies around the globe (European Food Safety Authority (EFSA), Food and Drug Administration (FDA), Food Safety Commission of Japan (FSCJ), World Health Organization (WHO), etc.). According to the WHO, more than 200 diseases are transmitted by food and the vast majority of the population will contract a foodborne disease at some point in their lifetime. For example, in the U.S. 48 million people (one in six) suffer a foodborne disease each year. Of these, 128,000 are hospitalized and 3000 die from such diseases [1,2]. Moreover, the real numbers are higher as many cases of foodborne disease go undetected and are not recorded as such, due to the difficulty in establishing a causal relationship between food contamination and illness or death. This highlights the importance of making sure that the food we consume is not contaminated with potentially harmful elements at any point along the food chain. Because food can become contaminated at any point along the global supply chain during production, distribution, and preparation–consumption, each individual along this chain, from producer to consumer, has a role to play in ensuring that the food we eat does not cause disease.

Furthermore, if we are to prevent such disease, we must gain a deeper understanding of the origin of foodborne illness and the way to control it.

The origin of foodborne illness could be bacteria, virus, parasite, mold, contaminants, metals, allergens, pesticides, natural toxins, etc., that can contaminate food and cause disease. In general, most food poisoning is caused by bacteria, viruses, and parasites as opposed to toxic substances. Nonetheless, there are cases of food poisoning that can be linked to chemical or natural toxins. From among these toxins, the FDA and EFSA pay particular attention to aflatoxins, mycotoxins, histamine, etc. Of these, it is worth noting that histamine, a biogenic amine, is present in most foods but in greater abundance in fish and fishery products. This biogenic amine is the main component in "scombrid poisoning" or "histamine poisoning" since these intoxications are related to the consumption of fish of the *Scombridae* and *Scomberesocidae* families (tuna, mackerel, bonito, bluefish, etc.) containing high levels of histamine. These species contain high levels of the free amino acid histidine in their muscle tissue, which is decarboxylated to histamine. However, other non scombroid species also contain high levels of free histamine in their muscle tissue [3,4], which is why this illness came to be known as "histamine poisoning". There have been recent cases involving vacuum-packed salmon. The most common symptoms of histamine poisoning are due to the effects it has on different systems (cardiovascular, gastrointestinal, respiratory, etc.) producing low blood pressure, skin irritation, headaches, edemas, and rashes typical of allergic reactions [5,6]. Furthermore, histamine plays a role in the health problem known as histaminosis or histamine intolerance associated with the increase of histamine in plasma [4]. It is also important to point out that histamine is a mediator of allergic disorders. Biogenic amines are released by mast cell degranulation (in response to an allergic reaction) and the consumption of foods containing histamine can have the same effect. Since food allergy symptoms are similar to those of histamine poisoning (food intolerance), physicians occasionally make a faulty diagnosis. For all these reasons, histamine is the biogenic amine (BA) causing major concerns in clinical and food chemistry. However, we would note that apparently histamine is not the only agent causing scombroid poisoning [7–12]. Other amines, such as putrescine and cadaverine, are also associated with this illness, although both seem to have much lower pharmacological activity on their own but enhance the toxicity of histamine and decrease the catabolism of this amine when they interact with amine oxidases, thus favoring intestinal absorption and hindering histamine detoxification [13,14].

Another important biogenic amine related to food poisoning is tyramine. In this case, intoxication is known as the "cheese reaction" as it is associated with the consumption of foods with high concentrations of tyramine, mainly associated with the consumption of cheese [10,14–17]. However, high levels of tyramine have also been observed in meat and meat products [14,18–21]. As in the case of histamine, this illness came to be known as "tyramine reaction" because of the main compound involved. Typical symptoms of tyramine poisoning are migraines, headaches, and increased blood pressure, since tyramine sparks the release of noradrenaline from the sympathetic nervous system [5,6,10].

Other Bas, such as spermidine or spermine, have also been associated with food allergies [6,22,23]. Tyramine and β-phenylethylamine are suspected of triggering hypertensive crises in certain patients and of producing dietary-induced migraines. Although tryptamine has toxic effects on humans (causing blood pressure to increase, thus leading to hypertension), the maximum amount of tryptamine permitted in sausages is not regulated in some countries [23]. It is worth noting an additional toxicological risk associated with BAs, mainly secondary BAs (putrescine and cadaverine), which are involved in other kinds of food poisoning, such as the formation of nitrosamines, that are believed to be cancer causing compounds [24,25]. This risk is greatest in meat products with high biogenic amine levels and which contain nitrite and nitrate salts used as curing agents, and also with heat treated products, as these factors favor interaction between BAs and nitrites to form nitrosamines [25,26]. However, under normal circumstances, the human body possesses detoxification systems to take care of these BAs, mainly in the intestine through the action of monoamine oxidase (MAO; CE 1.4.3.4), diamine oxidase (DAO; CE 1.4.3.6), and polyamine oxidase (PAO; CE 1.5.3.11). However, in certain cases this mechanism can be hindered by a variety of factors or circumstances, and BAs could accumulate in the body and cause serious toxicological problems and a high risk of poisoning [4,14]. Factors that could alter the detoxification mechanism include the consumption of amine oxidase inhibitors (mono and diamine oxidase inhibitors (MAOI/DAOI)), alcohol, immune deficiency of the consumer, gastrointestinal disorders, large amounts of BA, for example in the case of spoiled or fermented foods, etc. [5,13,27]. When calculating BA intake, one must consider that foods are not typically consumed in isolation but rather in the context of a meal where several foods are eaten simultaneously (meat, fish, cheese, wine, vegetables, etc.). Therefore, the aggregate amount of BA consumed would be the sum of all the amines from the different foods rather than one food considered individually. The potential toxicological effect would be the sum of the amines in all of the different foods, the synergies between them, and the other personal factors mentioned above. The role of various substances that enhance the toxicity of BA and the existence of synergic effects have been demonstrated. For example, in Europe approximately 20% of the population regularly takes MAOI and/or DAOI antidepressant drugs. In such circumstances, not even low amounts of biogenic amines can be metabolized efficiently, the result being increased sensitivity to BAs [14]. Some authors [28,29] have suggested that ripened meat products ("chorizo", "salchichón", "salami", etc.) contain enough tyramine to poison people taking MAOI even with low levels of tyramine (in the 6–9 mg/kg range). The consumption of 100 g of any of these products would interact with MAOI, while in the absence of MAOI none of these processed meats would be toxic if ingested in normal amounts, always depending of course on individual susceptibility. A new generation of MAOI has been developed that diminishes this sensitivity. The ingestion of even small amounts of tyramine has been known to cause severe migraines with intracranial hemorrhaging in patients treated with classic MAOIs, while tyramine between 50 and 150 mg is better tolerated by patients treated with a new generation of MAOIs, i.e., the so-called RIMA (reversible MAO-A inhibitor) [4,30]. The market is currently offering pharmaceutical preparations based on the DAO enzyme for the treatment of migraines whose fundamental function is to mitigate deficiencies of this enzyme (DAO), thus favoring the metabolism of histamine. It is very difficult to establish toxicity parameters for BAs considering the number of factors that affect their toxicity.

2. Biogenic Amines and Quality Control of Food Products

It is important to control and monitor biogenic amines not only for toxicological and health reasons as mentioned above, but also because they may play an important role as quality and/or acceptability indicators in some foods, and managing this quality is also a way to guarantee and ensure food safety. Food quality refers to main characteristics having to do with safety, nutrition, availability, convenience, integrity, and freshness [31].

BAs have been frequently employed as quality indexes in various foods (meat, fish, wines, etc.) to signal their degree of freshness and/or deterioration and also to control the processing and development of food and beverages. Individual BAs, such as histamine, tyramine, cadaverine, or a combination of various amines (putrescine–cadaverine, spermidine–spermine, etc.), have likewise been used as a quality index [19,26,32–40]. Also, different BA-based quality indexes have been proposed, such as the traditional one developed by Miet and Karmas [32] used as an indicator of the decomposition of fish. This index is based on the increase in putrescine, cadaverine, and histamine levels and the decrease in spermidine and spermine levels throughout the fish storage process. Scores of 0 and 1 are indicative of good quality fish, between 1 and 10 are tolerable, and a score of over 10 indicates decomposition of the product. However, in the case of other foods, such as cheese, meat, and meat products, this index has not yielded good results mainly because it does not include levels of tyramine, the main biogenic amine in these products. An alternative biogenic amine index (BAI) has been proposed for meat that consists of the sum of putrescine, cadaverine, histamine, and tyramine [40,41]. Hernández-Jover et al. [41] also suggested quality ranges for the index: BAI <5 mg/kg indicating good quality fresh meat, between 5 and 20 mg/kg for acceptable meat but with signs of initial spoilage, between 20 and 50 mg/kg for low quality meat, and >50 mg/kg for spoiled meat. However, the usefulness of BAs as a quality index depends on many factors, mainly concerning the nature of the product (fresh, canned, modified atmosphere, fermented, etc.). For example, BA indexes have proven to be more satisfactory in fresh meat and meat products and heat-treated products than in fermented products [40]. This is at least partly because biogenic amine concentrations vary much more in fermented products than in fresh and cooked meat products owing to the number of different factors involved in their processing (ripening, maturation, starter, additives, etc.) [13,14,20,21,42–44]. Therefore, establishing a biogenic amine index that reliably predicts product quality is no simple matter. It is important to note that sometimes foods with toxic levels of BAs, such as histamine or tyramine often appear organoleptically "normal". This could be the case of tuna, salmon, or fermented chorizo where unacceptable and toxic levels of histamine are undetectable prior to consumption and therefore consumers are unable to reject products based on sensorial parameters. This is another important reason to control these compounds.

3. Biogenic Amines in Food

Biogenic amines are compounds that are commonly found in food and beverages such as meat, fish, cheese, vegetables, wine, etc. The most important BAs found in food are histamine, tyramine, putrescine, cadaverine, β-phenylethylamine, agmatine, tryptamine, serotonin (SRT), spermidine, and spermine. These dietary amines are classified according to their chemical structure as aromatic amines (histamine, tyramine, serotonin, phenylethylamine, and tryptamine), aliphatic diamines (putrescine and cadaverine), and aliphatic polyamines (agmatine, spermidine, and spermine) [33,45]. In terms of origin or synthesis, they are classified as polyamines when they are endogenous and formed naturally by animals, plants, and microorganisms, which play an important role in physiological functions (neurotransmitter, psychoactive, vasoactive, regulating gene expression, cell growth and differentiation, gastric secretions, immune response, inflammatory processes, etc.), and biogenic amines, when formed mainly by the decarboxylation of free amino acids (FAAs) from the action of decarboxylase enzymes, which are mainly of microbial origin (Figure 1).

BA formation is influenced by numerous factors (Figure 1) that can be divided into three groups: raw materials (composition, pH, ion strength, etc.), microorganisms (decarboxylase activity is attributed chiefly to *Enterobacteriaceae*, *Pseudomonadaceae*, *Micrococcaceae*, lactic acid bacteria, etc.), and processing and storage conditions (fresh, cured, fermented, refrigerated, modified atmosphere, etc.) [9,14,17,43,46–48]. These factors do not act in isolation but rather have combined effects that determine the final concentration of BAs in food. Therefore, to ensure food quality from the perspective of BAs, it is vital to use suitable raw materials to limit the presence of BAs in the end product and hence assure better quality. It should be noted that these BAs are thermostable. In other words, once these biogenic amines are produced they are very difficult to destroy by subsequent processing (pasteurization, cooking, etc.) meaning that if they are present in the raw material or product, they will still be present in the final product.

In the case of factors such as microorganisms, it is necessary to control not only the microbial load in the product but also the type of microbiota constituting that load (bacterial species and strain), that in turn depends on factors associated with the raw material and processing and storage conditions [40]. These conditions directly or indirectly affect substrate and enzyme concentrations and determine the presence of other compounds or conditions that modulate (favor or not) decarboxylase activity (pH, temperature, co-factors, etc.). Therefore, there are many factors to be considered, especially in connection with the technology applied (thermal treatments, additives, fermentation, refrigeration, packaging, etc.). Hence, suitable raw materials are not enough to limit BA formation. Processing conditions must also be optimized as they are responsible for the specific profile of the biogenic amine in the different products. For example, fermentation generally promotes BAs, and in fact, this is the group of meat products with the greatest amount and diversity of these compounds. This has to do with several factors, such as the raw material, temperature of the medium (assuring conditions favorable to starter growth), the presence and concentration of additives (sugar, salt, antimicrobial agents, etc.), the microorganisms present, etc. The large quantities of microorganisms in these products, accompanied by proteolysis, gives rise to high concentrations of the amino acids constituting the nutrients required by the bacteria and the substrate on which decarboxylase enzymes work. In some cases, the presence of BAs in fermented products has been attributed to the poor quality of raw materials and defective processing.

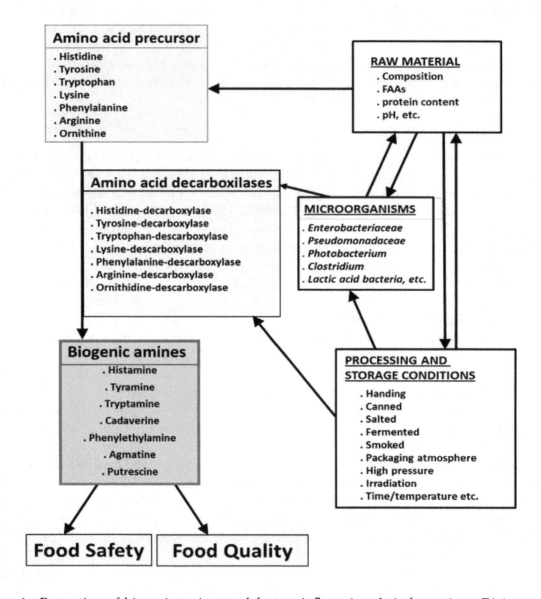

Figure 1. Formation of biogenic amines and factors influencing their formation. FAAs are free amino acids.

The storage temperature of final products is also one of the critical factors in the formation of BAs. Freezing temperatures inhibit microbial growth and therefore the production of biogenic amines. In contrast, higher chilled storage temperatures (>5 °C in fresh meat or fish) or poor temperature control foster the growth of microorganisms in products, which results in an increase in proteolysis in muscle tissue and an increase in decarboxylase enzymes and activity. Hence, low storage temperatures can make for improved quality and longer shelf-life of products. However, an increase in BAs is also related to processing and packaging conditions (modified atmosphere, vacuum, high hydrostatic pressure, irradiation, cooking products, etc.) that have an important influence on microbial flora. Controlling all of these factors improves the quality and shelf-life of food [14,34,48–50]. Today's lifestyle and global markets have led to the massive consumption of food and with this the development of new production and conservation systems and a complex food chain, that in many cases requires a deeper knowledge of how these foods are handled and forces us to face new challenges and problems in supplying safe foods.

4. Legislation Concerning Biogenic Amines in Food and Beverages

While it is very difficult to establish BA toxicity ranges owing to the many factors involved as described in the foregoing, given the dual importance of BAs (quality and health implications), efforts are being made to control BAs in food products and all countries have enacted legislation in this respect [4]. However, specific legislation only covers histamine in fishery products and no criteria have been established for other BAs or other food products, such as meat, dairy, or other products, despite the presence of important levels of BA in all types of food and the potential health risk in certain sectors of society where these products are consumed. However, in general the same legislation applicable to fish is applied to these products [19,22,40,44,51,52]. European Commission Regulations (2073/2005, 144/2007, 365/2010) set food safety criteria for histamine in fish. This legislation applies to particular fish species within the *Scombridae*, *Clupeidae*, *Eugraulidae*, *Coryphenidae*, *Pomatomidae*, and *Scomberesocidae* families throughout their shelf life with a sampling plan comprising nine units, two of which may be between 100–200 mg/kg of histamine and none above the limit of 200 mg/kg. This legislation also covers histamine levels in the processing (brine, enzyme maturation, curing, etc.) of these species with a sampling plan comprising nine units, two of which may be between 200–400 mg/kg of histamine and none above the limit of 400 mg/kg. The Australian and New Zealand standard codex feature similar levels between 100 mg/kg and none may exceed the limit of 200 mg/kg. In the U.S. the Food and Drug Administration [3] has set histamine limits in food in general at 50 mg/kg. This legislation is more advanced than its counterpart in the EU insofar as it applies to all food products.

Notwithstanding the difficulties and limitations in determining the real risk of toxicity for consumers posed by BAs in food, we should be aware that this legislation has its limitations. It is designed for one single biogenic amine (histamine) that, while admittedly one of the most important amines from a toxicological point of view, is not the only cause of toxicity. Limits should also be established for other amines, particularly tyramine, that have toxic effects, while also bearing in mind the other factors contributing to toxicity such as toxicity enhancers (individual susceptibility, consumption of MAOI, synergies resulting from the consumption of different foods during the same meal, etc.), with a view to establishing more restrictive legislation in certain cases. Although these aspects are truly difficult to address, they should be studied and included in future regulations to guarantee food safety and consumer health.

5. Analytical Determination of Biogenic Amines

As noted above, from the point of view of food safety and to assess the potential toxic effect of BAs, it is important to control and determine which BAs should be addressed. A number of swift and accurate analytical methods have been developed to determine BA levels in different foods and they were collected in various reviews [36,53–58]. These methods range from the more traditional colorimetric and fluorometric methods focused mainly on determining histamine individually, as is also the case with fast commercial kits based on the Elisa enzyme immunoassay to detect histamine in fish, to methods allowing for the simultaneous determination of several BAs (preferable) using chromatography methods such as: gas chromatography (CG) and gas chromatographic–mass spectrometry, high-performance liquid chromatography (HPLC), HPLC-tandem mass spectrometry, flow injection analysis (FIA), capillary electrophoresis, etc. (Table 1). Of all of these methods, HPLC is the most popular and frequently reported for the separation and quantification of BAs. This is the specific analytical method in European Commission (EC) [4]. This procedure offers high resolution, sensitivity, and versatility, and sample treatments are generally simple. Moreover, it offers the advantage of analyzing several BAs simultaneously. The HPLC method involves the first phase of BA extraction from the products and a second phase of determination. The extraction of BA is conducted using different solvents, such as hydrochloric acid, trichloroacetic acid, perchloric acid, methanol, etc., for the extraction procedure depending on the type of matrix (Table 1). The complexity of these matrices is a critical consideration for the adequate recovery of all BAs and to prevent interference with other compounds in the samples. This phase is also necessary in many other methods (Table 1). Chromatographic determination by HPLC is generally used pre- and post-column with reverse phase or ion exchange columns. Depending on the type of column employed, different derivate reagents are used to increase the sensitivity of the determination since BAs have low volatility and lack chromophores. The reagents commonly used in the literature are: dansyl chloride, ortho-ophtaldehyde (OPA), benzoyl chloride, p-phenyldiazonium sulfonate, 3-(4-fluorobenzoyl)-2-quinolinecarboxaldehyde, methanesulfonic acid, etc. Of these, OPA and dansyl chloride are the ones most widely used. The type of derivatization reagent used has implications for detection systems: UV/Vis, diode array, and fluorescence detector (Table 1). Important advances in analytical methods have paved the way for the use of more routine methods, such as flow injection analysis (FIA), which has been successfully used to determine BAs. This methodology offers a number of advantages, such as easy control of the chemical reaction, rapid reaction in the system, all reagent additions are performed automatically, etc. Moreover, FIA methods have been extensively used in combination with mass spectroscopy and with immobilized enzymes and electrodes or reactors using several different enzymes (amine oxidase, peroxidase, histaminase, etc.) to determine BAs in various elements by means of amperiometry or chemiluminescence. This has marked a major step forward in biosensor-assisted FIA determination of BAs [53].

Table 1. Methods and conditions for the determination of biogenic amines in food samples.

Analyte	Method/ Equipment	Sample	Extraction Solvents	Separation Technique	Derivatization Reagents	Detection System	Time of Analysis (min)	LOD	Ref
HIS	Fluorometric	Seafood, meats, cheeses, sauerkraut, etc.	MeOH	—	OPA	PF	—	0.02 mg/100 g	[59]
HIS	Fluorometric	Fish (fresh, dry, salted, frozen, brine, etc.)	MeOH	Ion exchange resin	OPA	PF	—	–	[60]
HIS	Colorimetric	Fish (tuna, mackerel),	NaCl solution	—	p-phenyldiazonium sulfonate	UV/vis	—	1 mg/100 g	[61]
HIS	ELISA immunoassay	Fish, wine			—	UV/vis	10	–	[62]
HIS, Tyr, Cad, Put, Phe, Trp, Spd, Spm	TLC	Cod, squid, MRS, TSB	TCA	—	Dansyl chloride	PF	—	5 ng–10 ng (1 mg/L per 10 µL spotted)	[63, 64]
HIS, Cad, Put, Spm	IEC	Tuna fish	MSA, HCL, PCA, PB	IonPac CS17	MSA	EDC	20	0.15–0.50 mg/kg	[65]
HIS, Tyr, Cad, Put, Spd	HPLC	Milk (cow, goat)	PCA	ODS2-C18	Benzoyl Chloride	UV/vis	13	0.03–1.30 mg/L	[66]
His, Tyr, Phe, Try, Cad, Put Spm Spd	HPLC	Sausages, cheese	PCA, HCL	Eclipse XDB-C18	Dansyl chloride, Fluorenylmethoxy-carbonyl chloride, Benzoyl chloride, Dansyl chloride	UV/vis	25–50	0.03–0.38 mg/kg,	[67]
HIS, Tyr, Cad, Put, Phe, Trp, Spd, Spm	HPLC or UHPLC	Meat, beer, wine, rice, mushroom, sausage, juice, oil, peanut butter, fish, shrimp sauce, etc.	PCA TCA HCL	ODS2-C-18, Nova-Pak C18, Zorbax XDB C18	Dansyl chloride	UV/vis	6–30	0.01–0.10 mg/kg 4.43–6.96 µg/L	[68–71]
HIS, Tyr, Cad, Put, Phe, Spd, Spm, Ser, Met, Etm	HPLC or UHPLC	Wines, meat, beverages, coffee	TCA	Phenomenex Luna 5u RP-18 Kromasil	Dansyl chloride	DAD	35	0.5 mg/kg	[72, 73]
Trp, Phe, Put, Cad, HIS, Tyr, Spm, Spd	HPLC	Chicken carcasses	PCA	C18	Dansyl chloride	FLD	32	0.05–25 µg/mL	[74]
HIS, Met, Etm, Tyr, Phe, Put, Cad	HPLC	Wine	–	Nova-Pak C18	OPA	FLD	42	0.006–0.057 mg/L	[62]
HIS, Tyr, Cad, Put, Phe, Agm, Trp, Spd, Spm	HPLC or UHPLC	Meat, fish, squid, prawn	TCA PCA	Cation exchange-Capcell Pak MG-C18	OPA	FLD	25–55	0.05–0.2 mg/L 0.2–2.0 µg/L	[75–77]

Table 1. *Cont.*

Analyte	Method/ Equipment	Sample	Extraction Solvents	Separation Technique	Derivatization Reagents	Detection System	Time of Analysis (min)	LOD	Ref
Met, Etm, HIS, Tym, Trp, Phe, Put, Cad	HPLC	Canned tuna fish	TCA	Inertsil ODS-3	Naphthalene-2,3-dicarboxaldehyde	FLD	50	2.5–330 mg/kg	[78]
HIS, Tyr, Phe, Ser, Trp, Oct, Dopa, Cad, Put, Agm, Spd, Spm	HPLC	Wine, cider, spinach hazelnut, banana, potato, milk, chocolate, meat	PCA	Nova-Pak C18	OPA	FLD	55–60	0.03–0.06 mg/L 0.07–0.2 mg/L ≤1.5 mg/kg	[41, 79, 80]
HIS, Tyr, Phe, Ser, Trp, Oct, Dopa, Cad, Put, Agm, Spd, Spm	UHPLC	Wine, fish, cheese, sausage	PCA	Acquity BEH C18	OPA	FLD	7	0.2–0.3 mg/L	[81]
Put, HIS, Cad, Phe, Tyr, Spd, Spm	HPLC-	Beer, cheese, fish, sausage, shrimp	TCA	Hypersil BDS C18	EAC	FLD	6	0.27–0.69 ng/mL	[82]
HIS, Tyr, Cad, Put, Phe, Agm, Trp, Spd, Spm, Ser, Oct, Dopa	HPLC	Wines	—	A Zorbax C18	NQS	DAD	45	0.2–3 mg/L,	[83]
Met, HIS, Put, Cad, Tyr, Spm, Spd, Trip, Phe, Etm	HPLC	Wine, beer	PVP	Inertsil ODS-3 column	Dansyl chloride	DAD–APCI-MS	35	0.008–40.0 mg/L	[84]
His, Tyr, Spd, Spm, Cad, Put, Agm	HPLC	Fish, cheese, meat, vegetable	HCL, TCA, MeOH	LiChrospher RP 18,	OPA	FLD–DAD	20	0.5–8.5 mg/kg	[85]
HIS, Cad, Agm, Tyr, Put, Phe	HPLC	Beer, wines	BB	Gemini C-18	p-toluenesulfonyl chloride	MS	22	0.023–12 µg/dm³	[86]
HIS, Tyr, Phe	HPLC	Cheese	HCL	Luna C18	—	MS	11	0.05–0.25 mg/kg	[87]
Cad, Put, HIS	GC	Cheese, fish	—	OV-225	Perfluoropropionyl derivatives	ECD	20	<1.1 µg/g	[88]
Etm, HIS, Put, Spm, Trp, Tyr, Phe, Met, Prp	GC	Wine	MeOH, CHCl3 (DLLME)	ZB-5MS capillary column	IBCF, PCF	MS-MS	25	<4.1 µg/L.	[89]
Cad, Put HIS	GC	Apple juice	DLLME	CC-DB-5	—	MS	8	0.06–2.20 µg/L	[90]
Put, Cad, HIS, Phe, Tyr	GC	Alcoholic beverages	Toluene	CC-HP-5MS	Isobutyl chloroformate	MS	12	1–10 µg/L	[91]
HIS	FIA	Mackerel, mahi-mahi	MeOH	—	OPA	FLD	—	0.8–6 mg/kg	[92, 93]
HIS	FIA	Cider, wine	—	Anion exchange mini-column	OPA	FLD	—	30–101 µg/L	[94]
HIS, Tyr, Put, Cad, Agm, Spm	FIA	Tuna	Water	Electrode-Biosensor (AO, HmDH)	OPA	APMD	—	100 pmol	[95, 96]

Table 1. *Cont.*

Analyte	Method/ Equipment	Sample	Extraction Solvents	Separation Technique	Derivatization Reagents	Detection System	Time of Analysis (min)	LOD	Ref
HIS, Phe	CE-FIA	Standard solutions	Water	—	—	MS	22	0.018–0.09 µg/mL	[97]
Put, Cad, Spm, Spd, Trp, Tyr, HIS	CE	Sauerkraut	PCA	Silica capillary	Benzoyl chloride	UV/vis	35	0.2–0.7 mg/L	[98]
HIS, Tyr, Phe, Put, Cad, Spm, Spd	CE	Soy sauce, fish, wine	TCA	Silica capillary	FBQCA	LIFD	14	0.4–10 nM	[99]
Put, Cad, Spd, Spm	CE	Fresh milk	PCA	Ag/AgCl electrode	—	APMD	27	100–400 nM	[100]
Put, HIS, Try, Phe, Spd	CE	Oyster	PCA	capillary column-Ag/AgCl	—	ECHL	30	9.2×10^{-4}–9.6×10^{-2} µg/mL	[101]
HIS, Tyr	CE	Meat, cheese, fish, vegetable	HCL, MeOH, TCA	—	—	DAD	9	2–6 mg/kg	[85]
Spm, Spd, Put, Cad, HIS, Phe, Trp, Tyr	CE	Beer, wine	—	Electrophoretic separation	—	MS	10	1–2 µg/L	[102]

Agm: agmatine, AO: amine oxidase, APMD: amperometric detection, BB: borate-buffer, Cad: cadaverine, CC: capillary column, CE: capillary electrophoresis, CHCl3: chloroform, CHMD: chemiluminescence detector, DAD: diode-array detector, DAD–APCI-MS: diode array detection–atmospheric pressure chemical ionization mass spectrometry system, DLLME: dispersive liquid microextraction, DOPA: dopamine, EAC: ethyl-acridine-sulfonyl chloride, EB: electrode-biosensor, ECD: electron capture detector, ECHL: electrochemiluminescence, EDC: electrochemical detector-conductivity, Etm: ethylamine, FBQCA: 3-(4-fluorobenzoyl)-2-quinolinecarboxaldehyde, FLD: fluorescence detection, HCL: chloridric acid, HIS: histamine, HmDH: histamine dehydrogenase, HPLC: high-performance liquid chromatography or high pressure liquid chromatography, HS-SPME: head space solid phase microextraction; IBCF: isobutyl cholorofromate; IBUT: isobutyl amine, IEC: ion-exchange chromatography, LIFD: laser-induced fluorescence detection, LLE: liquid-liquid extraction; LOD: limits of the detection, MAS: methanesulfonic acid, MeOH: methanol, Met: methylamine, MRS: Man, Rogosa and Sharpe Broth, MSA: methanesulphonic acid, MS: mass spectrometry, NQS: 1,2-naphthoquinone-4-sulfonate, Oct: octopamine, OPA: *o*-phthaldialdehyde, PB: phosphate buffer, PCA: percloric acid, PCF: propyl chloroformate, PF: photofluorometer, Phe: β-phenylethylamine, Put: putrescine, PVP: polyvinylpyrrolidone, Ser: serotonin, Spd: spermidine, Spm: spermine, TCA: tricloroacetic acid, TLC: thin layer chromatography, Trp: tryptamine, TSB: tryptic soy broth, Tyr: tyramine, UHPLC: ultra-high performance liquid chromatography, UV: ultraviolet.

6. Conclusions

There are many reasons to prevent the accumulation of biogenic amines in food products, mainly related to their utility as food quality indicators and their potential implications for consumer health. Controlling these compounds implies a deep understanding of the formation, monitoring, and reduction of biogenic amines during the processing and storage of food, and even of the effects of biogenic amines in consumers after the digestion of foods containing different levels of these compounds. Moreover, it is important to have quick, reliable, and precise analytical techniques to determine not only histamine and tyramine levels individually, but also to analyze other biogenic amines (putrescine, cadaverine, β-phenylethylamine, etc.) with implications for health and metabolic processes.

Such control of biogenic amines would benefit public authorities, industry, and consumers as it would help put higher quality products with fewer health implications on the market. However, guaranteeing the quality and safety of food requires a commitment not only from public institutions but also from production sectors, commercial processors, and ultimately from consumers who must play an important and active role in achieving food safety.

7. Future Trends and Perspectives

There are many lines of research looking into BAs in food and there are also many possibilities to be explored with regard to this subject from the technological, microbiological, analytical, and toxicological points of view.

Work should focus on determining the real risk of toxicity for consumers posed by BAs in food and should not be limited to a single amine or food product but should rather cover all the amines involved and in all foods consumed. Attention should also be given to the other factors contributing to toxicity, such as toxicity enhancers (individual susceptibility, consumption of MAOI, synergies resulting from the consumption of foods, etc.). Although these aspects are truly difficult to address, they should be studied and included in future regulations to guarantee food safety and consumer health.

Another important reason to control these compounds is the fact that often foods with toxic levels of BAs, such as histamine or tyramine, appear organoleptically 'normal' and consumers are unable to reject products based on sensorial parameters.

Moreover, today's market is trending towards the development of new products with new ingredients and new processing technologies, which create new conditions that could either favor or reduce the formation of biogenic amines. This is the case, for example, of the effect of decarboxylase enzymes responsible for their formation and the factors that modulate this activity. Therefore the implications of these new factors must be taken into account in new projects.

Important research efforts should continue in the field of analysis and determination of these BAs, always focused on the simultaneous determination of all of them, and on the different matrices, in order to solve the problems of extraction and interference of complex matrices. Also, advances need to be made in the search for more accurate, swift, simple, and unified determination methods that can easily be transferred to laboratories, industry, and the public administration.

Consequently, all research efforts should focus on the overarching goal of food safety and on providing the authorities with the tools they need to conduct swift checks of these compounds to reduce risk to consumers.

Author Contributions: C.R.-C. and A.M.H. designed and wrote the paper.

Acknowledgments: The authors wish to express heartfelt thanks to Francisco Jimenez-Colmenero for his welcome and generous support during his long, illustrious, and productive scientific career. Also thanks to Mehdi Triki for work performed during his stay in our lab.

References

1. FDA (Food and Drug Administration). Food Safety Modernization Act (FSMA). Available online: https://www.fda.gov/food/guidanceregulation/fsma/ (accessed on 18 September 2018).

2. CDCP (Centers for Disease Control and Prevention). Available online: https://www.cdc.gov/foodsafety/index.html (accessed on 25 September 2018).

3. FDA (Food and Drug Administration). Fish and Fishery Products Hazards and Controls Guidance - Fourth Edition. Available online: https://www.fda.gov/Food/GuidanceRegulation/GuidanceDocumentsRegulatoryInformation/Seafood/ucm2018426.htm (accessed on 25 October 2018).

4. EFSA. Scientific Opinion on risk based control of biogenic amine formation in fermented foods. *EFSA J.* **2011**, *9*, 2393. [CrossRef]

5. Bardócz, S. Polyamines in food and their consequences for food quality and human health. *Trends Food Sci. Technol.* **1995**, *6*, 341–346. [CrossRef]

6. Kalač, P. Health effects and occurrence of dietary polyamines: A review for the period 2005-mid 2013. *Food Chem.* **2014**, *161*, 27–39. [CrossRef] [PubMed]

7. Taylor, S.L.; Eitenmiller, R.R. Histamine food poisoning: Toxicology and clinical aspects. *Crit. Rev. Toxicol.* **1986**, *17*, 91–128. [CrossRef] [PubMed]

8. Lehane, L.; Olley, J. Histamine fish poisoning revisited. *Int. J. Food Microbiol.* **2000**, *58*, 1–37. [CrossRef]

9. Kim, M.K.; Mah, J.H.; Hwang, H.J. Biogenic amine formation and bacterial contribution in fish, squid and shellfish. *Food Chem.* **2009**, *116*, 87–95. [CrossRef]

10. Pegg, A.E. Toxicity of polyamines and their metabolic products. *Chem. Res. Toxicol.* **2013**, *26*, 1782–1800. [CrossRef]

11. Kovacova-Hanuskova, E.; Buday, T.; Gavliakova, S.; Plevkova, J. Histamine, histamine intoxication and intolerance. *Allergologia et Immunopathologia* **2015**, *43*, 498–506. [CrossRef]

12. Prester, L. Biogenic amines in fish, fish products and shellfish: A review. *Food Addit. Contam. Part A Chem. Anal. Control Expo Risk Assess.* **2011**, *28*, 1547–1560. [CrossRef]

13. Halász, A.; Baráth, Á.; Simon-Sarkadi, L.; Holzapfel, W. Biogenic amines and their production by microorganisms in food. *Trends Food Sci. Technol.* **1994**, *5*, 42–49. [CrossRef]

14. Ruiz-Capillas, C.; Jiménez-Colmenero, F. Biogenic amines in meat and meat products. *Crit. Rev. Food Sci. Nutr.* **2004**, *44*, 489–499. [CrossRef] [PubMed]

15. Karovičová, J.; Kohajdová, Z. Biogenic amines in food. *Chem. Papers* **2005**, *59*, 70–79.

16. Linares, D.M.; Martín, M.C.; Ladero, V.; Alvarez, M.A.; Fernández, M. Biogenic amines in dairy products. *Crit. Rev. Food Sci. Nutr.* **2011**, *51*, 691–703. [CrossRef] [PubMed]

17. Benkerroum, N. Biogenic Amines in Dairy Products: Origin, Incidence, and Control Means. *Compr. Rev. Food Sci. Food Saf.* **2016**, *15*, 801–826. [CrossRef]

18. Rice, S.L.; Eitenmiller, R.R.; Koehler, P.E. Biologically active amines in food: A review. *J. Milk Food Technol.* **1976**, *39*, 353–358. [CrossRef]

19. Hernández-Jover, T.; Izquierdo-Pulido, M.; Veciana-Nogués, M.T.; Vidal-Carou, M.C. Biogenic Amine Sources in Cooked Cured Shoulder Pork. *J. Agric. Food Chem.* **1996**, *44*, 3097–3101. [CrossRef]

20. Suzzi, G.; Gardini, F. Biogenic amines in dry fermented sausages: A review. *Int. J. Food Microbiol.* **2003**, *88*, 41–54. [CrossRef]

21. Stadnik, J.; Dolatowski, Z.J. Biogenic amines in meat and fermented meat products. *Acta Sci. Pol. Technol. Aliment* **2010**, *9*, 251–263.

22. Kalač, P.; Krausová, P. A review of dietary polyamines: Formation, implications for growth and health and occurrence in foods. *Food Chem.* **2005**, *90*, 219–230. [CrossRef]

23. Shalaby, A.R. Significance of biogenic amines to food safety and human health. *Food Res. Int.* **1996**, *29*, 675–690. [CrossRef]

24. Al Bulushi, I.; Poole, S.; Deeth, H.C.; Dykes, G.A. Biogenic amines in fish: Roles in intoxication, spoilage, and nitrosamine formation-A review. *Crit. Rev. Food Sci. Nutr.* **2009**, *49*, 369–377. [CrossRef] [PubMed]

25. De Mey, E.; De Klerck, K.; De Maere, H.; Dewulf, L.; Derdelinckx, G.; Peeters, M.C.; Fraeye, I.; Vander Heyden, Y.; Paelinck, H. The occurrence of N-nitrosamines, residual nitrite and biogenic amines in commercial dry fermented sausages and evaluation of their occasional relation. *Meat Sci.* **2014**, *96*, 821–828. [CrossRef] [PubMed]

26. Ruiz-Capillas, C.; Carballo, J.; Jiménez Colmenero, F. Biogenic amines in pressurized vacuum-packaged cooked sliced ham under different chilled storage conditions. *Meat Sci.* **2007**, *75*, 397–405. [CrossRef] [PubMed]

27. Alvarez, M.A.; Moreno-Arribas, M.V. The problem of biogenic amines in fermented foods and the use of potential biogenic amine-degrading microorganisms as a solution. *Trends Food Sci. Technol.* **2014**, *39*, 146–155. [CrossRef]

28. Vidal-Carou, M.C.; Izquierdo-Pulido, M.L.; Martín-Morro, M.C.; Mariné, F. Histamine and tyramine in meat products: Relationship with meat spoilage. *Food Chem.* **1990**, *37*, 239–249. [CrossRef]

29. Santos, C.; Jalón, M.; Marine, A. Contenido de tiramina en alimentos de origen animal. I. Carne, derivados cárnicos y productos relacionados. *Rev. Agroquim Technol. Aliment.* **1985**, *25*, 362–368.

30. McCabe-Sellers, B.J.; Staggs, C.G.; Bogle, M.L. Tyramine in foods and monoamine oxidase inhibitor drugs: A crossroad where medicine, nutrition, pharmacy, and food industry converge. *J. Food Composit. Anal.* **2006**, *19*, S58–S65. [CrossRef]

31. Herrero, A.M. Raman spectroscopy a promising technique for quality assessment of meat and fish: A review. *Food Chem.* **2008**, *107*, 1642–1651. [CrossRef]

32. Mietz, J.L.; Karmas, E. Polyamine and histamine content of rockfish, salmon, lobster, and shrimp as an indicator of decomposition. *J. Assoc. Off. Anal. Chem. (USA)* **1978**, *61*, 139–145.

33. Smith, T.A. Amines in food. *Food Chem.* **1980**, *6*, 169–200. [CrossRef]

34. Ruiz-Capillas, C.; Moral, A. Production of biogenic amines and their potential use as quality control indices for hake (*Merluccius merluccius*, L.) stored in ice. *J. Food Sci.* **2001**, *66*, 1030–1032. [CrossRef]

35. Rokka, M.; Eerola, S.; Smolander, M.; Alakomi, H.-L.; Ahvenainen, R. Monitoring of the quality of modified atmosphere packaged broiler chicken cuts stored in different temperature conditions: B. Biogenic amines as quality-indicating metabolites. *Food Control* **2004**, *15*, 601–607. [CrossRef]

36. Ruiz-Capillas, C.; Jiménez-Colmenero, F. Biogenic amines in seafood products. In *Handbook of Seafood and Seafood Products Analysis*; Leo, M.L., Nollet, F.T., Eds.; CRC Press Taylor & Francis Group: Boca Raton, FL, USA, 2009; pp. 833–850.

37. Galgano, F.; Favati, F.; Bonadio, M.; Lorusso, V.; Romano, P. Role of biogenic amines as index of freshness in beef meat packed with different biopolymeric materials. *Food Res. Int.* **2009**, *42*, 1147–1152. [CrossRef]

38. Vinci, G.; Antonelli, M.L. Biogenic amines: Quality index of freshness in red and white meat. *Food Control* **2002**, *13*, 519–524. [CrossRef]

39. Kalač, P.; Křížek, M. A review of biogenic amines and polyamines in beer. *J. Inst. Brewing* **2003**, *109*, 123–128. [CrossRef]

40. Triki, M.; Herrero, A.M.; Jiménez-Colmenero, F.; Ruiz-Capillas, C. Quality Assessment of Fresh Meat from Several Species Based on Free Amino Acid and Biogenic Amine Contents during Chilled Storage. *Foods* **2018**, *7*, 132–148. [CrossRef] [PubMed]

41. Hernández-Jover, T.; Izquierdo-Pulido, M.; Veciana-Nogués, M.T.; Vidal-Carou, M.C. Ion-Pair High-Performance Liquid Chromatographic Determination of Biogenic Amines in Meat and Meat Products. *J. Agric. Food Chem.* **1996**, *44*, 2710–2715. [CrossRef]

42. Latorre-Moratalla, M.L.; Veciana-Nogués, T.; Bover-Cid, S.; Garriga, M.; Aymerich, T.; Zanardi, E.; Ianieri, A.; Fraqueza, M.J.; Patarata, L.; Drosinos, E.H.; et al. Biogenic amines in traditional fermented sausages produced in selected European countries. *Food Chem.* **2008**, *107*, 912–921. [CrossRef]

43. Gardini, F.; Özogul, Y.; Suzzi, G.; Tabanelli, G.; Özogul, F. Technological factors affecting biogenic amine content in foods: A review. *Frontiers in Microbiology* **2016**, *7*. [CrossRef]

44. Eerola, H.S.; Roig Sagués, A.X.; Hirvi, T.K. Biogenic amines in Finnish dry sausages. *J. Food Saf.* **1998**, *18*, 127–138. [CrossRef]

45. Silla Santos, M.H. Biogenic amines: Their importance in foods. *Int. J. Food Microbiol.* **1996**, *29*, 213–231. [CrossRef]

46. Bodmer, S.; Imark, C.; Kneubühl, M. Biogenic amines in foods: Histamine and food processing. *Inflamm. Res.* **1999**, *48*, 296–300. [CrossRef]

47. Komprda, T.; Smělá, D.; Pechová, P.; Kalhotka, L.; Štencl, J.; Klejdus, B. Effect of starter culture, spice mix and storage time and temperature on biogenic amine content of dry fermented sausages. *Meat Sci.* **2004**, *67*, 607–616. [CrossRef] [PubMed]

48. Roig-Roig-Sagués, A.X.; Ruiz-Capillas, C.; Espinosa, D.; Hernández, M. The decarboxylating bacteria present in foodstuffs and the effect of emerging technologies on their formation. In *Biological Aspects of Biogenic Amines, Polyamines and Conjugates*; Dandrifosse, G., Ed.; Transworld Research Network: Kerala, India, 2009.

49. Naila, A.; Flint, S.; Fletcher, G.; Bremer, P.; Meerdink, G. Control of biogenic amines in food - existing and emerging approaches. *J. Food Sci.* **2010**, *75*, R139–R150. [CrossRef] [PubMed]

50. Kim, J.H.; Ahn, H.J.; Lee, J.W.; Park, H.J.; Ryu, G.H.; Kang, I.J.; Byun, M.W. Effects of gamma irradiation on the biogenic amines in pepperoni with different packaging conditions. *Food Chem.* **2005**, *89*, 199–205. [CrossRef]

51. Ten Brink, B.; Damink, C.; Joosten, H.M.; Huis in 't Veld, J.H. Occurrence and formation of biologically active amines in foods. *Int. J. Food Microbiol.* **1990**, *11*, 73–84. [CrossRef]

52. Bover-Cid, S.; Miguélez-Arrizado, M.J.; Vidal-Carou, M.C. Biogenic amine accumulation in ripened sausages affected by the addition of sodium sulphite. *Meat Sci.* **2001**, *59*, 391–396. [CrossRef]

53. Ruiz-Capillas, C.; Herrero, A.M.; Jiménez-Colmenero, F. Determination of biogenic amines. In *Flow Injection Analysis of Food Additives*; Ruiz-Capillas, C., Nollet, L.M.L., Eds.; CRC Press Taylor & Francis Group: Boca Raton, FL, USA, 2015; pp. 675–690.

54. Rivoira, L.; Zorz, M.; Martelanc, M.; Budal, S.; Carena, D.; Franko, M.; Bruzzoniti, M.C. Novel approaches for the determination of biogenic amines in food samples. *Stud. u. Babes-Bol. Chem.* **2017**, *62*, 103–122. [CrossRef]

55. Önal, A. A review: Current analytical methods for the determination of biogenic amines in foods. *Food Chem.* **2007**, *103*, 1475–1486. [CrossRef]

56. Mohammed, G.I.; Bashammakh, A.S.; Alsibaai, A.A.; Alwael, H.; El-Shahawi, M.S. A critical overview on the chemistry, clean-up and recent advances in analysis of biogenic amines in foodstuffs. *Trends Anal. Chem.* **2016**, *78*, 84–94. [CrossRef]

57. Ordóñez, J.L.; Troncoso, A.M.; García-Parrilla, M.D.C.; Callejón, R.M. Recent trends in the determination of biogenic amines in fermented beverages–A review. *Anal. Chim. Acta* **2016**, *939*, 10–25. [CrossRef] [PubMed]

58. Papageorgiou, M.; Lambropoulou, D.; Morrison, C.; Kłodzińska, E.; Namieśnik, J.; Płotka-Wasylka, J. Literature update of analytical methods for biogenic amines determination in food and beverages. *Trends Anal. Chem.* **2018**, *98*, 128–142. [CrossRef]

59. Taylor, S.L.; Lieber, E.R.; Leatherwood, M. A simplified method for histamine analysis of foods. *J. Food Sci.* **1978**, *43*, 247–250. [CrossRef]

60. AOAC. Histamine in seafood: Fluorometric method Sec. 35.1.32, Method 977.13. In *Official Methods of Analysis of AOAC International*; Cunniff, P.A., Ed.; AOAC International: Gaithersburg, MD, USA, 1995; pp. 6–17.

61. Patange, S.B.; Mukundan, M.K.; Kumar, K.A. A simple and rapid method for colorimetric determination of histamine in fish flesh. *Food Control* **2005**, *16*, 465–472. [CrossRef]

62. Marcobal, A.; Polo, M.C.; Martín-Álvarez, P.J.; Moreno-Arribas, M.V. Biogenic amine content of red Spanish wines: Comparison of a direct ELISA and an HPLC method for the determination of histamine in wines. *Food Res. Int.* **2005**, *38*, 387–394. [CrossRef]

63. Lapa-Guimarães, J.; Pickova, J. New solvent systems for thin-layer chromatographic determination of nine biogenic amines in fish and squid. *J. Chromatogr.* **2004**, *1045*, (1–2). [CrossRef]

64. Latorre-Moratalla, M.L.; Bover-Cid, S.; Veciana-Nogués, T.; Vidal-Carou, M.C. Thin-layer chromatography for the identification and semi-quantification of biogenic amines produced by bacteria. *J. Chromatogr.* **2009**, *1216*, 4128–4132. [CrossRef] [PubMed]

65. Cinquina, A.L.; Calì, A.; Longo, F.; De Santis, L.; Severoni, A.; Abballe, F. Determination of biogenic amines in fish tissues by ion-exchange chromatography with conductivity detection. *J. Chromatogr.* **2004**, *1032*, 73–77. [CrossRef]

66. Costa, M.P.; Balthazar, C.F.; Rodrigues, B.L.; Lazaro, C.A.; Silva, A.C.O.; Cruz, A.G.; Conte Junior, C.A. Determination of biogenic amines by high-performance liquid chromatography (HPLC-DAD) in probiotic cow's and goat's fermented milks and acceptance. *Food Sci. Nutr.* **2015**, *3*, 172–178. [CrossRef] [PubMed]

67. Liu, S.J.; Xu, J.J.; Ma, C.L.; Guo, C.F. A comparative analysis of derivatization strategies for the determination of biogenic amines in sausage and cheese by HPLC. *Food Chem.* **2018**, *266*, 275–283. [CrossRef]

68. Eerola, S.; Hinkkanen, R.; Lindfors, E.; Hirvi, T. Liquid chromatographic determination of biogenic amines in dry sausages. *J. AOAC Int.* **1993**, *76*, 575–577. [PubMed]

69. Yoon, H.; Park, J.H.; Choi, A.; Hwang, H.J.; Mah, J.H. Validation of an HPLC analytical method for determination of biogenic amines in agricultural products and monitoring of biogenic amines in Korean fermented agricultural products. *Toxicol. Res.* **2015**, *31*, 299–305. [CrossRef] [PubMed]

70. Dadáková, E.; Křížek, M.; Pelikánová, T. Determination of biogenic amines in foods using ultra-performance liquid chromatography (UPLC). *Food Chem.* **2009**, *116*, 365–370. [CrossRef]

71. Saaid, M.; Saad, B.; Hashim, N.H.; Mohamed Ali, A.S.; Saleh, M.I. Determination of biogenic amines in selected Malaysian food. *Food Chem.* **2009**, *113*, 1356–1362. [CrossRef]

72. Anli, R.E.; Vural, N.; Yilmaz, S.; Vural, Ỳ.H. The determination of biogenic amines in Turkish red wines. *J. Food Compos. Anal.* **2004**, *17*, 53–62. [CrossRef]

73. Casal, S.; Oliveira, M.B.P.P.; Ferreira, M.A. Determination of biogenic amines in coffee by an optimized liquid chromatographic method. *J. Liq. Chromatogr. Relat. Technol.* **2002**, *25*, 2535–2549. [CrossRef]

74. Tamim, N.M.; Bennett, L.W.; Shellem, T.A.; Doerr, J.A. High-performance liquid chromatographic determination of biogenic amines in poultry carcasses. *J. Agric. Food Chem.* **2002**, *50*, 5012–5015. [CrossRef]

75. Triki, M.; Jiménez-Colmenero, F.; Herrero, A.M.; Ruiz-Capillas, C. Optimisation of a chromatographic procedure for determining biogenic amine concentrations in meat and meat products employing a cation-exchange column with a post-column system. *Food Chem.* **2012**, *130*, 1066–1073. [CrossRef]

76. Sánchez, J.A.; Ruiz-Capillas, C. Application of the simplex method for optimization of chromatographic analysis of biogenic amines in fish. *Eur. Food Res. Technol.* **2012**, *234*, 285–294. [CrossRef]

77. Zhao, Q.X.; Xu, J.; Xue, C.H.; Sheng, W.J.; Gao, R.C.; Xue, Y.; Li, Z.J. Determination of biogenic amines in squid and white prawn by high-performance liquid chromatography with postcolumn derivatization. *J. Agric. Food Chem.* **2007**, *55*, 3083–3088. [CrossRef]

78. Zotou, A.; Notou, M. Enhancing Fluorescence LC Analysis of Biogenic Amines in Fish Tissues by Precolumn Derivatization with Naphthalene-2,3-dicarboxaldehyde. *Food Anal. Method.* **2013**, *6*, 89–99. [CrossRef]

79. Vidal-Carou, M.C.; Lahoz-Portolés, F.; Bover-Cid, S.; Mariné-Font, A. Ion-pair high-performance liquid chromatographic determination of biogenic amines and polyamines in wine and other alcoholic beverages. *J. Chromatogr.* **2003**, *998*, 235–241. [CrossRef]

80. Lavizzari, T.; Teresa Veciana-Nogués, M.; Bover-Cid, S.; Mariné-Font, A.; Carmen Vidal-Carou, M. Improved method for the determination of biogenic amines and polyamines in vegetable products by ion-pair high-performance liquid chromatography. *J. Chromatogr.* **2006**, *1129*, 67–72. [CrossRef] [PubMed]

81. Latorre-Moratalla, M.L.; Bosch-Fusté, J.; Lavizzari, T.; Bover-Cid, S.; Veciana-Nogués, M.T.; Vidal-Carou, M.C. Validation of an ultra high pressure liquid chromatographic method for the determination of biologically active amines in food. *J. Chromatogr.* **2009**, *1216*, 7715–7720. [CrossRef] [PubMed]

82. Li, G.; Dong, L.; Wang, A.; Wang, W.; Hu, N.; You, J. Simultaneous determination of biogenic amines and estrogens in foodstuff by an improved HPLC method combining with fluorescence labeling. *LWT Food Sci. Technol.* **2014**, *55*, 355–361. [CrossRef]

83. Hlabangana, L.; Hernández-Cassou, S.; Saurina, J. Determination of biogenic amines in wines by ion-pair liquid chromatography and post-column derivatization with 1,2-naphthoquinone-4-sulphonate. *J. Chromatogr.* **2006**, *1130*, 130–136. [CrossRef] [PubMed]

84. Loukou, Z.; Zotou, A. Determination of biogenic amines as dansyl derivatives in alcoholic beverages by high-performance liquid chromatography with fluorimetric detection and characterization of the dansylated amines by liquid chromatography-atmospheric pressure chemical ionization mass spectrometry. *J. Chromatogr.* **2003**, *996*, 103–113. [CrossRef]

85. Lange, J.; Thomas, K.; Wittmann, C. Comparison of a capillary electrophoresis method with high-performance liquid chromatography for the determination of biogenic amines in various food samples. *J. Chromatogr. B Analyt. Technol. Biomed. Life Sci.* **2002**, *779*, 229–239. [CrossRef]

86. Nalazek-Rudnicka, K.; Wasik, A. Development and validation of an LC–MS/MS method for the determination of biogenic amines in wines and beers. *Monatshefte fur Chemie* **2017**, *148*, 1685–1696. [CrossRef]

87. Calbiani, F.; Careri, M.; Elviri, L.; Mangia, A.; Pistarà, L.; Zagnoni, I. Rapid assay for analyzing biogenic amines in cheese: Matrix solid-phase dispersion followed by liquid chromatography-electrospray-tandem mass spectrometry. *J. Agric. Food Chem.* **2005**, *53*, 3779–3783. [CrossRef]

88. Staruszkiewicz, W.F., Jr.; Bond, J.F. Gas chromatographic determination of cadaverine, putrescine, and histamine in foods. *J. Assoc. Off. Anal. Chem.* **1981**, *64*, 584–591. [PubMed]

89. Płotka-Wasylka, J.; Simeonov, V.; Namieśnik, J. An in situ derivatization - dispersive liquid-liquid microextraction combined with gas-chromatography - mass spectrometry for determining biogenic amines in home-made fermented alcoholic drinks. *J. Chromatogr.* **2016**, *1453*, 10–18. [CrossRef] [PubMed]

90. Cunha, S.C.; Faria, M.A.; Fernandes, J.O. Gas chromatography-mass spectrometry assessment of amines in port wine and grape juice after fast chloroformate extraction/derivatization. *J. Agric. Food Chem.* **2011**, *59*, 8742–8753. [CrossRef] [PubMed]

91. Fernandes, J.O.; Judas, I.C.; Oliveira, M.B.; Ferreira, I.M.P.L.V.; Ferreira, M.A. A GC-MS method for quantitation of histamine and other biogenic amines in beer. *Chromatographia* **2001**, *53*, S327–S331. [CrossRef]

92. Hungerford, J.M.; Walker, K.D.; Wekell, M.M.; LaRose, J.E.; Throm, H.R. Selective Determination of Histamine by Flow Injection Analysis. *Anal. Chem.* **1990**, *62*, 1971–1976. [CrossRef] [PubMed]

93. Hungerford, J.M.; Hollingworth, T.A.; Wekell, M.M. Automated kinetics-enhanced flow-injection method for histamine in regulatory laboratories: Rapid screening and suitability requirements. *Anal. Chim. Acta* **2001**, *438*, 123–129. [CrossRef]

94. Del Campo, G.; Gallego, B.; Berregi, I. Fluorimetric determination of histamine in wine and cider by using an anion-exchange column-FIA system and factorial design study. *Talanta* **2006**, *68*, 1126–1134. [CrossRef] [PubMed]

95. Niculescu, M.; Frébort, I.; Peč, P.; Galuszka, P.; Mattiasson, B.; Csöregi, E. Amine oxidase based amperometric biosensors for histamine detection. *Electroanalysis* **2000**, *12*, 369–375. [CrossRef]

96. Takagi, K.; Shikata, S. Flow injection determination of histamine with a histamine dehydrogenase-based electrode. *Anal. Chim. Acta* **2004**, *505*, 189–193. [CrossRef]

97. Santos, B.; Simonet, B.M.; Ríos, A.; Valcárcel, M. Direct automatic determination of biogenic amines in wine by flow injection-capillary electrophoresis-mass spectrometry. *Electrophoresis* **2004**, *25*, 3427–3433. [CrossRef] [PubMed]

98. Křížek, M.; Pelikánová, T. Determination of seven biogenic amines in foods by micellar electrokinetic capillary chromatography. *J. Chromatogr.* **1998**, *815*, 243–250. [CrossRef]

99. Zhang, N.; Wang, H.; Zhang, Z.X.; Deng, Y.H.; Zhang, H.S. Sensitive determination of biogenic amines by capillary electrophoresis with a new fluorogenic reagent 3-(4-fluorobenzoyl)-2-quinolinecarboxaldehyde. *Talanta* **2008**, *76*, 791–797. [CrossRef] [PubMed]

100. Sun, X.; Yang, X.; Wang, E. Determination of biogenic amines by capillary electrophoresis with pulsed amperometric detection. *J. Chromatogr.* **2003**, *1005*, 189–195. [CrossRef]

101. An, D.; Chen, Z.; Zheng, J.; Chen, S.; Wang, L.; Huang, Z.; Weng, L. Determination of biogenic amines in oysters by capillary electrophoresis coupled with electrochemiluminescence. *Food Chem.* **2015**, *168*, 1–6. [CrossRef] [PubMed]

102. Daniel, D.; dos Santos, V.B.; Vidal, D.T.R.; do Lago, C.L. Determination of biogenic amines in beer and wine by capillary electrophoresis-tandem mass spectrometry. *J. Chromatogr.* **2015**, *1416*, 121–128. [CrossRef] [PubMed]

7

Effects of Soaking and Fermentation Time on Biogenic Amines Content of *Maesil* (*Prunus Mume*) Extract

So Hee Yoon [1], Eunmi Koh [2], Bogyoung Choi [2] and BoKyung Moon [1,*]

[1]   Department of Food and Nutrition, Chung-Ang University, Gyeonggi-do 17546, Korea; kisingserpe@naver.com

[2]   Major of Food & Nutrition, Division of Applied Food System, Seoul Women's University, Seoul 01797, Korea; kohem7@swu.ac.kr (E.K.); 5326460@hanmail.net (B.C.)

*   Correspondence: bkmoon@cau.ac.kr

Abstract: *Maesil* extract, a fruit-juice concentrate derived from *Prunus mume* prepared by fermenting with sugar, is widely used with increasing popularity in Korea. Biogenic amines in *maesil* extract were extracted with 0.4 M perchloric acid, derivatized with dansyl chloride, and detected using high-performance liquid chromatography. Among 18 home-made *maesil* extracts collected from different regions, total biogenic amine content varied from 2.53 to 241.73 mg/L. To elucidate the effects of soaking and fermentation time on biogenic amine content in *maesil* extract, *maesil* was soaked in brown sugar for 90 days and the liquid obtained was further fermented for 180 days at 15 and 25 °C, respectively. The main biogenic amines extracted were putrescine and spermidine and the total biogenic amine content was higher at 25 °C than at 15 °C. Soaking at 15 and 25 °C increased the total biogenic amines content from 14.14 to 34.98 mg/L and 37.33 to 69.05 mg/L, respectively, whereas a 180 day fermentation decreased the content from 31.66 to 13.59 mg/L and 116.82 to 57.05 mg/L, respectively. Biogenic amine content was correlated with total amino acid content (particularly, arginine content). Based on these results, we have considered that biogenic amine synthesis can be reduced during *maesil* extract production by controlling temperature and fermentation time.

Keywords: biogenic amine; *maesil*; amino acids; soaking; fermentation; temperature

---

1. Introduction

Maesil (*Prunus mume*) known as Japanese *Ume* has been used not only as a food but also as a medicine on account of its various functionalities [1–3]. As the seed of *maesil* has a toxic substance called amygdalin [4], *maesil* has been processed into various products such as alcoholic beverage, juice, pickle or extract rather than eaten raw [5]. *Maesil* extract is a fruit-juice concentrate produced by the fermentation of *maesil* and sugar. Recently, it has been increasingly used as a seasoning to impart sweetness and a unique flavor to foods [5–8]. Traditionally, *maesil* extract is soaked for a long period (90 days) at room temperature and fermented naturally under different conditions in individual households. Therefore, uncontrolled fermentation can lead to the formation of biogenic amines, which are produced by molds and bacteria.

As biogenic amines are mainly produced by the microbial decarboxylation of free amino acids, they are easily found in fermented foods [9,10]. These biogenic amines have been reported to be abundant and they have been found in a wide range of food products, including fish products, soy sauce, *Chunjang* (traditional fermented soybean paste in Korea and China), and agricultural products [11–15]. As a high intake of biogenic amines can cause various detrimental effects such as migraine and gastrointestinal problems, their ingestion needs to be restricted [16,17]. Indeed, their content is currently regulated in certain food products. For example, the histamine content in fish products is regulated by the US Food and Drug Administration (FDA, 50 mg/kg) and the European

Union (100 mg/kg) in fish products [18]. The formation of biogenic amines is influenced by microbial flora and their growth as well as the fermentation conditions used in the production of fermented foods [19,20]. To date, however, studies on the changes in biogenic amines during fruit fermentation have mainly focused on wine [9]. Moreover, little research has been conducted on the fermentation of biogenic amines during the fermentation of other fruits.

Therefore, in this study, we tried to monitor the biogenic amine content of *maesil* extracts and determine the effect of fermentation conditions on the changes in biogenic amines in *maesil* extracts during fermentation. For this purpose, we (i) determined the content of biogenic amines content in 18 home-made *maesil* extracts collected from different households in Korea and (ii) monitored the content of biogenic amines during the fermentation of *maesil* extracts at two different temperatures, 15 and 25 °C, over a period of 9 months.

2. Materials and Methods

2.1. Chemicals

Biogenic amine standards (histamine dihydrochloride (HIS), tryptamine hydrochloride (TRP), 2-phenylethylamine (2-PHE), putrescine dihydrochloride (PUT), cadaverine dihydrochloride (CAD), tyramine hydrochloride (TYR), spermidine trihydrochloride (SPD), and spermine tetrahydrochloride (SPM)) and dansyl chloride were obtained from Sigma-Aldrich Chemical Co. (St. Louis, MO, USA). Perchloric acid, 25% ammonium hydroxide solution, sodium hydroxide sodium hydrogen carbonate, and diethyl ether were acquired from Daejung Chemical Co. (Siheung, Korea). Acetone and acetonitrile (High-performance liquid chromatography (HPLC) grade) were purchased from Tedia Co. (Fairfield, OH, USA). Compound mixtures of amino acids, borate buffer, *o*-phthalaldehyde (OPA) and 9-fluorenylmethoxycarbonyl chloride (FMOC-Cl) were obtained from Agilent Technologies (Andover, MA, USA).

2.2. Preparation of Food Samples

During the period from 2010 to 2014, we collected samples 18 *maesil* extracts from different households in Korea for analysis of biogenic amines content. We also prepared our own *maesil* extract, following the method of Choi and Koh (2016) [5], and the process of preparation is shown in Figure 1. *Maesil* fruits obtained from a local market were washed with pure water, and drained at room temperature (23 ± 1 °C). To the 400 g of *maesil*, we added 400 g of brown sugar and the mixture was then placed in 1 L clear plastic jars, which were maintained in incubators set at 15 °C and 25 °C, respectively. After the *maesil* fruits were taken out from the jar after 90 days of soaking, the obtained liquid (490 mL at 15 °C and 476 mL at 25 °C) was further fermented for the next 180 days in the same jar. Biogenic amines were analyzed at 30, 45, 75, and 90 days of soaking period and 30, 60, 120, 150, and 180 days of fermentation.

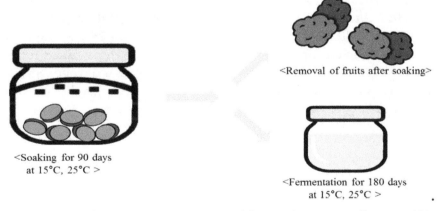

<Removal of fruits after soaking>

<Soaking for 90 days
at 15°C, 25°C >

<Fermentation for 180 days
at 15°C, 25°C >

Figure 1. The fermentation process used for producing *maesil* extract.888.

2.3. pH Measurement

To measure the pH, *maesil* extract (10 g) was mixed with 10 mL deionized water for 3 min and then filtered through Whatman paper No.2 filter paper (Advantec, Tokyo, Japan). The pH was measured using a pH meter (Beckman Coulter, FL, USA) following the method of Shukla et al. [21].

2.4. Amino Acids Analysis

Amino acids in the *maesil* extract were analyzed using an HPLC system (Dionex Ultimate 3000, Thermo Fisher Scientific, Waltham, MA, USA), equipped with a 1260 Infinity fluorescence detector (Agilent Technologies, Waldbronn, Germany), following the method described by Jajic et al. (2013) [22] with slight modifications. The samples were derivatized with OPA and FMOC via a programmed autosampler. After derivatization, samples (0.5 μL) were injected into an Inno-C$_{18}$ column (4.6 × 50 mm, 5 μm, Youngin Biochrom, Korea) at 40 °C. The fluorescence was detected at excitation and emission wavelengths of 340 and 450 mm, respectively for OPA, and at 266 and 305 nm, respectively, for FMOC. Primary and secondary amino acids were analyzed based on the OPA and FMOC derivatives, respectively. The mobile phase solvent A was 40 mM sodium phosphate (pH 7), and solvent B was a 10:45:45 (*v/v*) mixture of distilled water, acetonitrile, and methanol. The gradient program was run at a flow rate of 1.0 mL/min as follows: 5% B for 3 min; followed by elution with 5% to 55% B in 24 min; 55% to 90% B in 25 min; maintained 90% of B for next 6 min; and 90% to 5% B for 3.5 min, maintained for 0.5 min.

2.5. Biogenic Amine Analysis

2.5.1. Extraction of Biogenic Amines

Biogenic amines were extracted from the *maesil* extract using the method of Shukla et al. (2014) [21] with slight modifications. Briefly, 10 mL of 0.4 M perchloric acid solution was mixed with 5 g *maesil* extract, homogenized for 3 min, and then centrifuged at 3000× *g* for 10 min at 4 °C. The residue was re-extracted with 0.4 M perchloric acid solution (10 mL). After the supernatants were combined and 0.4 M perchloric acid solution was added to adjust the final volume to 50 mL. After filtering through Whatman filter paper No.1 (11 μm, Adventec, Tokyo, Japan), 1 mL of the extract was used for derivatization with dansyl chloride.

2.5.2. Derivatization of Biogenic Amines

Biogenic amines were derivatized following the methods described by Shukla et al. (2010) [23] and Frias et al. (2007) [24]. An extract sample (1 mL) or standard solution mixture (1 mL) was mixed with 200 μL 2 M sodium hydroxide; next, 300 μL of sodium hydrogen carbonate solution was added to saturate the solution. To the mixture, 1 mL of a dansyl chloride solution (10 mg/mL in acetone) was added and kept for 45 min at 40 °C. To stop the reaction, 100 μL of 25% ammonium hydroxide was added to the mixture and reacted for 30 min at 25 °C. Then, the derivatized biogenic amines were extracted twice with 1 mL of diethyl ether. Subsequent to drying in a nitrogen stream, the extract was redissolved in acetonitrile (1 mL) and filtered through a 0.22 μm polyvinylidene fluoride (PVDF) filter (Millipore Co., Bedford, MA, USA) for injection into the HPLC system.

2.5.3. HPLC Analysis of Biogenic Amines

Biogenic amines were analyzed using an HPLC system consisting of an Alliance 2695 separations module (Waters, Milford, MA, USA) and Ultra violet (UV)/Visible detector 2487 (Waters, Milford, MA, USA) with a Capcell Pak C18 column (4.6 × 250 mm i.d., 5 μm; Shiseido, Kyoto, Japan), thermostated at 30 °C, and detected at 210 nm [24–26]. The injection volume was 20 μL and the mobile phase consisted of solvent A (water) and B (acetonitrile) run at a flow rate of 0.8 mL/min with the following gradient elution program for 35 min: 65:35 (A:B, *v/v*), followed by 45% B for 5 min, elution with 45% to 65% B in 10.05 min, 65% to 80% B in 17.05 min, 80% to 90% B up to 26.25 min, and 90% to 35% B in 35 min.

2.6. Method Validation

The HPLC method for biogenic amines analysis was validated for linearity, limits of detection (LOD) and limits of quantification (LOQ), accuracy, and precision [22]. The linearity was evaluated using five concentrations (0.5, 1, 2, 5, and 10 mg/L) of each the biogenic amine standards (PUT, CAD, HIS, TRP, 2-PHE, TYR, SPD, and SPM) by constructing a calibration curve. The LOD and LOQ values were calculated using the following equations: LOD = 3.3 × (standard deviation (SD)/slope of calibration curve) and LOQ = 10 × (SD/slope of calibration curve). The accuracy of the method was verified by triplicate analysis of spiked samples at two different levels (5 and 10 mg/L) and expressed as % recovery. The recoveries were calculated by contrasting the peak area of measured concentration with the peak area of the spiked concentrations. To evaluate the precision, repeatability, inter-day, and intra-day were performed and expressed as the percentage relative standard deviation (RSD) of the peak area measurements. Repeatability was estimated by analysis of six consecutively injected samples. The inter-day precision was determined at two different levels, 5 and 10 mg/L, and the analyses were performed over a period of three consecutive days. The intra-day precision was determined by spiking five blank samples at concentrations levels of 5 and 10 mg/mL and the evaluation was based on the results obtained using the method operating over a single day under the same conditions.

2.7. Statistical Analysis

Quantitative data are expressed as the means ± SD of at least three measurements. Statistical analysis was performed using a one-way analysis of variance (ANOVA) and Duncan's multiple range test by SAS software, version 8.0 for Windows (SAS Institute, Cary, NC, USA). The probability value of $p < 0.05$ was considered statistically significant.

3. Results and Discussion

3.1. Method Validation

The results obtained from the different method validations are presented in Table 1. Standard curves for biogenic amines were constructed from triplicate analyses of five concentrations in the range 0.5–10 mg/L. With the exception of spermine (correlation coefficient (R^2) > 0.998), the linearity of the calibration curves for each biogenic amine was >0.999. The precision, expressed as %RSD, of inter-day variation was between 0.17% and 5.20%, and the RSD values for intra-day variation were between 0.07% and 6.46%. The LOD and LOQ of the biogenic amines ranged from 0.01 to 0.20 mg/L and 0.02 to 0.61 mg/L, respectively. The accuracy of the method with regard to recovery was between 89.4% and 110.8%.

Table 1. Summary results relating to validation of the high-performance liquid chromatography (HPLC) method used for biogenic amines.

Compound	R²	LOD (mg/L)	LOQ (mg/L)	Precision (%RSD)				Accuracy Recovery (%)	
				Inter-Day		Intra-Day			
				Low	High	Low	High	Low	High
HIS	1.0000	0.02	0.04	3.82	5.20	0.49	6.46	97.9	89.4
TRP	1.0000	0.01	0.02	0.66	0.19	0.10	0.09	101.8	105.3
2-PHE	1.0000	0.16	0.22	0.22	0.74	0.07	0.64	101.3	97.9
PUT	1.0000	0.01	0.02	0.45	0.34	0.25	0.76	99.8	104.0
CAD	1.0000	0.01	0.02	0.14	0.19	0.25	0.49	100.1	100.0
TYR	1.0000	0.07	0.10	0.27	0.17	0.09	0.22	99.7	99.3
SPM	0.9981	0.20	0.61	2.27	1.86	1.07	2.21	95.1	110.8
SPD	1.0000	0.04	0.08	0.14	0.40	0.16	0.11	107.1	99.6

LOD: Limits of detection; LOQ: Limits of quantification; RSD: Relative standard deviation HIS: Histamine; TRP: Tryptamine; PUT: Putrescine; 2-PHE: 2-phenylethylamine; CAD: Cadaverine; TYR: Tyramine; SPM: Spermine; SPD: Spermidine.

3.2. Content of Biogenic Amines in Home-Made Maesil Extract

Among the 18 home-made maesil extracts analyzed, the total content of biogenic amines ranged from 2.5 to 241.7 mg/L, the major individual biogenic amines were putrescine (not detectable (ND)-80.82 mg/L) and spermidine (ND-219.20 mg/L), followed by tryptamine (Table 2). Putrescine, histamine, tyramine, cadaverine, 2-phenylethylamine, spermidine, spermine, agmatine, and tryptamine are the main biogenic amines in wine [27]. Among these amines, putrescine has been reported to be generated from the raw material or by microbial decarboxylation [28]. In the case of wine, putrescine content has been found to be influenced by geographical region and grape variety [29]. Histamine and spermine detected in wine [27,29,30] are known to have toxicity or play a role in enhancing toxicity [11,31]. However, we were unable to detect either of these two amines in the 18 maesil extracts examined in the present study. These results imply that the amount and composition of biogenic amines may differ widely among different fruit-derived products and that these differences could be attributed to differences in manufacturing practice and fruit material.

Table 2. Biogenic amines content (mg/L) in 18 home-made *maesil* extracts prepared in individual households.

Sample	HIS	TRP	2-PHE	PUT	CAD	TYR	SPM	SPD	Total
A	ND	ND	ND	19.7 ± 0.53	2.9 ± 1.01	ND	ND	219.2 ± 6.30	241.8 ± 4.89
B	ND	3.0 ± 0.53	ND	12.4 ± 0.67	ND	ND	ND	44.5 ± 1.96	60.0 ± 1.64
C	ND	ND	ND	15.3 ± 0.74	ND	ND	ND	14.9 ± 0.91	30.2 ± 1.64
D	ND	1.9 ± 0.36	ND	8.8 ± 0.42	ND	ND	ND	25.6 ± 0.74	36.2 ± 0.18
E	ND	3.2 ± 0.47	ND	15.9 ± 1.69	ND	ND	ND	44.3 ± 2.31	63.4 ± 4.07
F	ND	ND	ND	13.1 ± 0.99	ND	ND	ND	83.5 ± 6.28	96.5 ± 6.21
G	ND	2.9 ± 0.65	ND	25.8 ± 2.33	17.3 ± 3.79	ND	ND	44.6 ± 3.11	90.6 ± 4.93
H	ND	5.7 ± 1.32	ND	ND	12.7 ± 0.42	ND	ND	35.9 ± 1.22	54.3 ± 2.09
I	ND	2.5 ± 0.48	ND	ND	ND	ND	ND	ND	2.5 ± 0.48
J	ND	3.5 ± 0.35	5.26 ± 1.47	ND	ND	ND	ND	ND	8.8 ± 1.19
K	ND	5.3 ± 0.45	ND	21.8 ± 0.70	ND	0.8 ± 0.02	ND	77.9 ± 1.81	105.8 ± 1.38
L	ND	ND	ND	ND	ND	1.0 ± 0.12	ND	ND	1.0 ± 0.12
M	ND	5.4 ± 0.38	ND	ND	ND	ND	ND	ND	5.4 ± 0.38
N	ND	ND	ND	17.5 ± 0.15	ND	ND	ND	71.8 ± 4.60	89.3 ± 4.45
O	ND	3.2 ± 0.35	ND	6.8 ± 0.26	ND	ND	ND	ND	9.9 ± 0.46
P	ND	ND	ND	80.8 ± 4.72	ND	ND	ND	17.1 ± 1.76	97.9 ± 4.29
Q	ND	4.1 ± 0.83	ND	5.9 ± 0.70	ND	0.7 ± 0.07	ND	10.1 ± 1.44	20.8 ± 3.10
R	ND	3.4 ± 0.45	ND	8.4 ± 0.97	ND	ND	ND	41.3 ± 1.41	53.1 ± 2.01

HIS: Histamine; TRP: Tryptamine; 2-PHE: 2-phenylethylamine; CAD: Cadaverine; TYR: Tyramine; PUT: Putrescine; SPM: Spermine; SPD: Spermidine; ND: Not detected.

The content of biogenic amines is known to be affected by fermentation conditions, including temperature, microorganisms, and the synthetic pathways of the biogenic amine formation [32–35]. In wine, cadaverine, histamine, putrescine, and tyramine are mainly detected, the content of which can vary depending on fermentation factors, storage, microbial decarboxylase activity, and vinification [27,30,36]. Marcobal et al. (2006) have reported that the content of biogenic amines in wine ranged from ND to 54.02 mg/L [31]. Garai et al. (2006) found that the main biogenic amine in commercial apple ciders was putrescine and that the total biogenic amine content ranged from ND to 23.26 mg/L [12]. In comparison, the results of this study indicate that the biogenic amine content in home-made *maesil* extracts is considerably higher than that reported in wine or apple ciders [12,27,30], thereby emphasizing the necessity to control biogenic amines productions during the fermentation of *maesil* extract.

3.3. Content of Biogenic Amines During Soaking and Fermentation

During the 90 day soaking of *maesil* examined in the present study, we found that the total biogenic amines content increased from 14.1 to 35.0 mg/L and 37.3 to 69.1 mg/L at 15 and 25 °C, respectively, indicating that the content was higher at the latter temperature throughout the soaking period (Figure 1a). Previous studies have reported that biogenic amines are generated via the catalytic activity of decarboxylase enzymes produced during the growth of microorganisms such as lactic acid bacteria [37], and thus, the increase in biogenic amines during the soaking period might be caused by microbial decarboxylase activity. At both incubation temperatures we assessed, the predominant biogenic amines detected in *maesil* extract were putrescine and spermidine (Figure 2b,c), and the latter comprised approximately 80% of the total biogenic amines.

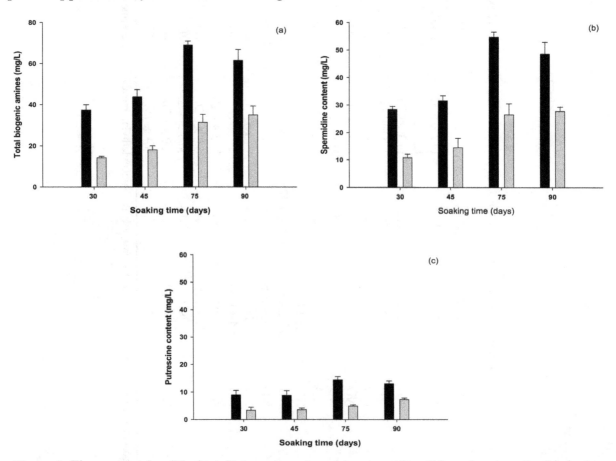

Figure 2. The content (mg/L) of total biogenic amines (**a**), spermidine (**b**), and putrescine (**c**) during soaking at 25 °C (■) and 15 °C (■).

After removing the *maesil* fruit from the sample jars at the end of the soaking period, the residual liquid was subsequently fermented for 180 days, during which, the content of biogenic amines decreased from 31.7 to 13.6 mg/L and 116.8 to 57.1 mg/L at 15 and 25 °C, respectively (Table 3). Generally, the extracts fermented at 25 °C exhibited biogenic amines content that was twice as high as that obtained at 15 °C. Moreover, at the end of the fermentation period, the total biogenic amines content at 15 °C was 23.8% of that at 25 °C. In addition to putrescine and spermidine, tryptamine was also detected at 0.33 mg/L when *maesil* was fermented at the higher temperature for 30 days. In this regard, Chong et al. (2011) have reported that temperature was the most important factor affecting biogenic amines formation [38], and Pinho et al. (2001) reported a higher increase in biogenic amines at a storage temperature of 21 °C than at 4 °C [35]. In addition, Kim et al. (2002) found that 25 °C was the optimum temperature for histamine production in fish muscles [39]. *Maesil* extract is typically produced by natural fermentation without controlling the temperature or starter culture. Moreover, it is sometimes consumed immediately after a 90 day soaking without subsequent fermentation. On the basis of the results obtained in this study, we recommend that, to yield a product with lower levels of biogenic amines, *maesil* extract should be fermented at a relatively low temperature and for a long period of time.

3.4. Effects of Processing Factors on Biogenic Amines Formation

The pathways implicated in the synthesis of biogenic amines can vary depending on the temperature, sugar content, precursors, and microorganisms involved in the fermentation of various food items [16,32,40]. Generally, putrescine is derived from the decarboxylation of arginine and ornithine or is already present in raw materials [28,32], whereas, spermidine is produced from arginine and ornithine or is converted from putrescine by spermidine synthase [16,31]. Poveda (2019) and Bardocz (1995) reported that most putrescine is either converted to spermidine or spermine, or is catabolized to succinate and other amino acids via succinate [28,31].

In the present study, to determine the effects of the factors influencing biogenic amines formation, we performed Pearson's correlation. Among these factors, we detected no significant correlation between biogenic amines content and pH, which had a narrow range (pH 2.9–3.3) during the fermentation.

Arena et al. (2008) reported a negative correlation between biogenic amines and sugar concentration and found that; the additions of glucose and fructose at 5 and 20 g/L reduced biogenic amines production by 82%–93% and 61%–99%, respectively [40]. Cid et al. (2008) reported that lower glucose concentration is associated with a high activity of ornithine-decarboxylase produced by *Lactobacillus* [41]. In our study, we found that, although there was a negative correlation between sugar content and biogenic amines content (Figure 3a), the relationship was not significant, which could be attributable to the narrow range of sugar content (61 to 81 °Brix) during fermentation.

Table 3. Biogenic amines content (mg/L) during fermentation at different temperatures.

Biogenic amine	Fermentation Days at 15 °C					Fermentation Days at 25 °C				
	30	60	120	150	180	30	60	120	150	180
TRP	ND	ND	ND	ND	ND	0.33 ± 0.09	ND	ND	ND	ND
PUT	8.1 ± 1.18 [a]	12.8 ± 1.15 [b]	13.8 ± 0.67 [b]	12.5 ± 1.95 [b]	7.4 ± 0.01 [a]	30.9 ± 0.77 [c]	23.9 ± 1.56 [b]	14.5 ± 0.76 [a]	20.0 ± 0.34 [b]	15.6 ± 4.03 [a]
CAD	ND	ND	ND	ND	ND	ND	ND	ND	ND	ND
SPD	23.6 ± 1.19 [b]	26.6 ± 2.32 [c]	22.2 ± 0.66 [b]	22.2 ± 0.65 [b]	6.2 ± 0.74 [a]	84.8 ± 0.68 [d]	74.9 ± 2.36 [c]	52.3 ± 3.25 [b]	46.5 ± 4.93 [a,b]	41.4 ± 2.58 [a]
Total	31.7 ± 2.15 [b]	39.4 ± 3.30 [d]	36.0 ± 1.33 [c,d]	34.7 ± 2.69 [b,c]	13.6 ± 0.75 [a]	116.8 ± 4.17 [d]	98.8 ± 0.93 [c]	66.8 ± 3.96 [b]	66.5 ± 4.59 [b]	57.0 ± 6.43 [a]

TRP: Tryptamine; PUT: Putrescine; CAD: Cadaverine; SPD: Spermidine; ND: Not detected. Each value expressed the mean of triplicates ± standard deviation (SD). Different superscript letters in the same row indicate significant difference ($p < 0.05$).

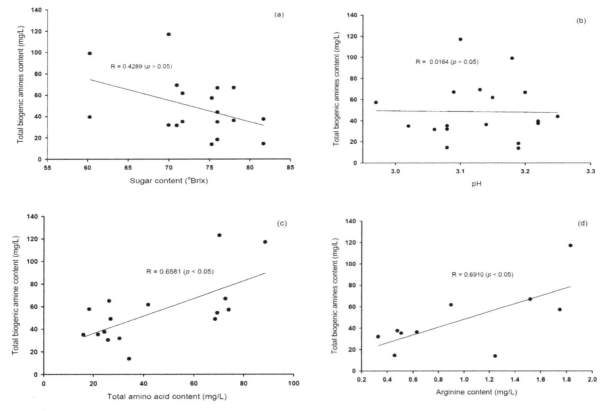

Figure 3. Correlations between sugar content and biogenic amine content (mg/L) (**a**); pH and biogenic amine content (mg/L) (**b**); the content (mg/L) of total amino acid and total biogenic amines (**c**); and the content (mg/L) of biogenic amines and arginine (**d**).

The total biogenic amines content showed a positive correlation with the total amounts of amino acids (R = 0.6581, $p < 0.05$), which could be explained by the fact that amino acids are precursors of biogenic amines [42]. We also detected a strong positive correlation between the amounts of putrescine and spermidine (R = 0.9277, $p < 0.01$; data not shown), consistent with the findings of Bardocz (1995) [31] and Nuriez et al. (2016) [16], which indicates that putrescine is a precursor of spermidine. However, apart from a positive correlation between arginine and total biogenic amines content (R = 0.6910, $p < 0.05$), we detected no correlation between individual biogenic amines and their respective precursor amino acid, which is consistent with the findings reported by Soufleros et al. (1998) [43]. Gezginc et al. (2013) reported that arginine serves as a precursor of putrescine, which can, in turn, be converted to spermidine [32]. Furthermore, it has been found that, in plants and some microorganisms, there are alternative pathways in which putrescine is generated from arginine via agmatine [33]. These results indicate that the fermentation of *maesil* extract at low temperature could reduce the production of biogenic amines. In addition, biogenic amine formation in *maesil* extract could be affected by the origin of *maesil*, the number of amino acids as well as the content of biogenic amine precursors.

4. Conclusions

The present study was conducted to evaluate the changes in biogenic amines formation and the relationship between biogenic amines and amino acids in *maesil* extract during the fermentation of this product. Although the consumption of *maesil* extract is currently increasing, there has, to date, been a lack of studies on the changes that biogenic amines undergo during *maesil* extract fermentation. The results of this study showed that the biogenic amines content in *maesil* extract is affected by both the inherent amino acids content and fermentation temperature and time. Moreover, the content of some biogenic amines may also be affected by the presence of other biogenic amines. We found that both amino acids and biogenic amines content was lower during fermentation at 15 °C than at 25 °C and decreased with increasing fermentation time. Accordingly, these observations indicate that employing protracted low-temperature fermentations could be an effective approach for reducing the production of biogenic amines in *maesil* extract. In further research, it will be necessary to study the types of microorganisms and formation on biogenic amines in *maesil* extract.

Author Contributions: Conceptualization, S.H.Y. and B.C.; methodology, S.H.Y. and B.C.; validation, S.H.Y., B.C., and E.K.; formal analysis, S.H.Y., B.C., and E.K.; investigation, S.H.Y. and B.C.; resources, B.M.; data curation, S.H.Y. and B.C.; writing—original draft preparation, S.H.Y. and B.C.; writing—review and editing, B.M.; supervision, B.M.; project administration, B.M.

Acknowledgments: This research did not receive any specific grant from funding agencies in the public, commercial, or not-for-profit sectors.

References

1.   Kang, M.Y.; Jeong, Y.H.; Eun, J.B. Physical and chemical characteristics of flesh and pomace of Japanese Apricots (*Prunus mume* Sieb. et Zucc). *Korean J. Food Sci. Technol.* **1999**, *31*, 1434–1439.

2.   Ko, M.S.; Yang, J.B. Antimicrobial activities of extracts of *Prunus mume* by sugar. *Korean Soc. Food Preser.* **2009**, *16*, 759–764.

3.   Miyazawa, M.; Yamada, T.; Utsunomiya, H. Suppressive effect of the SOS-inducing activity of chemical mutagen by citric acid esters from *Prunus mume* Sieb. et Zucc. using the *Salmonella typhimurium* Ta1535/Psk1002 Umu test. *Nat. Prod. Res.* **2003**, *17*, 319–323. [CrossRef] [PubMed]

4.   Bolarinwa, I.F.; Orfila, C.; Morgan, M.R.A. Amygdalin content of seeds, kernels and food products commercially-available in the UK. *Food Chem.* **2014**, *152*, 133–139. [CrossRef]

5.   Choi, B.G.; Koh, E. Changes of ethyl carbamate and its precursors in maesil (*Prunus mume*) extract curing one-year fermentation. *Food Chem.* **2016**, *209*, 318–322. [CrossRef]

6.   Kim, H.W.; Han, S.H.; Lee, S.W.; Suh, H.J. Effect of isomaltulose used for osmotic extraction of Prunus mume fruit juice substituting sucrose. *Food Sci. Technol.* **2018**, *27*, 1599–1605. [CrossRef]

7.   Ko, Y.J.; Jeong, D.Y.; Lee, J.O.; Park, M.H.; Kim, E.J.; Kim, J.W.; Kim, Y.S.; Ryu, C.H. The establishment of optimum fermentation conditions for *Prunus mume* vinegar and its quality evaluation. *Korean Soc. Food Sci. Nutr.* **2007**, *36*, 361–365. [CrossRef]

8.   Yan, X.T.; Lee, S.H.; Li, W.; Sun, Y.N.; Yang, S.Y.; Jang, H.D.; Kim, Y.H. Evaluation of the antioxidant and anti-osteoporosis activities of chemical constituents of the fruits of *Prunus mume*. *Food Chem.* **2014**, *156*, 408–415. [CrossRef]

9.   Henriquez-Aedo, K.; Galarce-Bustos, O.; Aqueveque, P.; Garcia, A.; Aranda, M. Dynamic of biogenic amines and precursor amino acids during cabernet sauvignon vinification. *LWT Food Sci. Technol.* **2018**, *97*, 238–244. [CrossRef]

10. Shalaby, A.R. Significance of biogenic amines to food safety and human health. *Food Res. Int.* **1996**, *29*, 675–690. [CrossRef]

11. Bai, X.; Byun, B.Y.; Mah, J.H. Formation and destruction of biogenic amines in Chunjang (a black soybean sauce) and Jajang (a blck soybean sauce). *Food Chem.* **2013**, *141*, 1026–1031. [CrossRef] [PubMed]

12. Garai, G.; Duenas, M.T.; Irastorza, A.; Martin-Alvarez, P.J.; Moreno-Arribas, M.V. Biogenic amines in natural ciders. *J. Food Prot.* **2006**, *69*, 3006–3012. [CrossRef] [PubMed]

13. Kim, J.H.; Park, H.J.; Kim, M.J.; Ahn, H.J.; Byun, M.W. Survey of biogenic amine contents in commercial soy sauce. *Korean J. Food Sci. Technol.* **2003**, *35*, 325–328.

14. Mohan, C.O.; Ravishankar, C.N.; Srinivasa Gopal, T.K.; Ashok Kumar, K.; Lalitha, K.V. Biogenic amines formation in seer fish (*Scomberomoruscommerson*) steaks packed with O_2 scavenger during chilled storage. *Food Res. Int.* **2009**, *42*, 411–416. [CrossRef]

15. Yoon, H.; Park, J.H.; Choi, A.; Hwang, H.J.; Mah, J.H. Validation of an HPLC analytical method for determination of biogenic amines in agricultural products and monitoring of biogenic amines in Korean fermented agricultural products. *Toxicol. Res.* **2015**, *31*, 299–305. [CrossRef] [PubMed]

16. Nuriez, M.; del Olmo, A.; Calzada, J. Biogenic amines. *Encycl. Food Health* **2016**, 416–423.
17. Zarei, M.; Najafzadeh, H.; Enayati, A.; Pashmforoush, M. Biogenic amines content of canned tuna fish marketed in Iran. *Am. Eurasian J. Toxicol. Sci.* **2011**, *3*, 190–193.
18. Nadeem, M.; Naveed, T.; Rehman, F.; Xu, Z. Determination of histamine in fish without derivatization by indirect reverse phase-HPLC method. *Microchem. J.* **2019**, *144*, 209–214. [CrossRef]
19. Carelli, D.; Centonze, D.; Palermo, C.; Quinto, M.; Rotunno, T. An interference free amperometric biosensor for the detection of biogenic amines in food products. *Biosens. Bioelectron.* **2007**, *23*, 64–647. [CrossRef]
20. Nout, M.J.R.; Ruikes, M.M.W.; Bouwmeester, H.M.; Belfaars, P.R. Effect of processing conditions on the formation of biogenic amines and ethyl carbamate in soybean Tempeh. *J. Food Saf.* **1993**, *13*, 293–303. [CrossRef]
21. Shukla, S.; Park, H.K.; Lee, J.S.; Kim, J.K.; Kim, M. Reduction of biogenic amines and aflatoxin in Doenjang samples fermented with various *Meju* as starter cultures. *Food Control.* **2014**, *42*, 181–187. [CrossRef]
22. Jajic, I.; Krstovic, S.; Glamocic, D.; Jaksic, S.; Abramovic, B. Validation of HPLC method for the determination of amino acids in feed. *J. Serb. Chem. Soc.* **2013**, *78*, 839–850. [CrossRef]
23. Shukla, S.; Park, H.K.; Kim, J.K.; Kim, M.H. Determination of biogenic amines in Korean traditional fermented soybean paste (Doenjang). *Food Chem. Toxicol.* **2010**, *48*, 1191–1195. [CrossRef] [PubMed]
24. Frias, J.; Martinez-Villaluenga, C.; Gulewicz, P.; Perez-Romero, A.; Pilarski, R.; Gulewicz, K.; Vidal-Valverde, C. Biogenic amines and HL 60 citotoxicity of alfalfa and fenugreek sprouts. *Food Chem.* **2007**, *105*, 959–967. [CrossRef]
25. Qiu, S.; Wang, Y.; Cheng, Y.; Liu, Y.; Yadav, M.P.; Yin, L. Reduction of biogenic amines in sufu by ethanol during ripening stage. *Food Chem.* **2018**, *239*, 1244–1252. [CrossRef]
26. Costa, M.P.; Balthazar, C.F.; Rodrigues, B.L.; Lazaro, C.A.; Silva, A.C.O.; Cruz, A.G.; Conte Junior, C.A. Determination of biogenic amines by high-performance liquid chromatography (HPLC-DAD) in probiotic cow's and goat's fermented milks and acceptance. *Food Sci. Nutr.* **2015**, *3*, 172–178. [CrossRef]
27. Jastrzebska, A.; Piasta, A.; Kowalska, S.; Krzeminski, M.; Szlyk, E. A new derivatization reagent for determination of biogenic amines in wines. *J. Food Compos. Anal.* **2016**, *48*, 111–119. [CrossRef]
28. Poveda, J.M. Biogenic amines and free amino acids in craft beers from the Spanish market: A statistical approach. *Food Control.* **2019**, *96*, 227–233. [CrossRef]
29. Landete, J.M.; Ferrer, S.; Polo, L.; Pardo, I. Biogenic amines in wines from three Spanish regions. *J. Agric. Food Chem.* **2005**, *53*, 1119–1124. [CrossRef]
30. Marcobal, A.; Martin-Alvarez, P.J.; Polo, M.C.; Moreno-Arribas, M.V. Formation of biogenic amines throughout the industrial manufacture of red wine. *J. Food Prot.* **2006**, *69*, 397–404. [CrossRef]
31. Bardocz, S. Polyamines in food and their consequences for food quality and human health. *Trends Food Sci. Technol.* **1995**, *6*, 341–346. [CrossRef]
32. Gezginc, Y.; Akyol, I.; Kuley, E.; Ozogul, F. Biogenic amines formation in *Streptococcus thermophiles* isolated from home-made natural yogurt. *Food Chem.* **2013**, *138*, 655–662. [CrossRef] [PubMed]
33. Karovicova, J.; Kohajdova, Z. Biogenic amines in food. *Chem. Pap.* **2005**, *59*, 70–79. [CrossRef]
34. Ozogul, F.; Hamed, I. The importance of lactic acid bacteria for the prevention of bacterial growth and their biogenic amines formation: A review. *Crit. Rev. Food Sci. Nutr.* **2018**, *58*, 1660–1670. [CrossRef]
35. Pinho, O.; Ferreira, I.M.P.L.V.O.; Mendes, E.; Oliveira, B.M.; Ferreira, M. Effect of temperature on evolution of free amino acid and biogenic amine contents during storage of Azeitao cheese. *Food Chem.* **2001**, *75*, 287–291. [CrossRef]
36. Santos, M.H.S. Biogenic amines: Their importance in foods. *Int. J. Food Microbiol.* **1996**, *29*, 213–231. [CrossRef]
37. Patel, M.A.; Ou, M.S.; Harbrucker, R.; Aldrich, H.C.; Buszko, M.L.; Ingram, L.O.; Shanmugam, K.T. Isolation and characterization of acid-tolerant, thermophilic bacteria for effective fermentation of biomass-derived sugars to lactic acid. *Appl. Environ. Micorobiol.* **2006**, *72*, 3228–3235. [CrossRef]
38. Chong, C.Y.; Abu Bakar, F.; Russly, A.R.; Jamilah, B.; Mahyudin, N.A. The effects of food processing on biogenic amines formation. *Int. Food Res. J.* **2011**, *18*, 867–876.
39. Kim, S.H.; Price, B.J.; Morrissey, M.T.; Field, K.G.; Wei, C.I.; An, H. Occurrence of histamine-forming bacteria in albacore and histamine accumulation in muscle at ambient temperature. *J. Food Sci.* **2002**, *67*, 1515–1521. [CrossRef]

40. Arena, M.E.; Landete, J.M.; Manca de Nadra, M.C.; Pardo, I.; Ferrer, S. Factors affecting the production of putrescine from agmatine by *Lactobacillus hilgardii* X1B isolated from wine. *J. Appl. Microbiol.* **2008**, *105*, 158–165. [CrossRef]

41. Cid, B.S.; Miguelez-Arrizado, M.J.; Becker, B.; Holzapfel, W.H.; Vidal-Carou, M.C. Amino acid decarboxylation by *Lactobacillus curvatus* CTC 273 affected by the pH and glucose availability. *Food Microbiol.* **2008**, *25*, 269–277. [PubMed]

42. Wang, Y.Q.; Ye, D.Q.; Zhu, B.Q.; Wu, G.F.; Duan, C.Q. Rapid HPLC analysis of amino acids and biogenic amines in wine during fermentation and evaluation of matrix effect. *Food Chem.* **2014**, *163*, 6–15. [CrossRef] [PubMed]

43. Soufleros, E.; Barrios, M.L.; Bertrand, A. Correlation between the content of biogenic amines and other wine compounds. *Am. J. Enol. Vitic.* **1998**, *49*, 266–278.

8

Biogenic Amine Production by Lactic Acid Bacteria

Federica Barbieri [1], **Chiara Montanari** [1], **Fausto Gardini** [1,2] **and Giulia Tabanelli** [1,2,*]

[1] Interdepartmental Center for Industrial Agri-Food Research, University of Bologna, 47521 Cesena, Italy; federica.barbieri16@unibo.it (F.B.); chiara.montanari8@unibo.it (C.M.); fausto.gardini@unibo.it (F.G.)

[2] Department of Agricultural and Food Sciences, University of Bologna, 40126 Bologna, Italy

[*] Correspondence: giulia.tabanelli2@unibo.it

Abstract: Lactic acid bacteria (LAB) are considered as the main biogenic amine (BA) producers in fermented foods. These compounds derive from amino acid decarboxylation through microbial activities and can cause toxic effects on humans, with symptoms (headache, heart palpitations, vomiting, diarrhea) depending also on individual sensitivity. Many studies have focused on the aminobiogenic potential of LAB associated with fermented foods, taking into consideration the conditions affecting BA accumulation and enzymes/genes involved in the biosynthetic mechanisms. This review describes in detail the different LAB (used as starter cultures to improve technological and sensorial properties, as well as those naturally occurring during ripening or in spontaneous fermentations) able to produce BAs in model or in real systems. The groups considered were enterococci, lactobacilli, streptococci, lactococci, pediococci, oenococci and, as minor producers, LAB belonging to *Leuconostoc* and *Weissella* genus. A deeper knowledge of this issue is important because decarboxylase activities are often related to strains rather than to species or genera. Moreover, this information can help to improve the selection of strains for further applications as starter or bioprotective cultures, in order to obtain high quality foods with reduced BA content.

Keywords: biogenic amines; decarboxylase enzymes; lactic acid bacteria; starter cultures

1. Biogenic Amine Toxicity and Physiological Role in Microorganisms

A large number of metabolites, exerting both beneficial and detrimental properties for human health, can be synthetized by microorganisms. Among these, amino acid derivatives produced during bacterial growth and fermentation can interact with human physiology in several ways, showing health-modulating potential [1]. This group includes bioactive compounds such as biogenic amines (BAs), which are responsible for adverse effects and are involved in several pathogenic syndromes [1]. In fact, ingestion of food containing high BA amounts is a risk for consumer health since these compounds can cause headache, heart palpitations, vomiting, diarrhea and hypertensive crises [2–4]. However, their toxic effect depends on the type of BA, on individual sensitivity or allergy and on the consumption of monoaminooxidase inhibitory drugs or ethanol, which interact with aminooxidase enzymatic systems responsible for the detoxification process of exogenous BAs [5,6].

Due to the severity of symptoms they may cause, histamine and tyramine are the most dangerous BAs and are responsible for symptomatology known as "scombroid fish poisoning" and "cheese reaction," respectively [3,7]. The "scombroid fish poisoning", often due to the consumption of fish such as tuna, sardines, anchovies, mackerel, etc., consists in flushing of face, neck and upper arms, oral numbness and/or burning, headache, heart palpitations, asthma attacks, hives, gastrointestinal symptoms, and difficulties in swallowing [8]. Tyramine intoxication is known as "cheese reaction" because this BA is the most frequently found in cheese and it can causes dietary-induced migraine, increased cardiac output, nausea, vomiting, respiratory disorders and elevated blood glucose [7,9]. As far as other BAs, the presence of high level of 2-phenylethylamine, putrescine, cadaverine, agmatine, spermine and spermidine can lead to toxicity. Moreover, they can potentiate the effects of histamine and tyramine toxicity by inhibiting their metabolizing enzymes [10].

Although the consumption of food containing large amounts of BAs can have toxicological consequences, there is no specific legislation regarding the presence of BAs in foods, with the exception of fishery products, for which the maximum acceptable level of histamine is defined [11]. However, recently, EFSA conducted a qualitative risk assessment concerning BA in fermented foods in the European Union, indicating concentrations that could induce adverse effects in consumers [12].

According to their chemical structures, BAs can be classified as aromatic (tyramine and 2-phenylethylamine), aliphatic (putrescine, cadaverine, spermine and spermidine) and heterocyclic (histamine and tryptamine) (see Table 1) and they are analogous to those naturally found in fresh food products, which exert a physiological role associated with cell growth and proliferation [13,14].

The exogenous BAs derive from bacterial decarboxylation of the corresponding amino acids through decarboxylase enzymes. Histamine and cadaverine can be formed by converting histidine and lysine via histidine decarboxylase (HDC) and via lysine decarboxylase (LDC), respectively. Tyrosine is converted in tyramine by tyrosine decarboxylase (TDC), which can act also on phenylalanine obtaining 2-phenylethylamine. This latter aromatic BA is produced by TDC with a lower efficiency with respect to tyramine and it is accumulated when tyrosine is almost completely depleted [15–17]. The formation of these BAs is based on one-step decarboxylation reactions of their respective amino acids and requires systems for amino acid active transport such as antiporter protein in exchange for the resulting BA. Putrescine can be accumulated with a single-step decarboxylation pathway by ornithine decarboxylase (ODC), common in Gram negative bacteria (such as enterobacteria and pseudomonads) or Lactic Acid Bacteria (LAB) deriving from wine environment [16,18–20]. However, this BA can also be formed through agmatinase pathway, which directly converts agmatine to urea and putrescine, or by agmatine deiminase (AgDI) pathway, common in LAB, which transforms arginine to agmatine by arginine decarboxylase. Subsequently, agmatine is converted to putrescine by the agmatine deiminase system, consisting of three enzymes: agmatine deiminase, putrescine carbamoyltransferase and carbamate kinase [21]. The biosynthesis of higher polyamines (spermine and spermidine) proceeds with complex pathways starting from putrescine released from ornithine or agmatine [22,23].

The decarboxylative pathways are activated for several physiological reasons. In fact, decarboxylation of amino acids is coupled with an electrogenic antiport system that can counteract intracellular acidification [24,25]. Therefore, BA accumulation can represent a cellular defense mechanism to withstand acid stress and it has been demonstrated that the transcription of many decarboxylase genes is induced by low pH and improves cell performances in acid conditions [9,17,19,20,26,27]. Moreover, the transfer of a net positive charge outside the cell can generate a proton motive force, leading to cell membrane energization and bringing supplementary energy. It has been demonstrated that the decarboxylase pathway can support the primary metabolism in environmental critical conditions [26–28]. This function can be particularly important for microorganisms lacking a respiratory chain, such as most LAB [29]. del Rio et al. [30] demonstrated that AgDI pathway promotes the growth of *Lactococcus lactis* after nutrient depletion.

Table 1. Biogenic amines, precursors, decarboxylase enzyme and their producers lactic acid bacteria found in fermented foods.

Biogenic Amine	Amino Acid Precursor	Classification	Decarboxylase Enzyme or Pathway in LAB	Lactic Acid Bacteria Producing Species	References
Histamine	Histidine	Heterocyclic	Histidine decarboxylase (HDC)	E. faecium, E. faecalis, L. sakei, L. curvatus, L. parabuchneri, L. buchneri, L. plantarum, L. brevis, L. casei, L. paracasei, L. vaginalis, L. reuteri, L. hilgardii, L. mali, L. rhamnosus, L. paracollinoides, L. rossiae, L. helveticus, S. thermophilus, O. oeni, P. parvulus, Leuc. mesenteroides, W. cibaria, W. confusa, W. paramesenteroides, T. muriaticus, T. halophilus	[24,31–46]
Tyramine	Tyrosine	Aromatic	Tyrosine decarboxylase (TDC)	E. faecium, E. faecalis, E. durans, E. hirae, E. casseliflavus, E. mundtii, L. sakei, L. curvatus, L. plantarum, L. brevis, L: buchneri, L. casei, L. paracasei, L. reuteri, L. hilgardii, L. homohiochii, L. delbrueckii subsp. bulgaricus, S. thermophilus, S. macedonicus, Lc. lactis, Leuc. mesenteroides, W. cibaria, W. confusa, W. paramesenteroides, W. viridescens, C. divergens, C. maltaromaticum, C. galliranum, T. halophilus, Sporolactobacillus sp.	[9,17,32,47–58]
2-phenylethylamine	Phenylalanine	Aromatic	Tyrosine decarboxylase (TDC)	E. faecium, E. faecalis, E. durans, E. hirae, E. casseliflavus, E. mundtii, L. brevis, Lc. lactis, Leuc. mesenteroides, C. divergens	[9,15-17,32,42,47,59]
Cadaverine	Lysine	Aliphatic	Lysine decarboxylase (LDC)	E. faecium, E. faecalis, L. curvatus, L. brevis, L. casei, L. paracasei, S. thermophilus, Pediococcus spp., Leuc. mesenteroides, T. halophilus	[32,47,56,60–63]
Putrescine	Arginine	Aliphatic	Ornithine decarboxylase (ODC)	E. faecium, E. faecalis, E. durans, E. hirae, E. casseliflavus, L. sakei, L. curvatus, L. buchneri, L. plantarum, L. brevis, L. paracasei, L. mali, L. rhamnosus, L. rossiae, L. homohiochii, Lc. lactis, S. thermophilus, S. mutans, P. parvolus, O. oeni, T. halophilus	[20,30,32,47,54,56,60,64–72]
	Agmatine	Aliphatic	Agmatine deiminase (AgDI)	E. faecalis, E. faecium, E. durans, E. hirae, E. mundtii, L. curvatus, L. plantarum, L. brevis, S. thermophilus, S. mutans, Lc. lactis, O. oeni, P. parvulus, P. pentosaceus, Leuc. mesenteroides, W. halotolerans, C. divergens, C. maltaromaticum, C. gallinarum	

Several Gram negative and Gram positive bacteria are able to produce BAs. Spoilage bacteria belonging to enterobacteria and pseudomonads can accumulate histamine, putrescine and cadaverine [47,64,73–75]. For this reason, BA content has been related to poor hygienic quality of non-fermented foods, being associated with a massive growth of decarboxylase positive spoilage microorganisms, and several authors proposed BA content as a microbial quality index [75,76]. Decarboxylase activity has been described also in Gram positive microbial groups, such as staphylococci, *Bacillus* spp. and, especially, LAB, considered the most efficient tyramine producers [9,48]. Moreover, the ability to produce histamine, cadaverine and putrescine by bacteria belonging to LAB have been reported [21,64,77]. According to some authors, also yeasts and moulds are implicated in BA accumulation, even if with a controversial role [78–80]. It is important to point out that the capability to produce BAs is generally a strain-specific characteristic, with strong variability in aminobiogenetic potential between different strains belonging to the same species.

2. Role of LAB in Fermented Food BA Content and Their Decarboxylase Clusters Genetic Organization

BA content in fermented foods is of great interest not only for its potential health concerns but also from an economic point of view. On the other hand, the presence of small concentrations of these compounds in fermented foods is unavoidable. In fact, the BA content in these products can range from concentrations below 20 mg/kg for alcoholic and no-alcoholic beverages, fermented vegetables and soy products, up to several hundred mg/kg for some sausages and cheeses [12]. The presence of different BAs is dependent on the precursor availability due to proteolysis during ripening. Moreover, the presence of decarboxylase positive non-starter microbiota, deriving from raw material and productive environment, often leads to high BA concentrations in fermented foods, especially in those obtained without the use of starter cultures [47,75,81,82]. In addition to precursor availability and the presence of BA producing microorganisms, the accumulation of these compounds depends on various intrinsic, environmental and technological factors, recently revised by Gardini et al. [83].

Decarboxylase activity is often expressed independently of cell viability and these enzymes maintain their activity after cell lysis also in harsh environmental conditions [31,49,84,85]. Moreover, once produced, BAs are stable to heat treatment, freezing, and smoking [86].

Dairy products, especially ripened cheeses, have been associated with foodborne intoxications due to their high content of BAs, such as tyramine, histamine, putrescine, and 2-phenylethylamine [32,47]. In any case, BA content varies between different types of cheeses and even among different sections of the same cheese [87]. In fermented meats the most prevalent BAs are tyramine, cadaverine, putrescine, and, with minor extent, histamine and their levels strongly vary among different types of products and [75,88–90]. The presence of these compounds in such products depends on low quality processing conditions favoring contamination and on the presence of autochthonous microbiota with decarboxylase potential [82]. Also in alcoholic beverages BAs (mainly histamine, tyramine, putrescine and cadaverine) can be formed through microbial activity during production and storage [60,61]. The presence of BAs has been reported also in fermented vegetables, such as sauerkraut or table olives, where the presence of aminobiogenic spoilage microorganisms can result in high putrescine, cadaverine and tyramine content [91,92]. Abundant amounts of histamine have been detected in fermented fish products [93].

Even if the production of diamines is usually attributed to spoiling Gram negative bacteria, such as enterobacteria and pseudomonads [82], LAB are considered mainly responsible for BA production in fermented foods [47,94]. Although starter cultures are accurately selected for the absence of decarboxylase activity, non-controlled autochthonous LAB involved in ripening process can contribute to BA accumulation. These non-starter LAB (NSLAB) consist mainly of mesophilic facultative or obligate heterofermentative bacteria, which exert a crucial role in maturation phenomena such as the development of flavor [95,96]. These bacteria show good adaptation to unfavorable growth conditions and can survive for long period after sugar depletion, thanks to their ability to obtain energy for growth and survival from other substrates, among which amino acids [97–100]. Moreover, the adaptation to some ecological niches has required the capability to resist acid stresses, activating bacterial mechanisms able to counteract low pH. In stress conditions occurring in fermented foods during ripening, NSLAB encode specific genetic mechanisms that lead to stress responses producing physiological changes among which decarboxylation reactions acquire important roles thanks to the maintenance of pH homeostasis [24,25,101,102]. In fact, expression (by transcriptional induction) and/or activation (by catalytic modulation) of amino acid decarboxylation systems in LAB are reported to be adaptive responses to energy depletion but also strategies to counteract acid stress [103]. The presence of the decarboxylase genes involved in the production of BAs are mostly strain dependent rather than species specific, highlighting the occurrence of horizontal gene transfer between strains as part of a mechanism of survival and adaptation to specific environments [16,33,50]. Recently, the genes belonging to BA biosynthetic pathways in LAB have been identified and the genetic organization of decarboxylase clusters has been reviewed [9,34,64,65]. Generally, enzymes responsible for specific amino acid decarboxylation are organized in clusters in which some genes are always present, i.e., the specific amino acid decarboxylase and the corresponding antiporter permease.

The first tyrosine decarboxylase locus (*tdc*) described in bacteria was found in *Enterococcus faecalis* JH2-2 [104]. This cluster has been annotated also in the genome sequence of other LAB [16,51,52,105–107]. Marcobal et al. [9] evidenced for all tyramine biosynthetic loci a high similarity in both gene sequence and organization, since this locus usually contains the genes encoding tyrosine decarboxylase (*tyrDC*), tyrosyl tRNA synthetase (*tyrS*, located upstream the *tyrDC* gene), putative tyrosine/tyramine permease (*tyrP*, located downstream the *tyrDC* gene) and a Na^+/H^+ antiporter (*nhaC*) [47]. The similar organization of different *tdc* clusters, their distribution, and their high similarity of sequence suggest a horizontal transfer of this cluster from a common source [106]. However, different strains can have different transcriptional organizations of the *tdc* gene cluster, as demonstrated by reverse transcription polymerase chain reaction (PCR) analyses. In fact, the four complete Open Reading Frame (ORF) can be co-transcribed [53] or *tyrS* can be transcribed independently and not included in the catabolic operon [27].

The LAB histidine decarboxylases belong to pyruvoil-dependent decarboxylases group and the encoding histidine decarboxylase gene (*hdcA*) has been identified in several LAB species [33–35,85,108–113]. The histidine decarboxylase gene clusters (*hdc*) of Gram positive bacteria usually comprise the decarboxylase gene *hdcA* and the histidine/histamine antiporter gene *hdcP*. Frequently, an *hdcB* gene, involved in the conversion of the histidine decarboxylase proenzyme to the active decarboxylase can be found [114]. Moreover, for lactobacilli, a histidyl-tRNA synthetase (*hisS*) gene has also been described [35]. The transcriptional studies demonstrated that these genes are located on an operon transcribed as a polycistronic mRNA. However, some authors demonstrated that the antiporter gene is transcribed as a monocistronic RNA and that transcriptional termination structures are present in the intergenic regions of histamine operon in *Lactobacillus buchneri* [111]. Rossi et al. [85] found that *hdcA* gene of *Streptococcus thermophilus* PRI60 was genetically different from the *hdcA* genes sequenced in other LAB, in agreement with the findings of Calles-Enríquez et al. [35], who reported that *hdc* cluster of *S. thermophilus* was more closely related to genera such as *Clostridium* and *Staphylococcus* than other LAB. Another interesting feature of *hdc* gene is its possibility to be located on a plasmid [34]. Lucas et al. [33,36] found that *Lactobacillus hilgardii* 0006, *Tetragenococcus muriaticus*, and *Oenococcus oeni* strains showed 99 to 100% identical *hdcA*- and *hdcB*-encoded proteins, highlighting the presence of a plasmid-encoded histidine decarboxylase system recently transferred horizontally between bacteria. Furthermore, they found that the *hdc* gene cluster, responsible for histamine production in *L. hilgardii* IOEB 0006, was located on an 80-kb plasmid that proved to be unstable. In fact, the capability to form histamine was lost in relation to the growth conditions.

Depending on the producer bacterium, genes/enzymes involved and the ecological niche from which it originates, two different metabolic routes have been described in LAB for the biosynthesis of putrescine [20,64,115]. The first is a decarboxylation system consisting of an ornithine decarboxylase (ODC) and an ornithine/putrescine exchanger. These enzymes are encoded by a gene cluster containing two adjacent genes: (i) *speC* encoding a biosynthetic/constitutive form of the ODC enzyme and (ii) *potE* encoding the transmembrane substrate/product exchanger protein [19,20,116]. Gram positive bacteria, however, have been infrequently reported to possess an ODC enzyme and putrescine-producing LAB strains via the ODC pathway are essentially, although not exclusively, derived from wine environment, belonging to the species *Lactobacillus saerimneri*, *Lactobacillus brevis* [19,20], *Lactobacillus mali* [18], and *O. oeni* [66]. In contrast, the agmatine deiminase (AgDI) pathway is relatively frequent in LAB and it is even considered a species trait in some enterococci [54]. This pathway consists of a more complex system, comprising AgDI, a putrescine transcarbamylase, a carbamate kinase, and an agmatine/putrescine exchanger [65,101]. Five genes are grouped in the agmatine deiminase cluster (*AgDI*): the regulator gene *aguR* and the metabolic genes *aguB, aguD, aguA* and *aguC (aguBDAC)*. Linares et al. [117] reported that *aguR* is constitutively transcribed from its promoter (P*aguR*) while the catabolic genes are co-transcribed in a single mRNA from the *aguB* promoter (P*aguB*) in a divergent orientation. These pathway genes were occasionally detected in a putative acid resistance locus in LAB species [101]. In this locus, the *AgDI* genes are found adjacent to the genes associated with the tyrosine decarboxylase pathway on the chromosome [53], suggesting the presence of genes for high-alkalinizing routes (such as amino acid decarboxylases) in LAB genome.

3. Main LAB Involved in BA Production in Fermented Foods

All fermented foods are subjected to the risk of BA contamination. Although LAB are considered GRAS (Generally Regarded As Safe) organisms, they can have the capability to produce toxic compounds as BAs. In particular, in fermented foods, NSLAB can accumulate BAs and strains of lactobacilli, enterococci, lactococci, pediococci, streptococci, and leuconostocs have been associated with high levels of these compounds [118]. Genetic studies have revealed that many of these strains harbor genes or operons coding for decarboxylating enzymes or other pathways implicated in BA biosynthesis [9,64].

Hereafter, the main LAB genera associated with fermented products and involved in BA production in vitro or in situ are described.

3.1. Enterococcus

The *Enterococcus* genus has not been yet classified as safe for human consumption since it neither is recommended for the Qualified Presumption of Safety (QPS) list nor have GRAS status. Most of the species harbor a series of virulence factors and antibiotic resistance and they have been associated with several infections, having the ability to mediate gene transfer with different genetic elements, including plasmids, phages and conjugative transposons [119,120]. The role of enterococci in fermented foods remains controversial. They show remarkable ecological adaptability and ability to grow in adverse conditions. Due to their tolerance to salt and low pH, they are highly adapted to several food systems and they are also involved in the fermentation process of traditional cheeses and dry sausages [121]. Moreover, some *Enterococcus* strains show probiotic features [122] or can improve sensorial properties of dairy products when added as adjunct starters, taking part to flavour generation through proteolytic and lipolytic activities and the accumulation of C4 metabolites such as diacetyl, acetoin or 2, 3-butanediol [123,124]. In addition, their ability to biosynthesize bacteriocins with a wide-range effectiveness on pathogenic and spoilage bacteria is known [125].

Nevertheless, enterococci presence in fermented foods has been associated with the production of BAs (mainly tyramine) and this activity has been reported for strains belonging to different species isolated from meat, cheese, fish, wine and human faeces [54,126–132]. However, not all the strains able to decarboxylate tyrosine were characterized by the same phenotypic potential in relation to the kinetics of tyramine accumulation [15] (Table 1).

Enterococci have been recognized as important part of the natural microbiota in many artisanal cheeses and, in some cases, they can predominate over lactobacilli and lactococci [133]. Usually, enterococci are not present in starter cultures and thus all species of this genus isolated from cheese samples represent contaminating microbial communities, and can include aminobiogenic strains. The most common species found in milk are *Enterococcus faecium*, *Enterococcus durans* and *E. faecalis* but, even if with minor extent, *Enterococcus casseliflavus* may also be isolated [123] and mostly of the strains belonging to these species and isolated from cheese have been identified as tyramine producers [132]. Several authors found a relation between the enterococci counts and the concentrations of tyramine [134–137] and putrescine [67] in dairy products.

Burdychova and Komprda [138] detected tyraminogenic isolates from cheese belonging to *E. durans, E. faecium, E. faecalis* and *E. casseliflavus* species. Rea et al. [139] studied the effect of six strains of *E. faecalis, E. faecium, E. durans* and *E. casseliflavus* species on tyramine production in Cheddar cheese during manufacturing and ripening and found that all strains, except *E. casseliflavus*, produced this BA, with *E. durans* responsible for the highest concentration after 9 months of ripening at 8 °C. Enterococcal strains isolated from an Italian cheese and from raw goat milk showed high decarboxylase activity with tyrosine and phenylalanine as substrates [59,136]. Kalhotka et al. [140] investigated the decarboxylase activity of enterococci isolated from goat milk and found that all the tested strains, identified as *Enterococcus mundtii, E. faecium* and *E. durans*, showed significant tyrosine and arginine decarboxylase activity, in relation to temperature and time of incubation. Martino et al. [141] studied safety features of four enterococcal strains isolated from a regional Argentinean cheese founding that these strains possessed *tdc* gene cluster, even if only two of four strains gave a positive result in Bover-Cid and Holzapfel decarboxylase screening medium [142]. These authors hypothesized the possibility that this pathway was not active, although all the strains possessed the complete decarboxylase cluster.

The presence of enterococci able to produce BAs is a relevant food issue also in meat products, despite their recognized role in the development of sensory properties of fermented products particularly in sausage [143,144]. In fact, enterococci are constituents of the natural microbiota of raw meat and of many fermented meat products [145], with *E. faecium* and *E. faecalis* being the predominant species, followed by *Enterococcus hirae, E. durans* and *E. mundtii* [122]. For this reason, dry fermented sausages can easily accumulate high levels of BAs, especially tyramine, putrescine and cadaverine [82]. In contrast, histamine is usually scarcely found in fermented sausages [146].

Landeta et al. [147] found that 79% of *E. faecium* strains isolated from Spanish dry-cured sausages were able to produce tyramine and that some strains were PCR-positive for the presence of the tyrosine decarboxylase gene, but were not able to accumulate this BA, due to the absence of gene expression. These results were in agreement with those obtained by Komprda et al. [37] who reported that 88% of enterococcal strains isolated in ripened fermented sausages and belonging to *E. faecium* and *E. faecalis* species, possessed *tdc* sequences. These authors found also that 71% of enterococcal isolates had *hdc* gene sequence, assuming that the decarboxylation pathway (producing proton motive force) gives the strains a competitive advantage in nutrient-depleted conditions and acidic environments, such as fermented sausages at the end of ripening. The potential of different indigenous enterococci to contribute to BA formation in spontaneously fermented game meat sausages has been reported also by Maksimovic et al. [148], who found that 100% of *E. durans* and about 7% of *E. casseliflavus* possessed *tdc* genes. Iacumin et al. [62] indicated enterococci able to accumulate large amount of BAs as responsible for spoilage in goose sausages produced in the north of Italy. In fact, despite the addition of starter, enterococci grew during ripening and produced a large amount of BAs. This ability was confirmed in vitro, since all the isolates (*n* = 100), belonging to the species *E. faecium* and *E. faecalis*, were able to decarboxylate amino acids and produce BAs. In particular, all the strains produced histamine, and 60 out of 70 *E. faecium* and 25 out 30 *E. faecalis* strains produced cadaverine and 10 isolates belonging to both species produced tyramine.

Enterococci has been reported as mainly responsible for tyramine accumulation in wine during malolactic fermentation, together with some *Lactobacillus* species [18,127,149,150]. These latter authors isolated *E. faecium* strains during malolactic fermentation of red wine and demonstrated that, although all the isolates harbored decarboxylase genes, only five strains were able to survive under the harsh conditions found in wine (high ethanol content and low pH), leading to a higher concentration of BAs in samples, including tyramine, histamine and 2-phenylethylamine.

E. faecium and *E. faecalis* have been considered responsible also for BA production in fermented soybean food [151] and in tofu [152].

Although aminobiogenic capability is reported to be strain dependent, Ladero et al. [54] suggested that tyramine and putrescine biosynthesis is a species level trait in E. faecalis. In fact, independently of the origin, several strains have been identified as BA producers. Moreover, PCR results demonstrated that the same genetic organization was present in all the tested strains and their decarboxylase clusters were independently located in the chromosome, with flanking regions showing within-species homogeneity.

In E. faecalis, putrescine is formed from agmatine by the AgDI pathway, which is repressed by carbon source, suggesting a role in the energy production [153]. Perez et al. [154] studied the possible co-regulation among TDC and AgDI pathways in E. faecalis. They investigated firstly the tyrosine effect on the tdc cluster transcription of E. faecalis by microarray experiment, highlighting, in the presence of tyrosine, an over-expression of tdcA, tdcP, and nhac-2 genes and a repression of tyrS. Bargossi et al. [15,155] have also demonstrated the same effect in other E. faecalis strains. Moreover, Perez et al. [154] showed that tyrosine induced putrescine biosynthesis genes, as confirmed by reverse transcription quantitative PCR (RT-qPCR) results. On the other hand, this effect was not observed in the mutant strain, which was unable to decarboxylate tyrosine and produce tyramine, showing that tdc cluster was involved in the tyrosine induction of putrescine biosynthesis.

Recently, some authors demonstrated that also E. mundtii possesses the capability to produce both tyramine and 2-phenilethylammine [107]. The genetic organization indicated that the tyramine-forming pathway in E. mundtii is similar to that found in phylogenetically closer enterococcal species, such as E. faecium, E. hirae and E. durans. The gene Na^+/H^+ antiporter (nhaC) that usually follows tyrP was missing. However, the analysis of the available data on E. mundtii genome revealed the presence of a further region that includes two genes encoding for an additional pyridoxal phosphate (PLP)-dependent decarboxylase and an amino acid permease, correlated with the tyrosine decarboxylating potential of this species.

In any case, tyramine is often accumulated by enterococci in high amounts already during the late exponential growth, before stationary phase, suggesting that this decarboxylation activity is not necessarily a response to starvation or nutrient depletion, and no competition between sugar catabolism and amino acid decarboxylation was observed [15,17]. In particular, these latter authors tested the ability to accumulate tyramine and 2-phenylethylamine by two strains of E. faecalis and two strains E. faecium in two culture media added or not with tyrosine. They demonstrated that, although all the tested enterococcal strains possessed a TDC pathway, they differed in BA accumulation level and in the expression rate of tdc gene, underlining the extremely variable decarboxylating potential of strains belonging to the same species, suggesting strain-dependent implications in food safety.

Environmental factors such as pH, temperature and NaCl concentrations can affect BA production in enterococci and several studies on decarboxylase activity of Enterococcus spp. in different conditions have been carried out. Gardini et al. [156] investigated the combined effects of temperature, pH and NaCl concentration on tyramine production by the strain E. faecalis EF37, finding that production of tyramine was mainly dependent on cell number. Moreover, these authors reported that this strain was able to accumulate also 2-phenylethylamine. A study regarding EF37 tyrDC expression revealed that stress could induce greater tyrosine decarboxylase activity, suggesting that suboptimal environmental conditions could lead to a higher tyrosine production, not necessarily associated with cell growth. This could be explained with the physiological role of this biochemical pathways associated with the survival of LAB in hostile environments [157]. Acidic conditions favored tyramine production in an E. durans BA-producing strain isolated from cheese [158] and in E. faecium [16,26], demonstrating the role of tyrosine decarboxylation in pH homeostasis. On the other hand, transcriptional studies of the tdc cluster in E. durans 655 showed a pH regulation of tyramine biosynthesis, being the gene expression quantification during the exponential phase induced by high concentrations of tyrosine, under acidic conditions [159].

Bargossi et al. [105] investigated the diversity of tyramine production capability of two E. faecalis and two E. faecium strains in buffered systems in relation to their genetic characteristics and to pH,

NaCl concentration and incubation temperature, comparing the results with those obtained with a purified tyrosine decarboxylase under the same conditions. They found that TDC activity was greatly heterogeneous within the enterococci, being *E. faecalis* EF37 the most efficient in tyramine accumulation. This heterogeneity depended on different genetic determinants, regulation mechanisms and environmental factors, above all incubation temperature.

A reduced transcription of genes involved in tyramine production was observed in the presence of 6.5% of NaCl in *E. faecalis* [160]. Also Bargossi et al. [105] showed that *E. faecalis* partially reduced its tyraminogenic potential passing from 0 to 5% of NaCl but the decarboxylation activity did not change significantly increasing NaCl concentration up to 15%. On the contrary, Liu et al. [161] demonstrated that NaCl stress can upregulate the expression of *tyrDC* and *tyrP* to improve the tyramine production of a single *E. faecalis* strain under certain conditions.

3.2. Lactobacillus

Lactobacilli are reported to be strong BA producers in different fermented foods [94].

In fermented sausages, beside to enterococci, the main tyramine producers among LAB are strains belonging to *Lactobacillus curvatus* species, which is, together with *Lactobacillus sakei*, the predominant *Lactobacillus* species in fermented meat products [162,163]. In fact, the majority of *L. curvatus* strains isolated from meat were reported to be tyramine producers [126]. However, Bover-Cid et al. [126] reported also some strains of *Lactobacillus paracasei*, *L. brevis*, and *L. sakei* isolated from pork-fermented meat as tyramine forming. Pereira et al. [164] demonstrated tyrosine and ornithine decarboxylase activities in *Lactobacillus homohiochii* and *L. curvatus* isolated from a Portuguese traditional dry fermented sausage.

Freiding et al. [165] screened *L. curvatus* strains from different origins, finding strain dependent tyrosine decarboxylase activity. Moreover, although *L. sakei* is usually described as non- aminogenic, histidine decarboxylase activity in one *L. sakei* strain has been evidenced [81,126,166].

Lactobacillus parabuchneri and *L. buchneri*, present as contaminants in fermented meat products, can produce histamine [167].

LAB populations were isolated from dry fermented sausages produced with different starters and using two spice mixtures in different process time by Kompdra et al. [37]. Tyrosine-decarboxylase and histidine-decarboxylase DNA sequence was identified in 44% and of 16% of lactobacilli isolates, respectively. In particular, several *Lactobacillus plantarum*, *L. brevis* and *Lactobacillus casei/paracasei* strains were identified as tyramine and histamine producers in the sausages analysed.

Although several microorganisms in cheese, including Gram negative bacteria, are able to produce BAs, *Lactobacillus* species, such as *Lactobacillus helveticus*, *L. buchneri* and *L. curvatus*, can be responsible for their accumulation in such products [38,138,168]. For example, specific strains of *L. buchneri* and *L. parabuchneri* harbor the histidine decarboxylase enzyme and can develop high levels of histamine, even at refrigerate temperature [39,169]. Wüthrich et al. [167] analysed several *L. parabuchneri* strains isolated from cheeses, finding some histamine positive among them. Moreover, these authors determined the complete genome of a histamine positive strain, showing that *hdc* gene cluster is located in a genomic island, transferred within the *L. parabuchneri* species. Diaz et al. [40] isolated, for the first time, 25 histamine-producing *Lactobacillus vaginalis* strains and sequenced *hdc* gene cluster and its flanking regions for a representative strain (*L. vaginalis* IPLA11050). These authors suggested that *hdc* locus was localized in the chromosome and, being the flanking regions the same in all histamine-producing *L. vaginalis* tested strains, histamine production has been suggested to be a species level trait. In addition, the organization of the examined genes was the same described for *Lactobacillus reuteri* [41], *L. buchneri* [111] and *L. hilgardii* [33] but differed to that of *S. thermophilus*.

L. brevis tyramine-producing strains have been isolated from cheeses by several authors [170,171] and this feature has been described as a strain-level trait (perhaps horizontally acquired) in *L. brevis*. Pachlová et al. [172] assessed the development of BA content in model cheese samples individually inoculated with two BA producing NSLAB strains of *L. curvatus* subsp. *curvatus* and *L. paracasei*, demonstrating the ability of these strains to accumulate tyramine up to 200 mg/kg in real dairy products during a 90 days ripening period. Yilmaz and Görkmen [173] demonstrated the capability to produce tyramine by a *L. plantarum* strain in yogurt and highlighted a possible indirect effect of *Lactobacillus delbrueckii* subsp. *bulgaricus* on accumulation of tyramine in the yoghurts, due to its synergistic interactions with tyraminogenic LAB strains.

It has been reported that non-starter *L. brevis* and *L. curvatus* are able to produce both tyramine and putrescine [174]. Although the ODC pathway has been described in several LAB, including strains of *L. brevis* [20], this pathway is not commonly used by dairy bacteria [64,67,175]. Ladero et al. [174] confirmed this aspect, showing that the detected putrescine-producing lactobacilli used AgDI pathway. Lucas et al. [101] found that *L. brevis* IOEB 9809 produced putrescine from agmatine but not from arginine, indicating the lack of a pathway converting arginine into agmatine. Moreover, it has been suggested that in *L. brevis* the AgDI genetic determinants are linked to those of the TDC pathway and are located in an acid resistance mechanism locus, probably acquired by horizontal gene transfer [20].

Several native lactobacilli, together with *O. oeni* and *Pediococcus parvulus* strains, are responsible for BA accumulation in wine. Their formation in wine depends on several conditions such as precursor amounts and the presence of specific decarboxylase-positive species and strains [176,177] and they are produced mainly during malolactic fermentation, particularly due to the presence of *L. brevis* and *L. hilgardii* [178–180]. Landete et al. [42] reported aminobiogenic potential of LAB isolated from wine samples, evidencing *L. mali* strains able to produce histamine, *L. brevis* strains able to accumulate tyramine and 2-phenylethylamine and a *L. hilgardii* strain showing histamine, tyramine, 2-phenylethylamine and putrescine production ability. On the other hand, the HDCs of *L. hilgardii* isolated from wine are well documented [33]. The enhancing effects of lower pH on histamine production (as responses to acidic stress) was observed in *L. brevis* [181]. Henríquez-Aedo et al. [182] reported that *Lactobacillus rhamnosus* was unexpectedly the predominant species in the vinification process of Chilean Cabernet Sauvignon wines and that it was mainly responsible for histamine accumulation in the products, presenting a significantly higher BA formation capability with respect to *O. oeni* isolated from the same samples. Arena and Manca de Nadra [21] studied a *L. plantarum* strain able to produce putrescine from arginine and ornithine while Moreno-Arribas et al. [183] found two wine strains of *L. buchneri* able to form putrescine via ornithine decarboxylase.

In wines, tyrosine decarboxylase has been associated with *Lactobacillus* spp., particularly *L. brevis* strains [84]. The same authors purified this pyridoxal 5P-phosphate dependent enzyme and showed it was highly substrate-specific for L-tyrosine and had an optimum pH of 5 [184]. The ability of a *L. plantarum* strain isolated from a red wine to produce tyramine from peptides containing tyrosine, especially during the late exponential growth phase, has been demonstrated [55] and *tdc* genes shared 98% identity with those in *L. brevis* consistent with horizontal gene transfer from *L. brevis* to *L. plantarum*. Arena et al. [185] assessed the expression of *L. brevis* IOEB 9809 *tdc* and *aguA1* genes during wine fermentation and evaluated the effect of substrate availability and pH on it, as well as on BA production, showing that the strain was able to produce both tyramine and putrescine. In addition, qRT-PCR analysis suggested a strong influence of substrate availability on the expression of BA pathway genes while less evident was pH influence. Afterwards, Lucas and Lonvaud-Funel [186] and Lucas et al. [53] reported for the same strain the complete *tdc* sequences, describing four complete genes (*tyrS, tyrDC, tyrP* and *nhaC*).

The BA production ability in different *Lactobacillus* strains, isolated from wine and cider, and their metabolic pathway were explored by Constantini et al. [68]. Their results demonstrated that most of the *L. brevis* analyzed harbor both *AgDI* and *tdc* genes and were tyramine and putrescine producers. Interestingly, these authors detected *hdc* genes in a *L. casei* strain isolated from cider.

Beer spoilage LAB showed several metabolic strategies to grow in nutrient poor environment, with acidic pH and hop presence among which BA production, contributing to energy supply and pH homeostasis, has been highlighted [187]. In particular, heterofermentative *L. brevis* strains accumulated tyramine and ornithine while *Lactobacillus lindneri* and *Lactobacillus paracollinoides* beer spoiling agent displayed ornithine and histamine production, respectively. Strains belonging to this latter species have been indicated as new potential histamine- and putrescine-producers in cider analysed by Ladero et al. [188]. Also Lorencová et al. [63] demonstrated that *L. brevis* strains isolated from beer can be a tyramine source in these products. The same authors showed the possibility to produce BAs by some probiotic strains belonging to *L. rhamnosus* species, opening a serious concern about the need to investigate the decarboxylation activity of probiotic or functional cultures before their use.

Recently, the capability to produce putrescine of a *Lactobacillus rossiae* strain, previously isolated from sourdough, has been reported [69]. This species is widely distributed in this fermented food [189] but the possibility of *L. rossiae* sourdough strains to produce BAs was not previously shown. In fact, only a strain of this species isolated from a wine starter has been described as histamine producer [190]. del Rio et al. [69] showed that *L. rossiae* strain accumulated this BA via the ODC pathway and the genetic organization and transcriptional analysis of the gene cluster identified the *odc* and *potE* genes forming an operon that is transcriptionally regulated by ornithine in a dose-dependent manner. Moreover, putrescine production via the ODC system improved the survival of *L. rossiae* by counteracting the cytoplasm acidification when the cells were subjected to acidic conditions, providing a biochemical defense mechanism against acidic environments. For this reason, this strain could easily produce putrescine during the fermentation process and the potential presence in sourdough of other BA-producing microorganisms cannot be ruled out.

The first description of ODC system in LAB was reported for *L. saerimneri* 30a [191]. Further investigation on this strain evidenced the presence of a unique genomic organization in which *odc* does not have an adjacent specific transporter gene but a three-component decarboxylase system with a lysine decarboxylase gene (*aadc*) and a promiscuous amino acid-amine transporter gene (*aat*), appearing atypical from those of other LAB [192].

3.3. Streptococcus

Although certain *Streptococcus* species are responsible for many disease (i.e., meningitis, bacterial pneumonia, endocarditis, necrotizing fasciitis etc.) many streptococcal species form part of the human microbiota and are important for fermented foods. In fact, *S. thermophilus* is often employed as selected starter culture and it is important for the dairy industry, since it is one of the principal components of many natural cultures used in fermented products such as hard cooked or pasta filata cheeses, yogurt and Cheddar [193]. This species is usually present in high numbers in the first steps of cheese-making and its relationships with BAs, which mainly accumulate during ripening, has been longer neglected. In fact, despite its wide industrial use, there are few papers regarding the decarboxylating potential of this species. Nevertheless, some BA producing strains have been identified and studied in recent years and several screenings on aminobiogenic potential of *Streptococcus* spp. strains have been performed. Ladero et al. [174] studied the BA producing ability of 137 strains of starter and NSLAB belonging to nine species of the genera *Lactobacillus*, *Lactococcus*, *Streptococcus* and *Leuconostoc* (all isolated from artisanal cheeses) in liquid media supplemented with the appropriate precursor amino acid by Ultra-High Performance Liquid Chromatography technique. Moreover, assessing the presence of key genes involved in the biosynthetic pathways of the target BA, they found that two *S. thermophilus* strains possessed *hdc* genes, although they were unable to synthesize histamine in broth. Also strains belonging to *Streptococcus macedonicus* species, isolated from Greek Kasseri cheese, showed tyramine production [194]. Some authors demonstrated that *Streptococcus mutans* expressed an agmatine deiminase system, encoded by the agmatine-inducible *aguBDAC* operon, which was induced in the presence of agmatine and was regulated by carbon catabolite repression. This metabolism was proposed to augment the acid resistance properties and pathogenic potential of *S. mutans*, etiological agent of dental caries and acid tolerance in oral biofilms [195,196].

Elsanhoty and Ramadan [197] reported the presence of *tdc*, *hdc* and *AgDI* genes in a *S. thermophilus* strain. Buňková et al. [48] studied BA production capability of selected technological important LAB belonging to *Lactococcus*, *Lactobacillus* and *Streptococcus* genera. Among these strains, one *S. thermophilus* of 11 was able to produce tyramine. Gezginc et al. [198] investigated the BA production capability of

S. thermophilus isolates in homemade natural yogurt, evidencing the presence of *hdcA* gene in several strains, although it was poorly correlated with histamine production in the decarboxylase medium. Yilmaz and Gökmen [173] investigated tyramine formation during yoghurt fermentation, focusing on interaction between a *S. thermophilus* strain and some *Lactobacillus* species The streptococci cells were able to produce tyramine depending on the fermentation conditions and synergistic interactions between *S. thermophilus* and *L. delbrueckii* subsp. *bulgaricus* were found in terms of BA accumulation.

The decarboxylating potential of the strain *S. thermophilus* NCFB2392 in lysine decarboxylase broth has been reported [199]. This strain was able to accumulate mostly putrescine, cadaverine and agmatine, and, when co-cultered with other BA producer Gram negative strains, had synergistic or antagonistic effect on BA concentrations. In fact, it caused 2-fold lower cadaverine production by *Salmonella* Paratyphi A and stimulated tyramine accumulation of *Escherichia coli*.

A high number of *S. thermophilus* strains have been investigated for tyramine production by La Gioia et al. [49]. Only the strain 1TT45, isolated from Taleggio cheese, demonstrated the capability to accumulate tyramine in broth. For this strain, a tyrosine decarboxylase (*tdc*A) gene was identified, with a nearly identical sequence to a *tdc*A of *L. curvatus*, indicated a horizontal gene transfer event. In the same work *tdc*A expression level and the production of tyramine were evaluated under different conditions during 7 days of incubation in skim milk. High transcript levels were evidenced only at the seventh day in presence of tyrosine, showing that the ability of *S. thermophilus* 1TT45 to form this BA depends on precursor availability in the culture medium, due to the incapability of this species to release peptides and free amino acids from milk proteins when grown in pure culture. On the other hand, the presence in cheeses of highly proteolytic LAB species would likely allow tyramine formation by *S. thermophilus*.

Calles-Enríquez et al. [35] observed two *S. thermophilus* strains able to produce histamine and reported their complete *hdc* gene cluster organization. This cluster began with the *hdcA* gene, was followed by a transporter (*hdcP*) and ended with the *hdcB* gene, located in the chromosome and orientated in the same direction.

The gene order of *hdcAPB* operon is similar to *Staphylococcus capitis* and *Clostridium perfringens*, which, however, lacks *hdcB* [200]. Transcriptional analysis of the *hdc* cluster revealed the maximum expression during the stationary growth phase, with high expression levels correlated with high histamine concentration. In the same work, also some factors affecting histamine biosynthesis and histidine-decarboxylating gene (*hdcA*) expression were studied. In particular, low temperature incubation determined lower levels of histamine in milk than in samples kept at 42 °C. This reduction was attributed to a reduction in the activity of the HDC enzyme itself rather than a reduction in gene expression or the presence of a lower cell number.

The occurrence of a histidine decarboxylase gene (*hdcA*) was demonstrated also in five among 83-screened *S. thermophilus* strains by Rossi et al. [85]. The sequence of the *hdcA* gene and closest flanking regions were determined for the strain PRI60, which produced the highest amounts of histamine. This strain synthesized HDC enzyme in milk even in the absence of histidine and it remained active also in cell-free extracts. Tabanelli et al. [201] continued the study of the histamine potential of PRI60 strain, testing histamine accumulation by cells or crude enzyme preparations with respect to factors related to dairy products, reporting a histamine concentration increase concomitantly with the cell growth. Moreover, HDC was mostly active at pH 4.5 and salt concentration up to 5% (w/v) did not affected enzyme activity. These authors evidenced enzyme thermal resistance up to temperature resembling low pasteurization, showing the risk of the presence of histaminogenic *S. thermophilus* strains in products from raw or mildly heat-treated milk. In fact, this strain was able to accumulate histamine in experimental cheeses, both when inoculated as starter or as cell-free crude enzyme preparations, highlighting that histamine formation by *S. thermophilus* in artisanal cheeses must not be overlooked, especially during typical production practices [31].

3.4. Lactococcus

Lactococci are among the most important LAB involved in the dairy industry and some species, including *Lc. lactis* subsp. *lactis*, *Lc. lactis* subsp. *lactis* biovar *diacetylactis* and *Lc. lactis* subsp. *cremoris*, play a critical role in the manufacture of many fermented dairy products [202]. As well as starters, they have been proposed as bioprotective cultures in the food industry and, for these reasons, safety criteria (such as BA production) should be evaluated.

Despite their QPS status (recognized by EFSA) and their generally regarded as safe (GRAS) status (recognized by FDA), some *Lc. lactis* have been reported to have aminobiogenic activity, both in vitro and in real systems. In fact, several strains of *Lc. lactis* subsp. *lactis* and *Lc. lactis* subsp. *cremoris* able to produce putrescine from agmatine via the AgDI pathway have been identified and these species are known to be, together with *E. faecalis*, *E. hirae*, *L. brevis* and *L. curvatus*, the main putrescine producers in dairy products [65,67,174]. The analysis of the *AgDI* cluster and their flanking regions revealed that the capability to produce putrescine via the AgDI pathway could be a specific characteristic that was lost during the adaptation to the milk environment by a process of reductive genome evolution [65]. The AgDI pathway increases the growth of *Lc. lactis* and causes the alkalinization of the culture medium, although it does not seem to be an acid stress resistance mechanism [30]. Linares et al. [203] investigated the role of *aguR* gene in putrescine formation in *Lc. lactis* subsp. *cremoris* CECT 8666, founding that it is essential for putrescine biosynthesis and it is transcribed independently of the polycistronic mRNA encoding the catabolic genes. Moreover, the transmembrane protein encoded by *aguR* can act as a transcription activator of the putrescine biosynthesis operon in response to the agmatine concentration. The same strain was tested in experimental Cabrales-like cheeses with different NaCl concentrations [204]. These authors evidenced that reducing the NaCl concentration of cheese led to increased putrescine accumulation and that NaCl was able to reduce the transcription of the *aguBDAC* operon, even if no effect on the transcription of *aguR* was recorded. The same authors investigated the effect of extracellular pH on putrescine biosynthesis and on the genetic regulation of the AgDI pathway of CECT 8666 strain. They showed increased putrescine biosynthesis at pH 5, when the transcription of the catabolic operon via the activation of the *aguBDAC* promoter P*aguB* was induced and a protection against acidic external conditions was reached through the counteraction of cytoplasm acidification [205].

It has been reported that *Lc. lactis* can produce other BAs in addition to putrescine. Martins Perin et al. [59] evaluated the BA production of bacteriocinogenic lactococci strains isolated from raw goat's milk reporting tyramine and 2-phenylethylamine accumulation capability in some of them. The decarboxylase activity of two aminobiogenic strains of *Lc. lactis* subsp. *cremoris* employed as starters in a model system of Dutch-type cheese was studied during a 90 day ripening period [206]. While in the control samples the amount of BAs was negligible, lactococci accumulated about 500 and 800 mg/kg of tyramine and putrescine, respectively. The putrescine decarboxylase activity observed in the model samples of cheeses with the inoculated strains was consistent with the results by Santos et al. [207].

3.5. Oenococcus and Pediococcus

Pediococcus spp. are often isolated from a large variety of plant materials and are involved in spontaneous fermentation of silage, sauerkraut, beans, cucumbers, olives, and cereals [208]. In addition, pediococci are also associated with fermented sausages and cheese, where selected strains of *Pediococcus pentosaceus* and *Pediococcus acidilactici* are often exploited as commercial starters, to control the development of undesired and pathogenic microbiota given their bacteriocinogenic features [209]. However, they can act also as spoilers in several fermented foods such as beer, wine and cider, causing turbidity, acidic off-tastes, adverse flavors and accumulating undesirable compounds [210].

O. oeni is a wine-associated LAB, considered the dominant species during the malolactic fermentation and possesses remarkable adaptability to harsh physico-chemical conditions. Several *O. oeni* strains have been described as BA producers and therefore many authors considered this species, together with other LAB species such as *L. hilgardii* and *Pediococcus*, as responsible for histamine presence in wine [33,36,43,108,176].

Lonvaud-Funel and Joyeux [44] isolated for the first time a strain of *O. oeni* able to produce histamine via histidine decarboxylase from a wine from the Bordeaux area. Subsequently, Coton et al. [108] purified and characterized this enzyme, concluding that it requires pyridoxal-5-phosphate as cofactor. Other authors have also shown that histamine-producing strains of *O. oeni* are frequent in wine [211] but this feature was strain dependent and in some strains no BA potential has been found [176]. For this reason, the role of *O. oeni* in BA accumulation in wine is still controversial [212]. As a possible explanation for these discrepancies, it has been suggested that the *hdc* genes are located on a large and possibly unstable plasmid and that culture collections will lose the capability to produce histamine in laboratory subcultures because of the loss of this unstable plasmid [36]. In the LAB isolated from wine by Landete et al. [42] mostly of *O. oeni* and *P. parvulus* harbored *hdc* genes and pediococci produced high level of histamine in synthetic medium, showing the highest histaminogenic potential within the tested genus. Nevertheless, in the same conditions *O. oeni* showed lower levels of histamine production, that were even lower when these strains were tested in wine samples. Berbegal et al. [213] showed the presence of both histamine producer and non-producer strains in a Spanish red Ribera del Duero wine and proposed a non histaminogenic strain to be used as starter reducing of 5-fold histamine content in inoculated wine than the non-inoculated control. It has been further demonstrated that, after one year, the barrel-ageing histamine concentrations were 3-fold lower in the inoculated vat than in the non-inoculated one.

Landete et al. [214] studied the influence of enological factors on the *hdc* expression and on HDC activity in *P. parvulus* and *O. oeni*. Gene expression was lowered by glucose, fructose, malic acid, and citric acid, whereas ethanol enhanced the HDC enzyme activity, so that the conditions normally occurring during malolactic fermentation and later on, could favor histamine production. On the other hand, Gardini et al. [177] evaluated the interactive effect of some variables on the BA production of *O. oeni*, demonstrating that high ethanol amounts and low concentration of pyridoxal-5-phosphate reduced their accumulation while higher pH enhanced BA concentrations. In addition, the SO_2 effect on tyramine accumulation depended also on other variables.

In cider, where a microbiological stabilization after malolactic fermentation is not performed, indigenous heterofermentative LAB constitute the predominant microbiota and several species such as oenococci and pediococci (beside to lactobacilli) are able to produce BAs [18,215]. Some pediococci strains isolated from wine and ciders showed the presence of decarboxylase genes, i.e., *AgDI* cluster in *P. parvulus* and *P. pentosaceus* [18,68,188]. In this latter work, also *O. oeni* strains had *AgDI* genes and the authors found a few discrepancies between phenotypic and genotypic data. On the other hand, the identification of an *odc* gene in a putrescine producer *O. oeni* strain has been reported by Marcobal et al. [70]. Later, the *odc* gene was also identified and sequenced in three *O. oeni* wine strains and in two *O. oeni* cider strains [216] and the sequencing of the complete *odc* gene from *O. oeni* and *L. brevis* showed an 83% identity [19].

Low production of cadaverine and tyramine was also found in a *Pediococcus* spp. strain isolated from beer [63] and Izquierdo-Pulido et al. [217] reported *Pediococcus* genus to be mainly responsible for tyramine accumulation in beer.

3.6. Other Genera: Weissella, Carnobacterium, Tetragenococcus, Leuconostoc, Sporolactobacillus

Leuconostocs are LAB associated with plants and decaying plant material, often detected in various fermented vegetable products but also in foods of animal origin [218]. Some species such as *Leuconostoc carnosum*, *Leuconostoc gasicomitatum*, and *Leuconostoc gelidum* have often been associated with food spoilage and some strains have been found to be decarboxylase positive. Although it is known that AgDI pathway has been demonstrated in *Leuconostoc mesenteroides* [18], some strains isolated from wine have been suggested to produce putrescine exclusively from arginine via the arginine deiminase pathway (ADI) pathway, given to the selective effect of the ecological niche on BA biosynthesis pathway [115,219]. Recently, based on current knowledge and QPS/GRAS/dairy (IDF) safety criteria guidelines, the safety of different LAB candidate antifungal bioprotective strains has been evaluated finding a tyramine-producer *Leuc. mesenteroides*. This result confirmed the importance to test decarboxylase activity before considering a candidate strain for use as a bioprotective agent in food products [220]. Dairy strains of leuconostocs have been associated with high levels of BAs in cheese and other dairy products and tyramine and 2-phenylethylamine production in *Leuconostoc* strains isolated from dairy products have been reported [32,221]. Moreno-Arribas et al. [183] found that *Leuc. mesenteroides* may also be responsible for tyramine production in wines and Landete et al. [42] isolated a wine strain belonging to this species able to produce histamine.

This genus can be also implicated in beer BA accumulation. In fact, some authors studied the occurrence of aminobiogenic strains during a craft brewing process, highlighting the presence of *Leuc. mesenteroides* possessing *tdc, hdc, odc* decarboxylase genes and able to produce tyramine in wort and beer [61]. *Leuc. mesenteroides* ssp. *mesenteroides* isolated from meat, fermented sausages and cheeses was able to form putrescine and cadaverine [56].

Weissella are heterofermentative LAB which occur in a wide range of habitats, i.e., milk, plants and as well as from a variety of fermented foods such as European sourdoughs and Asian and African traditional fermented vegetables. They can be involved in such traditional fermentations and some strains of *Weissella confusa* and *Weissalla cibaria* can produce copious amounts of dextran but strains of certain *Weissella* species are known as opportunistic pathogens involved in human infections [222]. Moreover, some *Weissella* strains have been demonstrated to be able to produce BAs in fermented foods. *Weissella viridescens* isolated from Tofu-misozuke, a traditional Japanese fermented food, resulted tyramine producers and several strains isolated from kimchi belonging to *W. cibaria*, *W. confusa* and *Weissella paramesenteroides* produced multiple BAs, including tyramine and histamine [45,152]. On the other hand, Pereira et al. [71] demonstrated that a *Weissella halotolerans* strain combines an ornithine decarboxylation pathway and an arginine deiminase pathway, leading to the accumulation of putrescine and producing a proton motive force.

Carnobacterium spp. can be found in vacuum or modified atmosphere packed, refrigerated raw or processed meat products and lightly preserved fish products, milk, and certain types of soft cheese [223]. Although *Carnobacterium divergens* and *Carnobacterium maltaromaticum* have been demonstrated to be bacteriocin producers, able to inhibit *Listeria monocytogenes*, some strains appear to display undesirable properties such as amino acid decarboxylation activities. In fact, these two species can produce tyramine while strains belonging to *C. divergens*, *Carnobacterium gallinarum*, *Carnobacterium maltaromaticum* and *C. mobile* can possess ADI pathway [57,104,223]. Curiel et al. [224] studied the BA production capability by LAB and enterobacteria isolated from fresh pork sausages and reported that all the tyramine-producer isolated strains were molecularly identified as *C. divergens*, whose abundance depended from the different packaging conditions. All these strains presented the *tdc* genes. Coton et al. [51] identified the gene encoding a putative tyrosine decarboxylase in *C. divergens*, evidencing the presence of three putative open reading frames (*tyrS, tyrDC* and amino acid transporter *PotE*) which showed the strongest homologies with *E. faecium* (94% identity, 98% homology) and *E. faecalis* (85% identity, 92% similarity) and exhibited conserved domains characteristic of the group II (PLP-dependent) decarboxylase family.

Other genera less frequent in fermented foods may also be involved in BA production.

Tetragenococcus is a halophilic facultative aerobic homofermentative coccus, which cannot be readily distinguished from members of the genus *Pediococcus* and which can play a role in halophilic fermentation processes such as the production of soy products, brined anchovies, fish sauce and fermented mustard or can constitute the dominant microbiota in concentrated sugar-concentrated juice [225]. Recently, the safety of 49 *Tetragenococcus halophilus* strains isolated from the Korean sauce doenjang has been assessed. The isolates produced higher tyramine level than reference strains and similar cadaverine, histamine, and putrescine production patterns [72]. A *T. muriaticus* strain isolated from fish sauce produced histamine during the late exponential growth phase, reaching a maximum production of this BA at 5–7% of NaCl, and was able to maintain a histidine decarboxylase activity also in the presence of 20% of salt [46].

Recently, a sporulating LAB, belonging to a novel species of *Sporolactobacillus* genus and isolated from cider must, harbored *tdc* gene, which showed the same organization as already described genes found in other tyramine-producing LAB. Moreover, genes showing the highest identities with mobile elements surrounded the *tdc* operon, suggesting that the tyramine-forming trait was acquired through horizontal gene transfer [58].

4. Conclusions

Biogenic amines can accumulate in high concentrations in fermented foods due to microbial activity and can cause toxic effects in consumers. LAB are considered mainly responsible for BA accumulation in these products and strains belonging to different species and genera, commonly found in fermented foods, have been characterized for their decarboxylase activities.

It is known that this decarboxylase activity provides cell advantages because it allows increasing the environmental pH and leads to the energization of membrane. The genetic clusters responsible for BA production have been described individually and they can show differences, within the same amine, that depend mainly on the species and the strain. Nevertheless, it is interesting to note, that the decarboxylation mechanisms constitute an important ecological tool which can favor strain competitiveness in stressful conditions (i.e., acid stress and nutritional stress) [20,26,27].

Even if differences between the chromosomic decarboxylase clusters are present, some interesting consideration can be drawn. The first is that the presence of these cluster are usually strain and not species dependent and can be regarded as genomic islands, as demonstrated for TDC cluster in *L. brevis* by Coton and Coton [50]. In addition, the genes encoding different decarboxylase pathways in several LAB species (*L. saerimneri*, *L. brevis*, and *O. oeni*) are clustered on the chromosome, acting as a genetic hotspot related to acid stress resistance [19,20,192].

Although the knowledge concerning the origin and factors involved in BA production in fermented foods is well documented, it is difficult to prevent the accumulation of these compounds since the fermentation conditions cannot be easily modified and the aminobiogenic ability is strain dependent. For these reasons, the selection of specific LAB starters lacking the pathways for BA accumulation and able to outgrow autochthonous microbiota under production conditions is essential to obtain high quality food with reduced contents of these toxic compounds. In fact, the inability of a strain to synthesize BAs has to be included as a selective criterion for starter cultures [226]. On the other hand, the metabolic heterogeneity observed in natural starter cultures could open a serious concern about the presence of aminobiogenic LAB strains. This risk could be avoided with the use of defined starters or selected autochthonous strain mixtures, chosen based on the absence of such activity and endowed with taylor-made metabolic and functional features for specific products. Nevertheless, when undefined cultures need to be used, strategies to prevent the presence and growth of aminobiogenic LAB should be actuated. Among them, the use of food microorganisms able to degrade BAs previously synthesized in the food matrix should be taken into consideration.

Author Contributions: Conceptualization, G.T. and F.G.; Literature data collection, G.T., F.B.; writing—original draft preparation, G.T., F.B.; writing—review and editing, C.M. and F.G.; supervision, F.G.

References

1.   Pessione, E.; Cirrincione, S. Bioactive molecules released in food by lactic acid bacteria: Encrypted peptides and biogenic amines. *Front. Microbiol.* **2016**, *7*, 876. [CrossRef] [PubMed]

2. Alvarez, M.A.; Moreno-Arribas, M.V. The problem of biogenic amines in fermented foods and the use of potential biogenic amine-degrading microorganisms as a solution. *Trends Food Sci. Technol.* **2014**, *39*, 146–155. [CrossRef]

3. Hungerford, J.M. Scombroid poisoning: A review. *Toxicon* **2010**, *56*, 231–243. [CrossRef] [PubMed]

4. Shalaby, A.R. Significance of biogenic amines to food safety and human health. *Food Res. Int.* **1996**, *29*, 675–690. [CrossRef]

5. Sathyanarayana Rao, T.S.; Yeragani, V.K. Hypertensive crisis and cheese. *Indian J. Psychiatry* **2009**, *51*, 65–66. [CrossRef] [PubMed]

6. Silla Santos, M.H. Biogenic amines: Their importance in foods. *Int. J. Food Microbiol.* **1996**, *29*, 213–231. [CrossRef]

7. McCabe-Sellers, B.; Staggs, C.G.; Bogle, M.L. Tyramine in foods and monoamine oxidase inhibitor drugs: A crossroad where medicine, nutrition, pharmacy, and food industry converge. *J. Food Comp. Anal.* **2006**, *19*, S58–S65. [CrossRef]

8. Knope, K.E.; Sloan-Gardner, T.S.; Stafford, R.J. Histamine fish poisoning in Australia, 2001 to 2013. *Commun. Dis. Intell. Q. Rep.* **2014**, *38*, E285–E293.

9. Marcobal, A.; de Las Rivas, B.; Landete, J.M.; Tabera, L.; Muñoz, R. Tyramine and phenylethylamine biosynthesis by food bacteria. *Crit. Rev. Food Sci. Nutr.* **2012**, *52*, 448–467. [CrossRef]

10. Pegg, A.E. Toxicity of polyamines and their metabolic products. *Chem. Res. Toxicol.* **2013**, *26*, 1782–1800. [CrossRef]

11. European Commission. Commission Regulation (EC) No. 2073/2005 of 15 November 2005 on microbiological criteria for foodstuffs. *Off. J. Eur. Union* **2005**, *50*, 1–26.

12. EFSA. Scientific opinion on risk based control of biogenic amine formation in fermented foods. *EFSA J.* **2011**, *9*, 2393–2486. [CrossRef]

13. Bover-Cid, S.; Latorre-Moratalla, M.L.; Veciana-Nogués, M.T.; Vidal-Carou, M.C. Biogenic amines. In *Encyclopedia of Food Safety*; Motarjemi, Y., Moy, G., Todd, E., Eds.; Academic Press: San Diego, CA, USA, 2014; pp. 381–391. ISBN 978-0-12-378613-5.

14. Halász, A.; Baráth, Á.; Simon-Sarkadi, L.; Holzapfel, W. Biogenic amines and their production by microorganisms in food. *Trends Food Sci. Technol.* **1994**, *5*, 42–49. [CrossRef]

15. Bargossi, E.; Tabanelli, G.; Montanari, C.; Lanciotti, R.; Gatto, V.; Gardini, F.; Torriani, S. Tyrosine decarboxylase activity of enterococci grown in media with different nutritional potential: Tyramine and 2-phenylethylamine accumulation and *tyrDC* gene expression. *Front. Microbiol.* **2015**, *6*, 259. [CrossRef] [PubMed]

16. Marcobal, A.; de las Rivas, B.; Muñoz, R. First genetic characterization of a bacterial b-phenylethylamine biosynthetic enzyme in *Enterococcus faecium* RM58. *FEMS Microbiol. Lett.* **2006**, *258*, 144–149. [CrossRef] [PubMed]

17. Pessione, E.; Pessione, A.; Lamberti, C.; Coïsson, D.J.; Riedel, K.; Mazzoli, R.; Bonetta, S.; Eberl, L.; Giunta, C. First evidence of a membrane-bound, tyramine and beta-phenylethylamine producing, tyrosine decarboxylase in *Enterococcus faecalis*: A two-dimensional electrophoresis proteomic study. *Proteomics* **2009**, *9*, 2695–2710. [CrossRef]

18. Coton, M.; Romano, A.; Spano, G.; Ziegler, K.; Vetrana, C.; Desmarais, C.; Coton, E. Occurrence of biogenic amine-forming lactic acid bacteria in wine and cider. *Food Microbiol.* **2010**, *27*, 1078–1085. [CrossRef]

19. Romano, A.; Trip, H.; Lonvaud-Funel, A.; Lolkema, J.S.; Lucas, P.M. Evidence of two functionally distinct ornithine decarboxylation systems in lactic acid bacteria. *Appl. Environ. Microbiol.* **2012**, *78*, 1953–1961. [CrossRef] [PubMed]

20. Romano, A.; Ladero, V.; Alvarez, M.A.; Lucas, P.M. Putrescine production via the ornithine decarboxylation pathway improves the acid stress survival of *Lactobacillus brevis* and is part of a horizontally transferred acid resistance locus. *Int. J. Food Microbiol.* **2014**, *175*, 14–19. [CrossRef]

21. Arena, M.E.; Manca de Nadra, M.C. Biogenic amine production by *Lactobacillus*. *J. Appl. Microbiol.* **2001**, *90*, 158–162. [CrossRef]

22. Bardócz, S. Polyamines in food and their consequences for food quality and human health. *Trends Food Sci. Technol.* **2005**, *6*, 341–346. [CrossRef]

23. Kalač, P.; Krausová, P. A review of dietary polyamines: Formation, implications for growth and health and occurrence in foods. *Food Chem.* **2005**, *90*, 219–230. [CrossRef]

24. Molenaar, D.; Bosscher, J.S.; Ten Brink, B.; Driessen, A.J.M.; Konings, W.N. Generation of a proton motive force by histidine decarboxylation and electrogenic histidine/histamine antiport in *Lactobacillus buchneri*. *J. Bacteriol.* **1993**, *175*, 2864–2870. [CrossRef]

25. Pessione, A.; Lamberti, C.; Pessione, E. Proteomics as a tool for studying energy metabolism in lactic acid bacteria. *Mol. BioSyst.* **2010**, *6*, 1419–1430. [CrossRef]

26. Pereira, C.I.; Matos, D.; Romão, M.V.S.; Barreto Crespo, M.T. Dual role for the tyrosine decarboxylation pathway in *Enterococcus faecium* E17: Response to an acid challenge and generation of a proton motive force. *Appl. Environ. Microbiol.* **2009**, *75*, 345–352. [CrossRef]

27. Perez, M.; Calles-Enríquez, M.; Nes, I.; Martin, M.C.; Fernández, M.; Ladero, V.; Alvarez, M.A. Tyramine biosynthesis is transcriptionally induced at low pH and improves the fitness of *Enterococcus faecalis* in acidic environments. *Appl. Microbiol. Biotechnol.* **2015**, *99*, 3547–3558. [CrossRef]

28. Konings, W.N. Microbial transport: Adaptations to natural environments. *Antonie Van Leeuwenhoek* **2006**, *90*, 325–342. [CrossRef]

29. Vido, K.; Le Bars, D.; Mistou, M.Y.; Anglade, P.; Gruss, A.; Gaudu, P. Proteome analyses of heme-dependent respiration in *Lactococcus lactis*: Involvement of the proteolytic system. *J. Bacteriol.* **2004**, *186*, 1648–1657. [CrossRef]

30. del Rio, B.; Linares, D.M.; Ladero, V.; Redruello, B.; Fernández, M.; Martin, M.C.; Alvarez, M.A. Putrescine production via the agmatine deiminase pathway increases the growth of *Lactococcus lactis* and causes the alkalinization of the culture medium. *Appl. Microbiol. Biotechnol.* **2015**, *99*, 897–905. [CrossRef]

31. Gardini, F.; Rossi, F.; Rizzotti, L.; Torriani, S.; Grazia, L.; Chiavari, C.; Coloretti, F.; Tabanelli, G. Role of *Streptococcus thermophilus* PRI60 in histamine accumulation in cheese. *Int. Dairy J.* **2012**, *27*, 71–76. [CrossRef]

32. Benkerroum, N. Biogenic amines in dairy products: Origin, incidence, and control means. *Compr. Rev. Food Sci. Food Saf.* **2016**, *15*, 801–826. [CrossRef]

33. Lucas, P.M.; Wolken, W.A.; Claisse, O.; Lolkema, J.S.; Lonvaud-Funel, A. Histamine-producing pathway encoded on an unstable plasmid in *Lactobacillus hilgardii* 0006. *Appl. Environ. Microbiol.* **2005**, *7*, 1417–1424. [CrossRef]

34. Landete, J.M.; de las Rivas, B.; Marcobal, A.; Muñoz, R. Updated molecular knowledge about histamine biosynthesis by bacteria. *Crit. Rev. Food Sci. Nutr.* **2008**, *48*, 697–714. [CrossRef]

35. Calles-Enríquez, M.; Eriksen, B.H.; Andersen, P.S.; Rattray, F.P.; Johansen, A.H.; Fernández, M.; Ladero, V.; Alvarez, M.A. Sequencing and transcriptional analysis of the *Streptococcus thermophilus* histamine biosynthesis gene cluster: Factors that affect differential *hdcA* expression. *Appl. Environ. Microbiol.* **2010**, *76*, 6231–6238. [CrossRef]

36. Lucas, P.M.; Claisse, O.; Lonvaud-Funel, A. High frequency of histamine-producing bacteria in the enological environmental and instability of the histidine decarboxylase production phenotype. *Appl. Environ. Microbiol.* **2008**, *74*, 811–817. [CrossRef]

37. Komprda, T.; Sládková, P.; Petirová, E.; Dohnal, V.; Burdychová, R. Tyrosine- and histidine-decarboxylase positive lactic acid bacteria and enterococci in dry fermented sausages. *Meat Sci.* **2010**, *86*, 870–877. [CrossRef]

38. Ladero, V.; Linares, D.M.; Fernández, M.; Alvarez, M.A. Real time quantitative PCR detection of histamine-producing lactic acid bacteria in cheese: Relation with histamine content. *Food Res. Int.* **2008**, *41*, 1015–1019. [CrossRef]

39. Diaz, M.; del Rio, B.; Sanchez-Llana, E.; Ladero, V.; Redruello, B.; Fernández, M.; Martin, M.C.; Alvarez, M.A. *Lactobacillus parabuchneri* produces histamine in refrigerated cheese at a temperature-dependent rate. *Int. J. Food Sci. Technol.* **2018**, *53*, 2342–2348. [CrossRef]

40. Diaz, M.; del Rio, B.; Ladero, V.; Redruello, B.; Fernández, M.; Martin, M.C.; Alvarez, M.A. Isolation and typification of histamine-producing *Lactobacillus vaginalis* strains from cheese. *Int. J. Food Microbiol.* **2015**, *215*, 117–123. [CrossRef]

41. Thomas, C.M.; Hong, T.; van Pijkeren, J.P.; Hemarajata, P.; Trinh, D.V.; Hu, W.; Britton, R.A.; Kalkum, M.; Versalovic, J. Histamine derived from probiotic *Lactobacillus reuteri* suppresses TNF via modulation of PKA and ERK signaling. *PLoS ONE* **2012**, *7*, e31951. [CrossRef]

42. Landete, J.M.; Ferrer, S.; Pardo, I. Biogenic amine production by lactic acid bacteria, acetic bacteria and yeast isolated from wine. *Food Control* **2007**, *18*, 1569–1574. [CrossRef]

43. Landete, J.M.; Ferrer, S.; Pardo, I. Which lactic acid bacteria are responsible for histamine production in wine? *J. App. Microbiol.* **2005**, *99*, 580–586. [CrossRef] [PubMed]

44. Lonvaud-Funel, A.; Joyeux, A. Histamine production by wine lactic acid bacteria: Isolation of a histamine-producing strain of *Leuconostoc oenos*. *J. Appl. Bacteriol.* **1994**, *77*, 401–407. [CrossRef]

45. Jeong, D.W.; Lee, J.H. Antibiotic resistance, hemolysis and biogenic amine production assessments of *Leuconostoc* and *Weissella* isolates for kimchi starter development. *LWT-Food Sci. Technol.* **2015**, *64*, 1078–1084. [CrossRef]

46. Kimura, B.; Konagaya, Y.; Fujii, T. Histamine formation by *Tetragenococcus muriaticus*, a halophilic lactic acid bacterium isolated from fish sauce. *Int. J. Food Microbiol.* **2001**, *70*, 71–77. [CrossRef]

47. Linares, D.M.; Martin, M.C.; Ladero, V.; Alvarez, M.A.; Fernández, M. Biogenic amines in dairy products. *Crit. Rev. Food Sci. Nutr.* **2011**, *51*, 691–703. [CrossRef] [PubMed]

48. Buňková, L.; Buňka, F.; Hlobilová, M.; Vakátková, Z.; Nováková, D.; Dráb, V. Tyramine production of technological important strains of *Lactobacillus*, *Lactococcus* and *Streptococcus*. *Eur. Food Res. Technol.* **2009**, *229*, 533–538. [CrossRef]

49. La Gioia, F.; Rizzotti, L.; Rossi, F.; Gardini, F.; Tabanelli, G.; Torriani, S. Identification of a tyrosine decarboxylase (*tdcA*) gene in *Streptococcus thermophilus* 1TT45: Analysis of its expression and tyramine production in milk. *Appl. Environ. Microbiol.* **2011**, *77*, 1140–1144. [CrossRef] [PubMed]

50. Coton, E.; Coton, M. Evidence of horizontal transfer as origin of strain to strain variation of the tyramine production trait in *Lactobacillus brevis*. *Food Microbiol.* **2009**, *26*, 52–57. [CrossRef] [PubMed]

51. Coton, M.; Coton, E.; Lucas, P.; Lonvaud, A. Identification of the gene encoding a putative tyrosine decarboxylase of *Carnobacterium divergens* 508. Development of molecular tools for the detection of tyramine-producing bacteria. *Food Microbiol.* **2004**, *21*, 125–130. [CrossRef]

52. Ladero, V.; Linares, D.M.; del Rio, B.; Fernández, M.; Martin, M.C.; Alvarez, M.A. Draft genome sequence of the tyramine producer *Enterococcus durans* strain IPLA 655. *Genome Announc.* **2013**, *1*, e00265-13. [CrossRef] [PubMed]

53. Lucas, P.; Landete, J.; Coton, M.; Coton, E.; Lonvaud-Funel, A. The tyrosine decarboxylase operon of *Lactobacillus brevis* IOEB 9809: Characterization and conservation in tyramine-producing bacteria. *FEMS Microbiol. Lett.* **2003**, *229*, 65–71. [CrossRef]

54. Ladero, V.; Fernández, M.; Calles-Enríquez, M.; Sánchez-Llana, E.; Cañedo, E.; Martin, M.C.; Alvarez, M.A. Is the production of the biogenic amines tyramine and putrescine a species-level trait in enterococci? *Food Microbiol.* **2012**, *30*, 132–138. [CrossRef] [PubMed]

55. Bonnin-Jusserand, M.; Grandvalet, C.; Rieu, A.; Weidmann, S.; Alexandre, H. Tyrosine-containing peptides are precursors of tyramine produced by *Lactobacillus plantarum* strain IR BL0076 isolated from wine. *BMC Microbiol.* **2012**, *12*, 199. [CrossRef] [PubMed]

56. Pircher, A.; Bauer, F.; Paulsen, P. Formation of cadaverine, histamine, putrescine and tyramine by bacteria isolated from meat, fermented sausages and cheeses. *Eur. Food Res. Technol.* **2007**, *226*, 225–231. [CrossRef]

57. Massona, F.; Johansson, G.; Montela, M.C. Tyramine production by a strain of *Carnobacterium divergens* inoculated in meat-fat mixture. *Meat Sci.* **1999**, *52*, 65–69. [CrossRef]

58. Coton, M.; Fernández, M.; Trip, H.; Ladero, V.; Mulder, N.L.; Lolkema, J.S.; Alvarez, M.A.; Coton, E. Characterization of the tyramine-producing pathway in *Sporolactobacillus* sp. P3J. *Microbiology* **2011**, *157*, 1841–1849. [CrossRef]

59. Martins Perin, L.; Belviso, S.; dal Bello, B.; Nero, L.A.; Cocolin, L. Technological properties and biogenic amines production by bacteriocinogenic lactococci and enterococci strains isolated from raw goat's milk. *J. Food Prot.* **2017**, *80*, 151–157. [CrossRef]

60. Guo, Y.Y.; Yang, Y.P.; Peng, Q.; Han, Y. Biogenic amines in wine: A review. *Int. J. Food Sci. Technol.* **2015**, *50*, 1523–1532. [CrossRef]

61. Poveda, J.M.; Ruiz, P.; Seseña, S.; Palop, M.L. Occurrence of biogenic amine-forming lactic acid bacteria during a craft brewing process. *LWT-Food Sci. Technol.* **2017**, *85*, 129–136. [CrossRef]

62. Iacumin, L.; Manzano, M.; Panseri, S.; Chiesa, L.; Comi, G. A new cause of spoilage in goose sausages. *Food Microbiol.* **2016**, *58*, 56–62. [CrossRef] [PubMed]

63. Lorencová, E.; Buňková, L.; Matoulková, D.; Dráb, V.; Pleva, P.; Kubáň, V.; Buňka, F. Production of biogenic amines by lactic acid bacteria and bifidobacteria isolated from dairy products and beer. *Int. J. Food Sci. Technol.* **2012**, *47*, 2086–2091. [CrossRef]

64. Wunderlichová, L.; Buňková, L.; Koutný, M.; Jančová, P.; Buňka, F. Formation, degradation, and detoxification of putrescine by foodborne bacteria: A review. *Compr. Rev. Food Sci. Food Saf.* **2014**, *13*, 1012–1033. [CrossRef]

65. Ladero, V.; Rattray, F.P.; Mayo, B.; Martin, M.C.; Fernández, M.; Alvarez, M.A. Sequencing and transcriptional analysis of the biosynthesis gene cluster of putrescine-producing *Lactococcus lactis*. *Appl. Environ. Microbiol.* **2011**, *77*, 6409–6418. [CrossRef] [PubMed]

66. Marcobal, A.; de las Rivas, B.; Moreno-Arribas, M.V.; Munoz, R. Evidence for horizontal gene transfer as origin of putrescine production in *Oenococcus oeni* RM83. *Appl. Environ. Microbiol.* **2006**, *72*, 7954–7958. [CrossRef]

67. Ladero, V.; Canedo, E.; Perez, M.; Cruz Martin, M.; Fernández, M.; Alvarez, M.A. Multiplex qPCR for the detection and quantification of putrescine-producing lactic acid bacteria in dairy products. *Food Control* **2012**, *27*, 307–313. [CrossRef]

68. Costantini, A.; Pietroniro, R.; Doria, F.; Pessione, E.; Garcia-Moruno, E. Putrescine production from different amino acid precursors by lactic acid bacteria from wine and cider. *Int. J. Food Microbiol.* **2013**, *165*, 11–17. [CrossRef]

69. del Rio, B.; Alvarez-Sieiro, P.; Redruello, B.; Martin, M.C.; Fernandez, M.; Ladero, V.; Alvarez, M.A. *Lactobacillus rossiae* strain isolated from sourdough produces putrescine from arginine. *Sci. Rep.* **2018**, *8*, 3989. [CrossRef]

70. Marcobal, A.; de Las Rivas, B.; Moreno-Arribas, M.V.; Muñoz, R. Identification of the ornithine decarboxylase gene in the putrescine-producer *Oenococcus oeni* BIFI-83. *FEMS Microbiol. Lett.* **2004**, *239*, 213–220. [CrossRef]

71. Pereira, C.I.; San Romão, M.V.; Lolkema, J.S.; Barreto Crespo, M.T. *Weissella halotolerans* W22 combines arginine deiminase and ornithine decarboxylation pathways and converts arginine to putrescine. *J. Appl. Microbiol.* **2009**, *107*, 1894–1902. [CrossRef]

72. Jeong, D.W.; Heo, S.; Le, J.H. Safety assessment of *Tetragenococcus halophilus* isolates from doenjang, a Korean high-salt-fermented soybean paste. *Food Microbiol.* **2017**, *62*, 92–98. [CrossRef] [PubMed]

73. Lorenzo, J.M.; Cachaldora, A.; Fonseca, S.; Gómez, M.; Franco, I.; Carballo, J. Production of biogenic amines "in vitro" in relation to the growth phase by *Enterobacteriaceae* species isolated from traditional sausages. *Meat Sci.* **2010**, *86*, 684–691. [CrossRef]

74. Morii, H.; Kasama, K. Activity of two histidine decarboxylases from *Photobacterium phosphoreum* at different temperatures, pHs, and NaCl concentrations. *J. Food Prot.* **2004**, *67*, 1736–1742. [CrossRef]

75. Ruiz-Capillas, C.; Jiménez-Colmenero, F. Biogenic amines in meat and meat products. *Crit. Rev. Food Sci. Nutr.* **2004**, *44*, 489–499. [CrossRef] [PubMed]

76. Özogul, F.; Özogul, Y. Biogenic amine content and biogenic amine quality indices of sardines (*Sardina pilchardus*) stored in modified atmosphere packaging and vacuum packaging. *Food Chem.* **2006**, *99*, 574–578. [CrossRef]

77. Ladero, V.; Sánchez-Llana, E.; Fernández, M.; Alvarez, M.A. Survival of biogenic amine-producing dairy LAB strains at pasteurisation conditions. *Int. J. Food Sci. Technol.* **2011**, *46*, 516–521. [CrossRef]

78. Gardini, F.; Tofalo, R.; Belletti, N.; Iucci, L.; Suzzi, G.; Torriani, S.; Guerzoni, M.E.; Lanciotti, R. Characterization of yeasts involved in the ripening of Pecorino Crotonese cheese. *Food Microbiol.* **2006**, *23*, 641–648. [CrossRef]

79. Qi, W.; Hou, L.H.; Guo, H.L.; Wang, C.L.; Fan, Z.C.; Liu, J.F.; Cao, X.H. Effect of salt-tolerant yeast of *Candida versatilis* and *Zygosaccharomyces rouxii* on the production of biogenic amines during soy sauce fermentation. *J. Sci. Food Agric.* **2014**, *94*, 1537–1542. [CrossRef]

80. Tristezza, M.; Vetrano, C.; Bleve, G.; Spano, G.; Capozzi, V.; Logrieco, A.; Mita, G.; Grieco, F. Biodiversity and safety aspects of yeast strains characterized from vineyards and spontaneous fermentations in the Apulia Region. *Food Microbiol.* **2013**, *36*, 335–342. [CrossRef] [PubMed]

81. Latorre-Moratalla, M.L.; Bover-Cid, S.; Talon, R.; Garriga, M.; Aymerich, T.; Zanardi, E.; Ianieri, A.; Fraqueza, M.J.; Elias, M.; Drosinos, E.H.; et al. Distribution of aminogenic activity among potential autochthonous starter cultures for dry fermented sausages. *J. Food Prot.* **2010**, *73*, 524–525. [CrossRef]

82. Suzzi, G.; Gardini, F. Biogenic amines in dry fermented sausages: A review. *Int. J. Food Microbiol.* **2003**, *88*, 41–54. [CrossRef]

83. Gardini, F.; Özogul, Y.; Suzzi, G.; Tabanelli, G.; Özogul, F. Technological factors affecting biogenic amine content in foods: A review. *Front. Microbiol.* **2016**, *7*, 1218. [CrossRef]

84. Moreno-Arribas, M.V.; Lonvaud-Funel, A. Tyrosine decarboxylase activity of Lactobacillus brevis IOEB 9809 isolated from wine and L. brevis ATCC 367. *FEMS Microbiol. Lett.* **1999**, *180*, 55–60. [CrossRef]

85. Rossi, F.; Gardini, F.; Rizzotti, L.; La Gioia, F.; Tabanelli, G.; Torriani, S. Quantitative analysis of histidine decarboxylase gene (*hdcA*) transcription and histamine production by *Streptococcus thermophilus* PRI60 under conditions relevant to cheese making. *Appl. Environ. Microbiol.* **2011**, *77*, 2817–2822. [CrossRef]

86. Becker, K.; Southwick, K.; Reardon, J.; Berg, R.; MacCormack, J.N. Histamine poisoning associated with eating tuna burgers. *JAMA* **2001**, *285*, 1327–1330. [CrossRef]

87. Novella-Rodríguez, S.; Veciana-Nogués, M.T.; Izquierdo-Pulido, M.; Vidal-Carou, M.C. Distribution of biogenic amines and polyamines in cheese. *J. Food Sci.* **2003**, *68*, 750–755. [CrossRef]

88. Jairath, G.; Singh, P.K.; Dabur, R.S.; Rani, M.; Chaudhari, M. Biogenic amines in meat and meat products and its public health significance: A review. *J. Food Sci. Technol.* **2015**, *52*, 6835–6846. [CrossRef]

89. Latorre-Moratalla, M.L.; Bover-Cid, S.; Bosch-Fusté, J.; Vidal-Carou, M.C. Influence of technological conditions of sausage fermentation on the aminogenic activity of *L. curvatus* CTC273. *Food Microbiol.* **2012**, *29*, 43–48. [CrossRef]

90. Ruiz-Capillas, C.; Pintado, T.; Jiménez-Colmenero, F. Biogenic amine formation in refrigerated fresh sausage "chorizo" keeps in modified atmosphere. *J. Food Biochem.* **2011**, *36*, 449–457. [CrossRef]

91. Medina-Pradas, E.; Arroyo-López, F.N. Presence of toxic microbial metabolites in table olives. *Front. Microbiol.* **2015**, *6*, 873. [CrossRef]

92. Rabie, M.A.; Siliha, H.; el-Saidy, S.; el-Badawy, A.A.; Malcata, F.X. Reduced biogenic amine contents in sauerkraut via addition of selected lactic acid bacteria. *Food Chem.* **2011**, *129*, 1778–1782. [CrossRef]

93. Prester, L. Biogenic amines in fish, fish products and shellfish: A review. *Food Addit. Contam.* **2011**, *28*, 1547–1560. [CrossRef]

94. Spano, G.; Russo, P.; Lonvaud-Funel, A.; Lucas, P.; Alexandre, H.; Grandvalet, C.; Coton, E.; Coton, M.; Barnavon, L.; Bach, B.; et al. Biogenic amines in fermented foods. *Eur. J. Clin. Nutr.* **2010**, *64*, 64–951. [CrossRef]

95. Gobbetti, M.; De Angelis, M.; Di Cagno, R.; Mancini, L.; Fox, P.F. Pros and cons for using non-starter lactic acid bacteria (NSLAB) as secondary/adjunct starters for cheese ripening. *Trends Food Sci. Technol.* **2015**, *45*, 167–178. [CrossRef]

96. Smid, E.J.; Kleerebezem, M. Production of aroma compounds in lactic fermentations. *Annu. Rev. Food Sci. Technol.* **2014**, *5*, 313–326. [CrossRef]

97. Cocconcelli, P.S.; Fontana, C. Starter cultures for meat fermentation. In *Handbook of Meat Processing*; Toldrà, F., Ed.; Wiley-Blackwell: Ames, IA, USA, 2010; pp. 199–218. ISBN 978-0-81-382089-7.

98. Montanari, C.; Barbieri, F.; Magnani, M.; Grazia, L.; Gardini, F.; Tabanelli, G. Phenotypic diversity of *Lactobacillus sakei* strains. *Front. Microbiol.* **2018**, *9*, 2003. [CrossRef]

99. Sgarbi, E.; Bottari, B.; Gatti, M.; Neviani, E. Investigation of the ability of dairy nonstarter lactic acid bacteria to grow using cell lysates of other lactic acid bacteria as the exclusive source of nutrients. *Int. J. Dairy Technol.* **2014**, *67*, 342–347. [CrossRef]

100. Skeie, S.; Kieronczyka, A.; Næssa, R.M.; Østliea, H. *Lactobacillus* adjuncts in cheese: Their influence on the degradation of citrate and serine during ripening of a washed curd cheese. *Int. Dairy J.* **2008**, *18*, 158–168. [CrossRef]

101. Lucas, P.M.; Blancato, V.S.; Claisse, O.; Magni, C.; Lolkema, J.S.; Lonvaud-Funel, A. Agmatine deiminase pathway genes in *Lactobacillus brevis* are linked to the tyrosine decarboxylation operon in a putative acid resistance locus. *Microbiology* **2007**, *153*, 2221–2230. [CrossRef]

102. Montanari, C.; Kamdem, S.L.S.; Serrazanetti, D.I.; Etoa, F.X.; Guerzoni, M.E. Synthesis of cyclopropane fatty acids in *Lactobacillus helveticus* and *Lactobacillus sanfranciscensis* and their cellular fatty acids changes following short term acid and cold stresses. *Food Microbiol.* **2010**, *27*, 493–502. [CrossRef]

103. Pessione, E. Lactic acid bacteria contribution to gut microbiota complexity: Lights and shadows. *Front. Cell. Infect. Microbiol.* **2012**, *2*, 86. [CrossRef]

104. Connil, N.; Plissoneau, L.; Onno, B.; Pilet, M.F.; Prevost, H.; Dousset, X. Growth of *Carnobacterium divergens* V41 and production of biogenic amines and divercin V41 in sterile cold-smoked salmon extract at varying temperatures, NaCl levels, and glucose concentrations. *J. Food Prot.* **2002**, *65*, 333–338. [CrossRef]

105. Bargossi, E.; Gardini, F.; Gatto, V.; Montanari, C.; Torriani, S.; Tabanelli, G. The capability of tyramine production and correlation between phenotypic and genetic characteristics of *Enterococcus faecium* and *Enterococcus faecalis* strains. *Front. Microbiol.* **2015**, *6*, 1371. [CrossRef]

106. Fernández, M.; Linares, D.M.; Alvarez, M.A. Sequencing of the tyrosine decarboxylase cluster of *Lactococcus lactis* IPLA 655 and the development of a PCR method for detecting tyrosine decarboxylating lactic acid bacteria. *J. Food Prot.* **2004**, *67*, 2521–2529. [CrossRef]

107. Gatto, V.; Tabanelli, G.; Montanari, C.; Prodomi, V.; Bargossi, E.; Torriani, S.; Gardini, F. Tyrosine decarboxylase activity of *Enterococcus mundtii*: New insights into phenotypic and genetic aspects. *Microb. Biotechnol.* **2016**, *9*, 801–813. [CrossRef]

108. Coton, E.; Rollan, G.C.; Lonvaud-Funel, A. Histidine decarboxylase of *Leuconostoc oenos* 9204: Purification, kinetic properties, cloning and nucleotide sequence of the *hdc* gene. *J. Appl. Microbiol.* **1998**, *84*, 143–151. [CrossRef]

109. Coton, E.; Coton, M. Multiplex PCR for colony direct detection of Gram-positive histamine- and tyramine-producing bacteria. *J. Microbiol. Methods* **2005**, *63*, 296–304. [CrossRef]

110. Konagaya, Y.; Kimura, B.; Ishida, M.; Fujii, T. Purification and properties of a histidine decarboxylase from *Tetragenococcus muriaticus*, a halophilic lactic acid bacterium. *J. Appl. Microbiol.* **2002**, *92*, 1136–1142. [CrossRef]

111. Martin, M.C.; Fernández, M.; Linares, D.M.; Alvarez, M.A. Sequencing, characterization and transcriptional analysis of the histidine decarboxylase operon of *Lactobacillus buchneri*. *Microbiology* **2005**, *151*, 1219–1228. [CrossRef]

112. Satomi, M.; Furushita, M.; Oikawa, H.; Yoshikawa-Takahashi, M.; Yano, Y. Analysis of a 30 kbp plasmid encoding histidine decarboxylase gene in *Tetragenococcus halophilus* isolated from fish sauce. *Int. J. Food Microbiol.* **2008**, *126*, 202–209. [CrossRef]

113. Vanderslice, P.; Copeland, W.C.; Robertus, J.D. Cloning and nucleotide sequence of wild type and a mutant histidine decarboxylase from *Lactobacillus* 30a. *J. Biol. Chem.* **1986**, *261*, 15186–15191.

114. Trip, H.; Mulder, N.L.; Rattray, F.P.; Lolkema, J.S. HdcB, a novel enzyme catalysing maturation of pyruvoyl-dependent histidine decarboxylase. *Mol. Microbiol.* **2011**, *79*, 861–871. [CrossRef]

115. Nannelli, F.; Claisse, O.; Gindreau, E.; de Revel, G.; Lonvaud-Funel, A.; Lucas, P.M. Determination of lactic acid bacteria producing biogenic amines in wine by quantitative PCR methods. *Lett. Appl. Microbiol.* **2008**, *47*, 594–599. [CrossRef]

116. Coton, E.; Mulder, N.; Coton, M.; Pochet, S.; Trip, H.; Lolkema, J.S. Origin of the putrescine-producing ability of the coagulase-negative bacterium *Staphylococcus epidermidis* 2015B. *Appl. Environ. Microbiol.* **2010**, *76*, 5570–5576. [CrossRef]

117. Linares, D.M.; Perez, M.; Ladero, V.; del Rio, B.; Redruello, B.; Martin, M.C.; Fernández, M.; Alvarez, M.A. An agmatine-inducible system for the expression of recombinant proteins in *Enterococcus faecalis*. *Microb. Cell Fact.* **2014**, *13*, 169. [CrossRef]

118. Özogul, F.; Hamed, I. The importance of lactic acid bacteria for the prevention of bacterial growth and their biogenic amines formation: A review. *Crit. Rev. Food Sci. Nutr.* **2018**, *58*, 1660–1670. [CrossRef]

119. Coburn, P.S.; Baghdayan, A.S.; Dolan, G.T.; Shankar, N. Horizontal transfer of virulence genes encoded on the *Enterococcus faecalis* pathogenicity island. *Mol. Microbiol.* **2007**, *63*, 530–544. [CrossRef]

120. Davis, I.J.; Roberts, A.P.; Ready, D.; Richards, H.; Wilson, M.; Mullany, P. Linkage of a novel mercury resistance operon with streptomycin resistance on a conjugative plasmid in *Enterococcus faecium*. *Plasmid* **2005**, *54*, 26–38. [CrossRef]

121. Foulquié Moreno, M.; Sarantinopoulos, P.; Tsakalidou, E.; De Vuyst, L. The role and application of enterococci in food and health. *Int. J. Food Microbiol.* **2006**, *106*, 1–24. [CrossRef]

122. Franz, C.M.; Huch, M.; Abriouel, H.; Holzapfel, W.; Gálvez, A. Enterococci as probiotics and their implications in food safety. *Int. J. Food Microbiol.* **2011**, *151*, 125–140. [CrossRef]

123. Giraffa, G. Functionality of enterococci in dairy products. *Int. J. Food Microbiol.* **2003**, *88*, 215–222. [CrossRef]

124. Martino, G.P.; Quintana, I.M.; Espariz, M.; Blancato, V.S.; Gallina Nizo, G.; Esteban, L.; Magni, C. Draft genome sequences of four *Enterococcus faecium* strains isolated from Argentine cheese. *Genome Announc.* **2016**, *4*, e01576-15. [CrossRef]

125. Hanchi, H.; Mottawea, W.; Sebei, K.; Hammami, R. The genus *Enterococcus*: Between probiotic potential and safety concerns-an update. *Front. Microbiol.* **2018**, *9*, 1791. [CrossRef]

126. Bover-Cid, S.; Hugas, M.; Izquierdo-Pulido, M.; Vidal-Carou, M.C. Amino acid decarboxylase activity of bacteria isolated from fermented pork sausages. *Int. J. Food Microbiol.* **2001**, *66*, 185–189. [CrossRef]

127. Capozzi, V.; Ladero, V.; Beneduce, L.; Fernández, M.; Alvarez, M.A.; Benoit, B.; Laurent, B.; Grieco, F.; Spano, G. Isolation and characterization of tyramine-producing *Enterococcus faecium* strains from red wine. *Food Microbiol.* **2011**, *28*, 434–439. [CrossRef]

128. Jiménez, E.; Ladero, V.; Chico, I.; Maldonado-Barragán, A.; López, M.; Martin, V.; Fernández, L.; Fernández, M.; Álvarez, M.A.; Torres, C.; Rodríguez, J.M. Antibiotic resistance, virulence determinants and production of biogenic amines among enterococci from ovine, feline, canine, porcine and human milk. *BMC Microbiol.* **2013**, *13*, 288. [CrossRef]

129. Ladero, V.; Fernández, M.; Alvarez, M.A. Isolation and identification of tyramine-producing enterococci from human fecal samples. *Can. J. Microbiol.* **2009**, *55*, 215–218. [CrossRef]

130. Ladero, V.; Martínez, N.; Cruz Martin, M.; Fernández, M.; Alvarez, M.A. qPCR for quantitative detection of tyramine-producing bacteria in dairy products. *Food Res. Int.* **2010**, *43*, 289–295. [CrossRef]

131. Muñoz-Atienza, E.; Landeta, G.; de Las Rivas, B.; Gómez-Sala, B.; Muñoz, R.; Hernández, P.E.; Cintas, L.M.; Herranz, C. Phenotypic and genetic evaluations of biogenic amine production by lactic acid bacteria isolated from fish and fish products. *Int. J. Food Microbiol.* **2011**, *146*, 212–216. [CrossRef]

132. Sarantinopoulos, P.; Andrighetto, C.; Georgalaki, M.D.; Rea, M.C.; Lombardi, A.; Cogan, T.M.; Kalantzopoulos, G.; Tsakalidou, E. Biochemical properties of enterococci relevant to their technological performance. *Int. Dairy J.* **2001**, *11*, 621–647. [CrossRef]

133. Suzzi, G.; Caruso, M.; Gardini, F.; Lombardi, A.; Vannini, L.; Guerzoni, M.E.; Andrighetto, C.; Lanorte, M.T. A survey of the enterococci isolated from an artisanal Italian goat's cheese (semicotto caprino). *J. Appl. Microbiol.* **2000**, *89*, 267–274. [CrossRef]

134. Bonetta, S.; Bonetta, S.; Carraro, E.; Coïsson, J.D.; Travaglia, F.; Arlorio, M. Detection of biogenic amine producer bacteria in a typical Italian goat cheese. *J. Food Prot.* **2008**, *71*, 205–209. [CrossRef]

135. Fernández, M.; Linares, D.M.; Del Rio, B.; Ladero, V.; Alvarez, M.A. HPLC quantification of biogenic amines in cheeses: Correlation with PCR-detection of tyramine-producing microorganisms. *J. Dairy Res.* **2007**, *74*, 276–282. [CrossRef]

136. Galgano, F.; Suzzi, G.; Favati, F.; Caruso, M.; Martuscelli, M.; Gardini, F.; Salzano, G. Biogenic amines during ripening in 'Semicotto Caprino' cheese: Role of enterococci. *J. Food Sci. Technol.* **2001**, *36*, 153–160. [CrossRef]

137. Joosten, H.M.L.J.; Northolt, M.D. Conditions allowing the formation of biogenic amines in cheese. 2. Decarboxylative properties of some non-starter bacteria. *Neth. Milk Dairy J.* **1987**, *41*, 259–280.

138. Burdychova, R.; Komprda, T. Biogenic amine-forming microbial communities in cheese. *FEMS Microbiol. Lett.* **2007**, *276*, 149–155. [CrossRef]

139. Rea, M.C.; Franz, C.M.A.P.; Holzapfel, W.H.; Cogan, T.M. Development of enterococci and production of tyramine during the manufacture and ripening of cheddar cheese. *Irish J. Agric. Food Res.* **2004**, *43*, 247–258.

140. Kalhotka, L.; Manga, I.; Přichystalová, J.; Hůlová, M.; Vyletělová, M.; Šustová, K. Decarboxylase activity test of the genus *Enterococcus* isolated from goat milk and cheese. *Acta Vet. Brno* **2012**, *81*, 145–151. [CrossRef]

141. Martino, G.P.; Espariz, M.; Gallina Nizo, G.; Esteban, L.; Blancato, V.S.; Magni, C. Safety assessment and functional properties of four enterococci strains isolated from regional Argentinean cheese. *Int. J. Food Microbiol.* **2018**, *277*, 1–9. [CrossRef]

142. Bover-Cid, S.; Holzapfel, W.H. Improved screening procedure for biogenic amine production by lactic acid bacteria. *Int. J. Food Microbiol.* **1999**, *53*, 33–41. [CrossRef]

143. Coloretti, F.; Chiavari, C.; Armaforte, E.; Carri, S.; Castagnetti, G.B. Combined use of starter cultures and preservatives to control production of biogenic amines and improve sensorial profile in low-acid salami. *J. Agric. Food Chem.* **2008**, *56*, 11238–11244. [CrossRef]

144. Hugas, M.; Garriga, M.; Aymerich, M. Functionality of enterococci in meat products. *Int. J. Food Microbiol.* **2003**, *88*, 223–233. [CrossRef]

145. Garriga, M.; Aymerich, T. The microbiology of fermentation and ripening. In *Handbook of Fermented Meat and Poultry*; Toldrá, F., Hui, Y.H., Astiasarán, I., Sebranek, J.G., Talon, R., Eds.; John Wiley & Sons: Chichester, UK, 2014; pp. 107–115. ISBN 978-1-118-52269-1.

146. Latorre-Moratalla, M.L.; Comas-Basté, O.; Bover-Cid, S.; Vidal-Carou, M.C. Tyramine and histamine risk assessment related to consumption of dry fermented sausages by the Spanish population. *Food Chem. Toxicol.* **2017**, *99*, 78–85. [CrossRef]

147. Landeta, G.; Curiel, J.A.; Carrascosa, A.V.; Muñoz, R.; de las Rivas, B. Technological and safety properties of lactic acid bacteria isolated from Spanish dry-cured sausages. *Meat Sci.* **2013**, *95*, 272–280. [CrossRef]

148. Maksimovic, A.Z.; Zunabovic-Pichler, M.; Kos, I.; Mayrhofer, S.; Hulak, N.; Domig, K.J.; Fuka, M.M. Microbiological hazards and potential of spontaneously fermented game meat sausages: A focus on lactic acid bacteria diversity. *LWT-Food Sci. Technol.* **2018**, *89*, 418–426. [CrossRef]

149. Marcobal, A.; de las Rivas, B.; García-Moruno, E.; Muñoz, R. The tyrosine decarboxylation test does not differentiate *Enterococcus faecalis* from *Enterococcus faecium*. *Syst. Appl. Microbiol.* **2004**, *27*, 423–426. [CrossRef]

150. Pérez-Martín, F.; Seseña, S.; Izquierdo-Pulido, M.; Llanos Palop, M. Are Enterococcus populations present during malolactic fermentation of red wine safe? *Food Microbiol.* **2014**, *42*, 95–101. [CrossRef]

151. Jeon, A.R.; Lee, J.H.; Mah, J.H. Biogenic amine formation and bacterial contribution in *Cheonggukjang*, a Korean traditional fermented soybean food. *LWT-Food Sci. Technol.* **2018**, *92*, 282–289. [CrossRef]

152. Takebe, Y.; Takizaki, M.; Tanaka, H.; Ohta, H.; Niidome, T.; Morimura, S. Evaluation of the biogenic amine-production ability of lactic acid bacteria from Tofu-misozuke. *Food Sci. Technol. Res.* **2016**, *22*, 673–678. [CrossRef]

153. Suárez, C.; Esparíz, M.; Blancato, V.S.; Magni, C. Expression of the agmatine deiminase pathway in *Enterococcus faecalis* is activated by the *AguR* regulator and repressed by CcpA and PTS^{Man} systems. *PLoS ONE* **2013**, *8*, e76170. [CrossRef]

154. Perez, M.; Victor Ladero, V.; del Rio, B.; Redruello, B.; de Jong, A.; Kuipers, O.; Kok, J.; Martin, M.C.; Fernández, M.; Alvarez, M.A. The relationship among tyrosine decarboxylase and agmatine deiminase pathways in *Enterococcus faecalis*. *Front. Microbiol.* **2017**, *8*, 2107. [CrossRef] [PubMed]

155. Bargossi, E.; Tabanelli, G.; Montanari, C.; Gatto, V.; Chinnici, F.; Gardini, F.; Torriani, S. Growth, biogenic amine production and *tyrDC* transcription of *Enterococcus faecalis* in synthetic medium containing defined amino acid concentrations. *J. Appl. Microbiol.* **2017**, *122*, 1078–1091. [CrossRef] [PubMed]

156. Gardini, F.; Martuscelli, M.; Caruso, M.C.; Galgano, F.; Crudele, M.A.; Favati, F.; Guerzoni, M.E.; Suzzi, G. Effects of pH, temperature and NaCl concentration on the growth kinetics, proteolytic activity and biogenic amine production of *Enterococcus faecalis*. *Int. J. Food Microbiol.* **2001**, *64*, 105–117. [CrossRef]

157. Torriani, S.; Gatto, V.; Sembeni, S.; Tofalo, R.; Suzzi, G.; Belletti, N.; Gardini, F.; Bover-Cid, S. Rapid detection and quantification of tyrosine decarboxylase gene (*tdc*) and its expression in Gram-positive bacteria associated with fermented foods using PCR-based methods. *J. Food Prot.* **2008**, *71*, 93–101. [CrossRef] [PubMed]

158. Fernández, M.; Linares, D.M.; Rodríguez, A.; Alvarez, M.A. Factors affecting tyramine production in *Enterococcus durans* IPLA 655. *Appl. Microbiol. Biotechnol.* **2007**, *73*, 1400–1406. [CrossRef]

159. Linares, D.M.; Fernández, M.; Martín, M.C.; Alvarez, M.A. Tyramine biosynthesis in *Enterococcus durans* is transcriptionally regulated by the extracellular pH and tyrosine concentration. *Microbiol. Biotechnol.* **2009**, *2*, 625–633. [CrossRef] [PubMed]

160. Solheim, M.; Leanti La Rosa, S.; Mathisen, T.; Snipen, L.G.; Nes, I.F.; Anders Brede, D. Transcriptomic and functional analysis of NaCl-induced stress in *Enterococcus faecalis*. *PLoS ONE* **2014**, *9*, e94571. [CrossRef]

161. Liu, F.; Wang, X.; Du, L.; Wang, D.; Zhu, Y.; Geng, Z.; Xu, X.; Xu, W. Effect of NaCl treatments on tyramine biosynthesis of *Enterococcus faecalis*. *J. Food. Prot.* **2015**, *78*, 940–945. [CrossRef]

162. Hugas, M.; Garriga, M.; Aymerich, T.; Monfort, J.M. Biochemical characterization of lactobacilli from dry fermented sausages. *Int. J. Food Microbiol.* **1993**, *18*, 107–113. [CrossRef]

163. Holck, A.; Axelsson, L.; McLeod, A.; Rode, T.M.; Heir, E. Health and safety considerations of fermented sausages. *J. Food Qual.* **2017**, 9753894. [CrossRef]

164. Pereira, C.I.; Barreto Crespo, M.T.; Romao, M.V.S. Evidence for proteolytic activity and biogenic amines production in *Lactobacillus curvatus* and *L. homohiochii*. *Int. J. Food Microbiol.* **2001**, *68*, 211–216. [CrossRef]

165. Freiding, S.; Gutsche, K.A.; Ehrmann, M.A.; Vogel, R.F. Genetic screening of *Lactobacillus sakei* and *Lactobacillus curvatus* strains for their peptidolytic system and amino acid metabolism, and comparison of their volatilomes in a model system. *Syst. Appl. Microbiol.* **2011**, *34*, 311–320. [CrossRef] [PubMed]

166. Latorre-Moratalla, M.L.; Bover-Cid, S.; Veciana-Nogués, M.T.; Vidal-Carou, M.C. Control of biogenic amines in fermented sausages: Role of starter cultures. *Front. Microbiol.* **2012**, *3*, 169. [CrossRef]

167. Wüthrich, D.; Berthoud, H.; Wechsler, D.; Eugster, E.; Irmler, S.; Bruggmann, R. The histidine decarboxylase gene cluster of *Lactobacillus parabuchneri* was gained by horizontal gene transfer and is mobile within the species. *Front. Microbiol.* **2017**, *8*, 218. [CrossRef] [PubMed]

168. Linares, D.M.; del Rio, B.; Redruello, B.; Fernández, M.; Martin, M.C.; Ladero, V.; Alvarez, M.A. The use of qPCR-based methods to identify and quantify food spoilage microorganisms. In *Novel Food Preservation and Microbial Assessment Techniques*; Boziaris, I.S., Ed.; CRC Press: Boca Raton, FL, USA, 2014; pp. 313–334. ISBN 978-1-46-658075-6.

169. Fröhlich-Wyder, M.T.; Guggisberg, D.; Badertscher, R.; Wechsler, D.; Wittwer, A.; Irmler, S. The effect of *Lactobacillus buchneri* and *Lactobacillus parabuchneri* on the eye formation of semi-hard cheese. *Int. Dairy J.* **2013**, *33*, 120–128. [CrossRef]

170. Bunková, L.; Bunka, F.; Mantlová, G.; Cablová, A.; Sedlácek, I.; Svec, P.; Pachlová, V.; Krácmar, S. The effect of ripening and storage conditions on the distribution of tyramine, putrescine and cadaverine in Edam-cheese. *Food Microbiol.* **2010**, *27*, 880–888. [CrossRef]

171. Komprda, T.; Burdychová, R.; Dohnal, V.; Cwiková, O.; Sládková, P.; Dvorácková, H. Tyramine production in Dutch-type semi-hard cheese from two different producers. *Food Microbiol.* **2008**, *25*, 219–227. [CrossRef] [PubMed]

172. Pachlová, V.; Buňková, L.; Flasarová, R.; Salek, R.N.; Dlabajová, A.; Butor, I.; Buňka, F. Biogenic amine production by nonstarter strains of *Lactobacillus curvatus* and *Lactobacillus paracasei* in the model system of Dutch-type cheese. *LWT-Food Sci. Technol.* **2018**, *97*, 730–735. [CrossRef]

173. Yılmaz, C.; Gökmen, V. Formation of tyramine in yoghurt during fermentation – Interaction between yoghurt starter bacteria and *Lactobacillus plantarum*. *Food Res. Int.* **2017**, *97*, 288–295. [CrossRef] [PubMed]

174. Ladero, V.; Martin, M.C.; Redruello, B.; Mayo, B.; Flórez, A.B.; Fernández, M.; Alvarez, M.A. Genetic and functional analysis of biogenic amine production capacity among starter and non-starter lactic acid bacteria isolated from artisanal cheeses. *Eur. Food Res. Technol.* **2015**, *241*, 377–383. [CrossRef]

175. Linares, D.M.; del Rio, B.; Ladero, V.; Martínez, N.; Fernández, M.; Martin, M.C.; Alvarez, M.A. Factors influencing biogenic amines accumulation in dairy products. *Front. Microbiol.* **2012**, *3*, 180. [CrossRef] [PubMed]

176. Costantini, A.; Cersosimo, M.; Del Prete, V.; Garcia-Moruno, E. Production of biogenic amines by lactic acid bacteria: Screening by PCR, thin-layer chromatography and high-performance liquid chromatography of strains isolated from wine and must. *J. Food Prot.* **2006**, *69*, 391–396. [CrossRef] [PubMed]

177. Gardini, F.; Zaccarelli, A.; Belletti, N.; Faustini, F.; Cavazza, A.; Martuscelli, M.; Mastrocola, D.; Suzzi, G. Factors influencing biogenic amine production by a strain of *Oenococcus oeni* in a model system. *Food Control* **2005**, *16*, 609–616. [CrossRef]

178. Ancín-Azpilicueta, C.; González-Marco, A.; Jiménez-Moreno, N. Current knowledge about the presence of amines in wine. *Crit. Rev. Food Sci. Nutr.* **2008**, *48*, 257–275. [CrossRef] [PubMed]

179. Lerm, E.; Engelbrecht, L.; du Toit, M. Malolactic fermentation: The ABC's of MLF. *S. Afr. J. Enol. Vitic.* **2010**, *31*, 186–212. [CrossRef]

180. Marcobal, A.; Martín-Álvarez, P.J.; Polo, C.; Muñoz, R.; Moreno-Arribas, M.V. Formation of biogenic amines throughout the industrial manufacture of red wine. *J. Food Prot.* **2006**, *69*, 397–404. [CrossRef]

181. Marcobal, A.; Martín-Álvarez, P.J.; Moreno-Arribas, M.V.; Muñoz, R. A multifactorial design for studying factors influencing growth and tyramine production of the lactic acid bacteria *Lactobacillus brevis* CECT 4669 and *Enterococcus faecium* BIFI-58. *Res. Microbiol.* **2006**, *157*, 417–424. [CrossRef]

182. Henríquez-Aedo, K.; Durán, D.; Garcia, A.; Hengst, M.B.; Aranda, M. Identification of biogenic amines-producing lactic acid bacteria isolated from spontaneous malolactic fermentation of chilean red wines. *LWT-Food Sci. Technol.* **2016**, *68*, 183–189. [CrossRef]

183. Moreno-Arribas, M.V.; Polo, M.C.; Jorganes, F.; Muñoz, R. Screening of biogenic amine production by lactic acid bacteria isolated from grape must and wine. *Int. J. Food Microbiol.* **2003**, *84*, 117–123. [CrossRef]

184. Moreno-Arribas, V.; Lonvaud-Funel, A. Purification and characterization of tyrosine decarboxylase of *Lactobacillus brevis* IOEB 9809 isolated from wine. *FEMS Microbiol. Lett.* **2001**, *195*, 103–107. [CrossRef]

185. Arena, M.P.; Romano, A.; Capozzi, V.; Beneduce, L.; Ghariani, M.; Grieco, F.; Lucas, P.; Spano, G. Expression of *Lactobacillus brevis* IOEB 9809 tyrosine decarboxylase and agmatine deiminase genes in wine correlates with substrate availability. *Lett. Appl. Microbiol.* **2001**, *53*, 395–402. [CrossRef] [PubMed]

186. Lucas, P.; Lonvaud-Funel, A. Purification and partial gene sequence of the tyrosine decarboxylase of *Lactobacillus brevis* IOEB 9809. *FEMS Microbiol. Lett.* **2002**, *211*, 85–89. [CrossRef] [PubMed]

187. Geissler, A.J.; Behr, J.; von Kamp, K.; Vogel, R.F. Metabolic strategies of beer spoilage lactic acid bacteria in beer. *Int. J. Food Microbiol.* **2016**, *216*, 60–68. [CrossRef] [PubMed]

188. Ladero, V.; Coton, M.; Fernández, M.; Buron, N.; Martín, M.C.; Guichard, H.; Coton, E.; Alvarez, M.A. Biogenic amines content in Spanish and French natural ciders: Application of qPCR for quantitative detection of biogenic amine-producers. *Food Microbiol.* **2011**, *28*, 554–561. [CrossRef] [PubMed]

189. Corsetti, A.; Settanni, L. Lactobacilli in sourdough fermentation. *Food Res. Int.* **2007**, *40*, 539–558. [CrossRef]

190. Costantini, A.; Vaudano, E.; Del Prete, V.; Danei, M.; Garcia-Moruno, E. Biogenic amine production by contaminating bacteria found in starter preparations used in winemaking. *J. Agric. Food Chem.* **2009**, *57*, 10664–10669. [CrossRef]

191. Rodwell, A.W. The occurrence and distribution of amino-acid decarboxylases within the genus *Lactobacillus*. *J. Gen. Microbiol.* **1953**, *8*, 224–232. [CrossRef]

192. Romano, A.; Trip, H.; Lolkema, J.S.; Lucas, P.M. Three-component lysine/ornithine decarboxylation system in *Lactobacillus saerimneri* 30a. *J. Bacteriol.* **2013**, *195*, 1249–1254. [CrossRef]

193. Delorme, C. Safety assessment of dairy microorganisms: *Streptococcus thermophiles*. *Int. J. Food Microbiol.* **2008**, *126*, 274–277. [CrossRef]

194. Georgalaki, M.D.; Sarantinopoulos, P.; Ferreira, E.S.; De Vuyst, L.; Kalantzopoulos, G.; Tsakalidou, E. Biochemical properties of *Streptococcus macedonicus* strains isolated from Greek Kasseri cheese. *J. Appl. Microbiol.* **2000**, *88*, 817–825. [CrossRef]

195. Griswold, A.R.; Jameson-Lee, M.; Burne, R.A. Regulation and physiologic significance of the agmatine deiminase system of *Streptococcus mutans* UA159. *J. Bacteriol.* **2006**, *188*, 834–841. [CrossRef] [PubMed]

196. Liu, Y.; Zeng, L.; Burne, R.A. *AguR* is required for induction of the *Streptococcus mutans* agmatine deiminase system by low pH and agmatine. *Appl. Environ. Microbiol.* **2009**, *75*, 2629–2637. [CrossRef] [PubMed]

197. Elsanhoty, R.M.; Ramadan, M.F. Genetic screening of biogenic amines production capacity from some lactic acid bacteria strains. *Food Control* **2016**, *68*, 220–228. [CrossRef]

198. Gezginc, Y.; Akyol, I.; Kuley, E.; Özogul, F. Biogenic amines formation in *Streptococcus thermophilus* isolated from home-made natural yogurt. *Food Chem.* **2013**, *138*, 655–662. [CrossRef] [PubMed]

199. Kuley, E.; Balıkcı, E.; Özoğul, I.; Gökdogan, S.; Ozoğul, F. Stimulation of cadaverine production by foodborne pathogens in the presence of *Lactobacillus*, *Lactococcus*, and *Streptococcus* spp. *J. Food Sci.* **2012**, *77*, M650–M658. [CrossRef] [PubMed]

200. de las Rivas, B.; Rodríguez, H.; Carrascosa, A.V.; Muñoz, R. Molecular cloning and functional characterization of a histidine decarboxylase from *Staphylococcus capitis*. *J. Appl. Microbiol.* **2008**, *104*, 194–203. [CrossRef] [PubMed]

201. Tabanelli, G.; Torriani, S.; Rossi, F.; Rizzotti, L.; Gardini, F. Effect of chemico-physical parameters on the histidine decarboxylase (HdcA) enzymatic activity in Streptococcus thermophilus PRI60. *J. Food Sci.* **2012**, *77*, M231–M237. [CrossRef]

202. Fox, P.F.; McSweeney, P.L.H.; Cogan, T.M.; Guinee, T.P. *Cheese: Chemistry, Physics and Microbiology*, 3rd ed.; Volume 1 General aspects; Elsevier Academic Press: London, UK, 2004; ISBN 978-0-12-263652-3.

203. Linares, D.M.; del Rio, B.; Redruello, B.; Ladero, V.; Martin, M.C.; de Jong, A.; Kuipers, O.P.; Fernández, M.; Alvarez, M.A. *AguR*, a transmembrane transcription activator of the putrescine biosynthesis operon in *Lactococcus lactis*, acts in response to the agmatine concentration. *Appl. Environ. Microbiol.* **2015**, *81*, 6145–6157. [CrossRef]

204. del Rio, B.; Redruello, B.; Ladero, V.; Fernández, M.; Martin, M.C.; Alvarez, M.A. Putrescine production by *Lactococcus lactis* subsp. *cremoris* CECT 8666 is reduced by NaCl via a decrease in bacterial growth and the repression of the genes involved in putrescine production. *Int. J. Food Microbiol.* **2016**, *232*, 1–6. [CrossRef]

205. del Rio, B.; Linares, D.; Ladero, V.; Redruello, B.; Fernández, M.; Martin, M.C.; Alvarez, M.A. Putrescine biosynthesis in *Lactococcus lactis* is transcriptionally activated at acidic pH and counteracts acidification of the cytosol. *J. Food Microbiol.* **2016**, *236*, 83–89. [CrossRef]

206. Flasarová, R.; Pachlová, V.; Buňková, L.; Menšíková, A.; Georgová, N.; Dráb, V.; Buňka, F. Biogenic amine production by *Lactococcus lactis* subsp. *cremoris* strains in the model system of Dutch-type cheese. *Food Chem.* **2016**, *194*, 68–75. [CrossRef] [PubMed]

207. Santos, W.C.; Souza, M.R.; Cerqueira, M.M.O.P.; Gloria, M.B.A. Bioactive amines formation in milk by *Lactococcus* in the presence or not of rennet and NaCl at 20 and 32 °C. *Food Chem.* **2003**, *81*, 595–606. [CrossRef]

208. Holzapfel, W.H.; Franz, C.M.A.P.; Ludwig, W.; Back, W.; Dicks, L.M.T. The genera *Pediococcus* and *Tetragenococcus*. In *The Prokaryotes*, 3rd ed.; Dworkin, M., Falkow, S., Rosenberg, E., Schleifer, K.H., Stackebrandt, E., Eds.; Springer: New York, NY, USA, 2006; Volume 4, pp. 229–266. ISBN 978-0-387-30744-2.

209. Hugas, M.; Monfort, J.M. Bacterial starter cultures for meat fermentation. *Food Chem.* **1997**, *59*, 547–554. [CrossRef]

210. Walling, E.; Gindreau, E.; Lonvaud-Funel, A. A putative glucan synthase gene *dps* detected in exopolysaccharide-producing *Pediococcus damnosus* and *Oenococcus oeni* strains isolated from wine and cider. *Int. J. Food Microbiol.* **2005**, *98*, 53–62. [CrossRef] [PubMed]

211. Guerrini, S.; Mangani, S.; Granchi, L.; Vincenzini, M. Biogenic amine production by *Oenococcus oeni*. *Curr. Microbiol.* **2002**, *44*, 374–378. [CrossRef] [PubMed]

212. Garcia-Moruno, E.; Muñoz, R. Does *Oenococcus oeni* produce histamine? *Int. J. Food Microbiol.* **2012**, *157*, 121–129. [CrossRef] [PubMed]

213. Berbegal, C.; Benavent-Gil, Y.; Navascués, E.; Calvo, A.; Albors, C.; Pardo, I.; Ferrer, S. Lowering histamine formation in a red Ribera del Duero wine (Spain) by using an indigenous *O. oeni* strain as a malolactic starter. *Int. J. Food Microbiol.* **2017**, *244*, 11–18. [CrossRef] [PubMed]

214. Landete, J.M.; Pardo, I.; Ferrer, S. Regulation of *hdc* expression and HDC activity by enological factors in lactic acid bacteria. *J. Appl. Microbiol.* **2008**, *105*, 1544–1551. [CrossRef]

215. Garai, G.; Dueñas, M.T.; Irastorza, A.; Moreno-Arribas, M.V. Biogenic amine production by lactic acid bacteria isolated from cider. *Lett. Appl. Microbiol.* **2007**, *45*, 473–478. [CrossRef]

216. Bonnin-Jusserand, M.; Grandvalet, C.; David, V.; Alexandre, H. Molecular cloning, heterologous expression, and characterization of Ornithine decarboxylase from *Oenococcus oeni*. *J. Food Prot.* **2011**, *74*, 1309–1314. [CrossRef]

217. Izquierdo-Pulido, M.; Mariné-Font, A.; Vidal-Carou, M.C. Effect of tyrosine on tyramine formation during beer fermentation. *Food Chem.* **2000**, *70*, 329–332. [CrossRef]

218. Huys, G.; Leisner, J.; Björkroth, J. The lesser LAB gods: *Pediococcus, Leuconostoc, Weissella, Carnobacterium*, and affiliated genera. In *Lactic Acid Bacteria: Microbiological and Functional Aspects*, 4th ed.; Lahtinen, S., Ouwehand, A.C., Salminen, S., von Wright, A., Eds.; CRC Press: Boca Raton, FL, USA, 2011; pp. 93–121. ISBN 978-1-43-983677-4.

219. Liu, S.; Pritchard, G.G.; Hardman, M.J.; Pilone, G.J. Occurrence of arginine deiminase pathway enzymes in arginine catabolism by wine lactic acid bacteria. *Appl. Environ. Microbiol.* **1995**, *61*, 310–316. [PubMed]

220. Coton, M.; Lebreton, M.; Marcia Leyva Salas, M.L.; Garnier, L.; Navarri, M.; Pawtowski, A.; Le Bla, G.; Valence, F.; Coton, E.; Mounier, J. Biogenic amine and antibiotic resistance profiles determined for lactic acid bacteria and a propionibacterium prior to use as antifungal bioprotective cultures. *Int. Dairy J.* **2018**, *85*, 21–26. [CrossRef]

221. González del Llano, D.; Cuesta, P.; Rodríguez, A. Biogenic amine production by wild lactococal and leuconostoc strains. *Lett. Appl. Microbiol.* **1998**, *26*, 270–274. [CrossRef]

222. Fusco, V.; Quero, G.M.; Cho, G.S.; Kabisch, J.; Meske, D.; Neve, H.; Bockelmann, W.; Franz, C.M.A.P. The genus *Weissella*: Taxonomy, ecology and biotechnological potential. *Front. Microbiol.* **2015**, *6*, 155. [CrossRef] [PubMed]

223. Leisner, J.J.; Laursen, B.G.; Prévost, H.; Drider, D.; Dalgaard, P. *Carnobacterium*: Positive and negative effects in the environment and in foods. *FEMS Microbiol. Rev.* **2007**, *31*, 592–613. [CrossRef] [PubMed]

224. Curiel, J.A.; Ruiz-Capillas, C.; de las Rivas, B.; Carrascosa, A.V.; Jiménez-Colmenero, F.; Muñoz, R. Production of biogenic amines by lactic acid bacteria and enterobacteria isolated from fresh pork sausages packaged in different atmospheres and kept under refrigeration. *Meat Sci.* **2011**, *88*, 368–373. [CrossRef]

225. Justé, A.; Lievens, B.; Frans, I.; Marsh, T.L.; Klingeberg, M.; Michiels, C.W.; Willems, K.A. Genetic and physiological diversity of *Tetragenococcus halophilus* strains isolated from sugar- and salt-rich environments. *Microbiology* **2008**, *154*, 2600–2610. [CrossRef]

226. Torriani, S.; Felis, G.E.; Fracchetti, F. Selection criteria and tools for malolactic starters development: An update. *Ann. Microbiol.* **2001**, *61*, 33–39. [CrossRef]

9

Identification of a Lactic Acid Bacteria to Degrade Biogenic Amines in Chinese Rice Wine and Its Enzymatic Mechanism

Tianjiao Niu [1,2], **Xing Li** [1], **Yongjie Guo** [2] and **Ying Ma** [1,*]

[1]   School of Chemistry and Chemical Engineering, Harbin Institute of Technology, Harbin 150090, China
[2]   Mengniu Hi-tech Dairy (Beijing) Co., Ltd., Beijing 101107, China
*   Correspondence: maying@hit.edu.cn

Abstract: A *L. plantarum*, CAU 3823, which can degrade 40% of biogenic amines (BAs) content in Chinese rice wine (CRW) at the end of post-fermentation, was selected and characterized in this work. It would be an optimal choice to add 10^6 cfu/mL of selected strain into the fermentation broth to decrease the BAs while keeping the character and quality of CRW. Nine amine oxidases were identified from the strain and separated using Sephadex column followed by LC-MS/MS analysis. The purified amine oxidase mixture showed a high monoamine oxidase activity of 19.8 U/mg, and more than 40% of BAs could be degraded. The biochemical characters of the amine oxidases were also studied. This work seeks to provide a better solution to degrade BAs in CRW prior to keeping the character and quality of CRW and a better understanding of the degradability of the strain to the BAs.

Keywords: biogenic amines; *L. plantarum*; amines oxidase; Chinese rice wine; industrial fermentation

1. Introduction

Biogenic amines (BAs) are low molecular weight organic compounds that have been identified as toxicological agents in various foods, such as fishery products, dairy, meat, wine, and so on [1,2]. The ingestion of foods containing relatively high concentrations of BAs could lead to several health hazards, such as headaches, hypotension, respiratory distress, heart palpitations and digestive problems, particularly when alcohol is present [3,4]. Histamine, which is well-known because of its implication in many food poisoning cases, has a potent vasodilatory action that could cause important drops in blood pressure [5]. Tyramine, as one of the vasoconstrictor amines, can provoke a release of noradrenaline resulting in an increase of arterial pressure [5]. Even though there are no accurate regulations for BAs, several countries including France, Germany and Australia have set regulations and limits for histamine and many wine importers in the EU require a BA analysis [4,6]. The presence of BAs is considered a marker of poor wine quality and bad winemaking practices [4,7].

BAs are synthesized in fermented food by decarboxylation of corresponding amino acids by microorganisms [1]. According to the previous studies, BAs could be formed by lactic acid bacteria in wine [8,9], Chinese rice wine [10] and Korean rice wine [1]. As a traditional alcoholic beverage, Chinese rice wine (CRW), which has been popular in China for thousands of years [11], has high nutritional values, and thus, it has been used as an ingredient in traditional Chinese medicine [12]. Since the brewing process of CRW is the typical open semisolid-state fermentation, lots of microorganisms (molds, yeast, bacteria) are brought in the glutinous rice with the addition of Chinese koji [3,13], and the system is favorable to BAs generation combining with the high amount of free amino acids [2]. The abundant bacteria in CRW, mainly originating from Chinese koji, the surroundings and the surfaces of the equipment, could be one of the main reasons for the formation of BAs [10].

Histamine, tyramine, putrescine, cadaverine and phenylethylamine are the most representative BAs detected in the wine [6]. Histamine and tyramine have been considered as the most toxic products in wine, and putrescine and cadaverine could potentiate these effects [4]. The formation of BAs was traditionally controlled by avoiding the growth of spoilage bacteria, decreasing the amino acid precursors and inoculating starter cultures with negative decarboxylase activity [6,7]. Driven by greater awareness of the importance of food quality and safety by consumers, the methods for degradation of BAs in fermented foods have been explored. Biological enzymatic degradation of BAs would be a safe and economic way while avoiding the production difficulties. Two *Lactobacillus plantarum* strains (named NDT 09 and NDT 16) isolated from red wine were able to degrade 22% of tyramine and 31% of putrescine, respectively [14]. Three different strains of *Brevibacterium linens* were utilized to eliminate tyramine and histamine in cheese [6], and the strain *K. varians* LTH 1540, it was also found, could degrade tyramine during sausage ripening [15]. Two lactic acid bacteria were used to degrade 50%–54% of histamine in fish silage [16]. However, the relationship between BAs degradation and microbiological enzymes of the strains has not been explored yet.

In this work, a *Lactobacillus plantarum* was obtained from CRW which could degrade BAs. The optimal industrial conditions of the selected strain were analyzed, and the microbiological amine oxidase enzymes were identified and biochemically characterized. Our results could receive considerable interest by providing a green industrial strategy to control the BAs contents in the rice wine and improve the safety consumption of the fermented foodstuffs.

2. Materials and Methods

2.1. Materials

Man Rogosa Sharpe agar (MRS) medium was obtained from Oxoid. Ltd. (Basingstoke, Hants, UK). The BA standards were purchased from Sigma-Aldrich (St. Louis, MO, USA). Bacterial genomic DNA extraction kit was obtained from Tiangen (Beijing, China). Ultra-pure water was obtained from a Millipore purification system (>18.3 MΩ·cm). Formic acid, methanol and acetonitrile used in the preparation of the mobile phase were of LC-MS grade. All other chemicals used were of analytical grade.

2.2. Strains Screening and Identification

Fermentation broths were collected at the later stage from a typical rice wine production process in Shaoxing (Zhejiang, China). The suspension was filtered through four layers of sterile gauze to remove the unliquefied rice and sealed in a sterile plastic bottle. One gram of fermentation broths was diluted 10-fold by a 0.85% NaCl solution and routinely subcultured 5 to 10 times on MRS medium to obtain purified clones. The screening medium designed was based on the method of Landete [17] to obtain the bacteria that could decrease biogenic amine content. These strains isolated were kept frozen at −20 °C in a sterilized mixture of culture medium and glycerol (50:50, *v/v*) according to the methods described by García-Ruiz [18], and further identified by 16S rRNA gene sequencing.

2.3. HPLC Determination of Biogenic Amines

Eight biogenic amines of Histamine (HIS), tyramine (TYR), putrescine (PUT), cadaverine (CAD), phenylethylamine (PHE), tryptamine (TRY), spermine (SPM) and spermidine (SPD) were analyzed according to the method of Callejon, Sendra [13] with slight modifications. The individual strains were cultured on MRS, and 10^7 cfu/mL were inoculated with the MRS liquid medium contaminated with 50 mg/L of each amine at pH 5.5. After 48 h incubation at 30 °C, the reaction was stopped by adding HCl. Samples were centrifuged at 8000 rpm for 15 min and the supernatant was pipetted into a screw-capped vial. The pre-column derivatization procedure using dansyl chloride as derivatization reagent was performed according to the report of Yongmei, Xin [12]. The samples were filtered through 0.22 μm millipore syringe filters and analyzed by RP-HPLC using on LC-20A HPLC system (Shimadzu, Kyoto, Japan) with an Agilent C18 column (250 mm × 4.6 mm, 300 A pores, 5 μm particles, Agilent Technologies, Inc., Santa Clara, CA, USA). The column temperature was kept at 30 °C and the detection wavelength was 254 nm with a flow rate of 1.0 mL/min by using water (A) and methanol (B) as eluents. The gradient elution program consisting of a linear gradient from 65% to 70% B in 7 min followed by from 70% to 80% B in 13 min and 3 min isocratic elution.

The percentage of BAs degradation was calculated based on the HPLC data as following,

$$\text{BAs degradation (\%)} = (C_{control} - C_{strain})/C_{control}$$

where $C_{control}$ was the concentration of the BAs in the control medium and C_{strain} was the concentration of the BAs in the medium incubated with the strain.

2.4. Bacterial Growth Analysis

The bacterial growth was measured according to the methods described by Cui [19]. Briefly, the isolated lactic acid bacteria (LAB) strains were diluted to 10^5 cfu/mL in MRS liquid medium, and the pH and optical density ($OD_{600\,nm}$) of medium was checked at 28 °C, 33 °C and 37 °C for 36 h, respectively.

2.5. The Bacterial Starter Application in Pilot Scale Fermentation

A pilot fermentation was performed according to the methods described by Zhang, Xue [10] with modifications (Figure 1). Glutinous rice (12 kg) was soaked at 18 °C for 20 h and steamed for 30 min. After naturally cooling to room temperature (about 25 °C), the steamed rice was transferred into a 33 L wide-mouth bottle to which 14.5 kg water, 1.5 kg Chinese koji (unique saccharifying agent including molds, yeasts and bacteria, obtained from COFCO Shaoxin wine Co., Ltd., Shaoxin, China) were added. The main fermentation was carried out at 33 °C for 4 days with intermittent oxygen filling, and post-fermentation was then carried out at 28 °C for 20 days. The isolated strain with 10^5 (low level), 10^6 (middle level) and 10^7 (high level) cfu/mL was added into the CRW at the main fermentation and post-fermentation stage, respectively. After filter pressing, clarification, wine frying and sterilization (90 °C for 3 min), finished Chinese rice wines were obtained. Ten milliliters of fermentation broths were taken from different fermentation stages, including addition of starter (AS); main fermentation (MF); post-fermentation 5d (PF5d); post-fermentation 10d (PF10d); and post-fermentation 20d (PF20d)), to analysis the changes in the BAs contents by using the HPLC method. According to the previous studies [20,21], pH, alcohol content, total sugar, total acid, non-sugar solid and amino acid nitrogen of CRW were analyzed by using official methods (Chinese National Standard GB/T 13662-2008). Sensory evaluation of CRW was conducted by 30 panelists (15 males and 15 females) who have professional training certificates.

The procedure was conducted in a sensory laboratory following GB/T 13662-2008 and ISO 4121. A total of 11 sensory attributes of appearance (color and turbidity), aroma (alcohol, fruit and cereal), taste (sweet, sour and bitter), mouthfeel (astringency, continuation and full body) and harmony were chosen to characterize the sensory properties using quantitative descriptive analysis involving a 0–9 ten-point linear scale (0: none; 1–2: very weak; 3–4: ordinary; 5–6: moderate; 7–8: strong; 9: very strong).

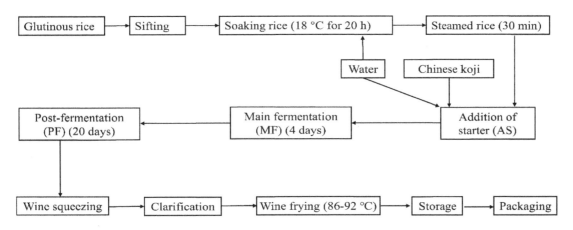

Figure 1. Diagram of the Chinese rice wine production process.

2.6. Separation of the Amine Oxidases

Cell-free extracts were obtained by using the method of Callejon [22]. The bacterial cells from a 1 L culture were collected by centrifugation at 10,000 rpm, 20 min at 4 °C and washed twice with 50 mM sodium phosphate buffer (PBS), pH 7.4. The samples were resuspended in PBS buffer containing 1 mM of phenyl methylsulfonyl fluoride (PMSF) as protease inhibitor. Cell-free extracts were obtained by disrupting the bacterial cells with 1 g of 106 μM diameter glass beads in a Mikro-dismenbrator® Sartorius: 10 cycles of 40 s, alternating 5 cycles of disruption with a cooling step of 5 min in ice. The samples were centrifuged at 13,000 rpm for 15 min (PrismR, Labnet, USA), and supernatants were saved at −20 °C until use. The protein content was determined by using the bicinchoninic acid assay kit (BCA, Solarbio, Beijing, China). Monoamine oxidase (MAO) assay kit and diamine oxidase (DAO) assay kit (Jiancheng Institute, Nanjing, China) were both used to determine the amine oxidase activity. The MAO assay kit was based on the ability of MAO to form H_2O_2 substrate, which could be determined by a fluorimetric method. The DAO assay kit was based on the oxidation of PUT to pyrroline plus NH_3 and H_2O_2, which can be determined by the fluorimetric method.

The cell-free extracts were further ultracentrifuged at 47,000 rpm for 1 h, and the supernatant was precipitated by 75% saturation of ammonium sulfate precipitation [22]. The protein was redissolved with 50 mM PBS and were loaded onto a Sephadex G-100 column (1.6 cm × 70 cm) followed by a linear gradient elution with a flow rate of 1 mL/min. The protein fraction was collected and measured at 280 nm by using a HD-93-1 spectrophotometer (Purkinje General Instrument Co. Ltd., Beijing, China). There fractions were collected (P1, P2 and P3, Supplement Figure S1), and were then concentrated and freeze-dried. The degradation ability of the fractions was further evaluated by incubation with 50 mg/L eight biogenic amines at pH 4.0, 33 °C for 2 h.

2.7. Identification of the Amine Oxidases

The fractions separated from the cell-free extracts were digested with trypsin (Promega, Madison, WI, USA) overnight at 37 °C and were identified by LC-MS/MS using the Easy nLC-1000 nano ultra-high-pressure system (Thermo Fisher Scientific, San Jose, CA, USA) coupling with a Q Exactive mass spectrometer (Thermo Fisher Scientific, San Jose, CA, USA). The peptide mixture was loaded onto a Zorbax 300SB-C18 peptide traps (Agilent Technologies, Wilmington, DE, USA) in buffer A (0.1% Formic acid) and separated with a linear gradient of 4%–50% buffer B (80% acetonitrile and 0.1% formic acid) for 50 min, 50%–100% B for 4 min, and held at 100% B for 6 min at a flow rate of 250 nL/min. The mass spectrometer was operated in positive ion mode. MS data was acquired using a data-dependent top10 method dynamically choosing the most abundant precursor ions from the survey scan for high-energy collisional dissociation (HCD) fragmentation and was searched by using MASCOT engine and Proteome Discoverer 1.3 against the local uniport_lactobocilluspiantarum database.

2.8. Enzymatic Properties of the Amine Oxidases

Effects of temperatures (15, 20, 25, 28, 30, 35, 40, 80 °C at pH 4.0 for 2 h), pH (3.0–5.0) at 30 °C for 2 h, and metal ions (0.2 mol/L, copper ion, ferrous ion, zinc ion, calcium ion and magnesium ion) at 30 °C for 2 h (pH 4.0) on the amine oxidase degradation activity were further investigated.

2.9. Statistical Analysis

All samples were prepared in three independent and each was analyzed in triplicate by the analysis of variance (ANOVA). The results were considered significant at $p \leq 0.05$ by the Duncan test.

3. Result

3.1. Strains Screening and Identification

A total of 61 strains were isolated from the five major stages (soaking rice, steamed rice, addition of starter, main fermentation and post-fermentation, Figure 1) of CRW fermentation. After screening their potentials to degrade/eliminate the contents of BAs, about 30% of strains were able to degrade BAs even though most of them degraded BAs to less than 10% extents (results not known). Only one strain drew attentions for more than 40% degradation efficiency of the BAs (Table 1). 16S rDNA sequencing identified that the strain had 100% similarity in 16S rDNA sequences to *Lactobacillus plantarum* CAU 3823 (GenBank accession no. MF424991.1). In the details, *Lactobacillus plantarum* CAU 3823 was a *L. plantarum* that exhibited the greatest potential for BAs degradation, as 56% degradation, for TRY, 41% for PHE, 42% for PUT, 43% for CAD, 40% for TYR, 45% for HIS, 44% for SPD and 43% for SPM, which should be considered in the further analysis.

Table 1. Percentage (%) of degradation of the biogenic amines by *Lactobacillus plantarum* CAU 3823 from Chinese rice wine [a].

Strains	Tryptamine	Phenylethylamine	Putrescine	Cadaverine	Tyramine	HISTAMINE	Spermidine	Spermine
Lactobacillus plantarum CAU 3823	55.95 ± 6.59	40.85 ± 9.87	41.82 ± 7.97	42.79 ± 7.76	40.12 ± 8.09	44.72 ± 7.56	43.51 ± 8.39	42.56 ± 8.41

[a] 10^7 cfu/mL of *Lactobacillus plantarum* CAU 3823 was incubated in the Man Rogosa Sharpe agar (MRS) liquid medium contaminated with 50 mg/L of each amine at pH 5.5 for 48 h.

3.2. The Bacterial Growth Ability

The growth ability of *L. plantarum* CAU 3823 at different temperatures (28 °C, 33 °C and 37 °C) was shown in Figure 2. *L. plantarum* CAU 3823 was able to grow at different temperatures, showing $OD_{600} > 1$ at main fermentation temperature (33 °C) for 9–25 h and post-fermentation temperature (28 °C) for 12~25 h. The maximum OD_{600} value of 1.4 was found at different temperatures at 25 h of growth, suggesting the good growth trends indicated that *L. plantarum* CAU 3823 could be used in industry producing CRW fermentation.

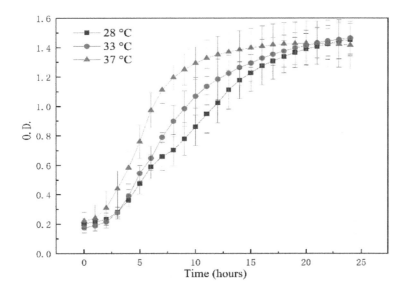

Figure 2. The growth ability of *L. plantarum* CAU 3823 at different temperatures (28 °C, 33 °C and 37 °C). Growth curves are representative of all determinations.

3.3. Changes in the BAs Induced by L. plantarum in Pilot Scale Fermentation

To investigate the capability to degrade BAs of *L. plantarum* CAU 3823 to the BAs in pilot scale fermentation of CRW, RP-HPLC was applied to quantify the contents of BAs in CRW incubation with various levels (10^5, 10^6 and 10^7 cfu/mL) of *L. plantarum* CAU 3823 as extra starter during fermentation, and the results are shown in Figure 3. Compared to control group, the total contents of BAs in CRW with *L. plantarum* CAU 3823 were significantly lower ($p < 0.05$) during the entire fermentation period (Figure 3A). The degradation percentages of BAs were 32%, 54% and 58%, respectively, at low, middle and high level of *L. plantarum* CAU 3823 at the main fermentation stage, suggesting the dose dependent manner. Total content of BAs was significantly reduced to 34%, 60% and 61% at low, middle and high levels of *L. plantarum* CAU 3823, respectively, at 5th day of post-fermentation, and similar degradation efficiency was obtained in the 10th day of post-fermentation and 20th day of post-fermentation, respectively.

The degrading abilities of *L. plantarum* CAU 3823 to TRY, PUT, HIS, CAD, PHE, SPD and SPM were also studied in Figure 3B–H, respectively. A marked decrease in the contents of BAs was observed during fermentation with the increasing of the strain content. As the most content of BAs detected in Chinese rice wine, TRY was degraded by *L. plantarum* CAU 3823 with the degradation rate of 39% at low level, 56% at middle level and 58% at high level strain at main fermentation; 41% at low level, 60% at middle level and 62% at high level strain at post-fermentation 5d; 51% at low level, 63% at middle level and 66% at high level strain at post-fermentation 10d; and 49% at low level, 57% at middle level and 61% at high level strain at post-fermentation 20d (Figure 3B). Similar degradation efficiency to PUT, HIS, CAD, PHE, SPD and SPM was also found as follows: PUT with 13% reduction at low level, 39% at middle level and 43% at high level strain; HIS with 4% at low level, 42% at middle level and 55% at high level; PHE with 45% at low level, 74% at middle level and 82% at high level; CAD with 38% at low level, 55% at middle level and 55% at high level; SPD with 23% at low level, 46% at middle level and 89% at high level; SPM with 25% at low level, 50% at middle level and 75%, respectively, at high level at the end of post-fermentation. Overall, more than 40% contents of BAs could be degraded incubation with *L. plantarum* CAU 3823 at the middle and high levels than the one at low level during fermentation.

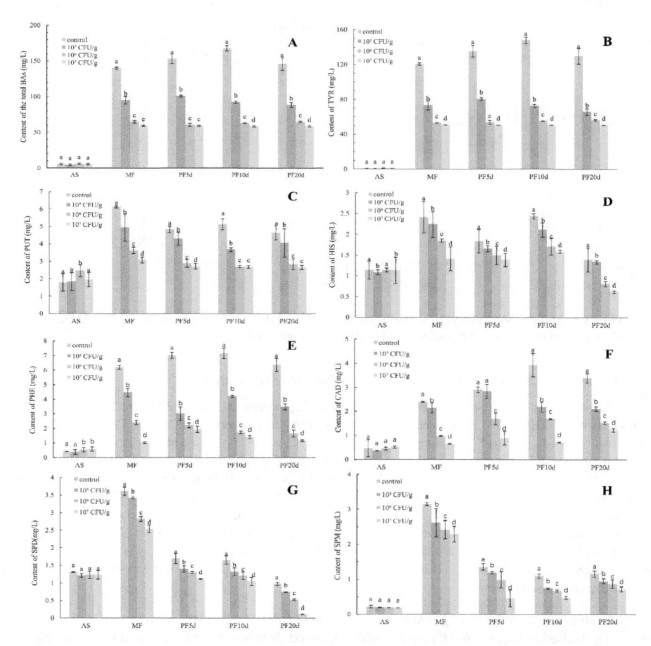

Figure 3. The contents of the total biogenic amines (BAs, **A**), tyramine (TYR, **B**), putrescine (PUT, **C**), Histamine (HIS, **D**) phenylethylamine (PHE, **E**), cadaverine (CAD, **F**), spermidine (SPD, **G**) and spermine (SPM, **H**) in Chinese rice wine adding different level of *L. plantarum* CAU 3823 at the post-fermentation and main fermentation stage during different fermentation stages (addition of starter (AS); main fermentation (MF); post-fermentation 5d (PF5d); post-fermentation 10d (PF10d); post-fermentation 20d (PF20d)).

3.4. Total Acid and pH in Pilot Scale Fermentation

As shown in Table 2, the changes in total acid and pH value of Chinese rice wine when different levels of *L. plantarum* CAU 3823 were added during fermentation were investigated, to evaluate the effect of this strain on the quality of CRW. At the initial stage of starter addition, there was no difference ($P > 0.05$) in lactic acid content and pH value among the four CRW samples. The total acid content of the CRW showed a slightly increase from 6.53 at low level to 6.86 g/L at middle level strain incubated with *L. plantarum* CAU 3823 at the end of post-fermentation, compared to the control group of 5.94 g/L. However, the total acid of CRW of 9.14 g/L incubated with high level of *L. plantarum* CAU 3823 indicated the over-acidification.

Table 2. Changes in the total acid and pH in the Chinese rice wine adding different level of *L. plantarum* CAU 3823 at the post-fermentation and main fermentation stage during different fermentation stages (addition of starter; main fermentation; post-fermentation 5d; post-fermentation 10d; post-fermentation 20d).

	Addition the Selected Strain (cfu/mL)	The Addition of Starter	Main Fermentation	Post-Fermentation 5d	Post-Fermentation 10d	Post-Fermentation 20d
Total acid (g/L)	Control	6.03 ± 0.22 [a]	3.81 ± 0.15 [a]	4.91 ± 0.07 [a]	5.55 ± 0.33 [a]	5.94 ± 0.20 [a]
	10^5 (low level)	6.17 ± 0.13 [a]	4.93 ± 0.13 [b]	5.41 ± 0.19 [b]	5.92 ± 0.19 [a]	6.53 ± 0.13 [b]
	10^6 (middle level)	5.92 ± 0.19 [a]	6.01 ± 0.14 [c]	6.51 ± 0.14 [c]	6.74 ± 0.44 [b]	6.86 ± 0.13 [d]
	10^7 (high level)	6.03 ± 0.15 [a]	6.04 ± 0.10 [c]	7.06 ± 0.09 [c]	8.01 ± 0.23 [c]	9.14 ± 0.45 [c]
pH	Control	6.33 ± 0.19 [a]	4.04 ± 0.12 [a]	4.19 ± 0.05 [a]	4.21 ± 0.03 [b]	4.14 ± 0.12 [a]
	10^5 (low level)	6.37 ± 0.28 [a]	4.00 ± 0.14 [a]	4.36 ± 0.12 [a]	4.12 ± 0.07 [a]	3.99 ± 0.16 [a]
	10^6 (middle level)	6.45 ± 0.22 [a]	3.84 ± 0.16 [a]	4.34 ± 0.08 [a]	4.45 ± 0.13 [b]	3.87 ± 0.12 [a]
	10^7 (high level)	6.43 ± 0.23 [a]	3.71 ± 0.04 [b]	4.24 ± 0.12 [a]	4.32 ± 0.12 [b]	3.63 ± 0.03 [b]

Presented data (mean ± standard deviation) are the mean values of three independent samples and each analyzed in triplicate. Values in a column with different superscripts differ significantly ($p < 0.05$).

3.5. Alcohol Content, Total Sugar, Non-Sugar Solid and Amino Acid Nitrogen in Pilot Scale Fermentation

The effects of *L. plantarum* CAU 3823 on the alcohol content, total sugar, non-sugar solid and amino acid nitrogen in the Chinese rice wine were analyzed after production process. As presented in Table 3, there was no notable change in alcohol, amino acid nitrogen and total sugar contents among the CRWs incubated with low and middle level of *L. plantarum* CAU 3823. The non-sugar solid was markedly higher ($p < 0.05$) when CRW was fermented involving with the selected strain.

Table 3. The alcohol content, amino acid nitrogen, total sugar and non-sugar solid in the Chinese rice wine after production process.

Addition the Selected Strain (cfu/mL)	Alcohol Content (% vol)	Amino Acid Nitrogen (g/L)	Total Sugar (g/L)	Non-Sugar Solid (g/L)
Control	11.52 ± 0.23 [a]	1.44 ± 0.11 [a]	31.98 ± 1.37 [a]	39.81 ± 0.33 [a]
10^5 (low level)	11.49 ± 0.35 [a]	1.28 ± 0.35 [a]	15.35 ± 2.34 [b]	62.34 ± 0.32 [c]
10^6 (middle level)	10.33 ± 0.41 [b]	0.82 ± 0.13 [b]	11.98 ± 3.25 [b]	71.52 ± 0.18 [d]
10^7 (high level)	9.29 ± 0.25 [c]	0.59 ± 0.02 [c]	10.97 ± 2.23 [b]	51.16 ± 0.25 [b]

Presented data (mean ± standard deviation) are the mean values of three independent samples and each analyzed in triplicate. Values in a column with different superscripts differ significantly ($p < 0.05$).

3.6. Sensory Evaluation

The sensory characteristics of CRW adding with different levels of the isolated strain were described by the 30 sensory panelists. As presented in Figure 4, CRW with high level of strain exhibited the lowest score (appearance 6, aroma 7, taste 6, mouthfeel 6 and harmony 6.2) among the four CRW samples. No significant difference was observed between the CRW incubated with middle level and low level strain compared to the control CRW ($p > 0.05$), indicating *L. plantarum* CAU 3823 with low and middle level would not have an influence on the sensory behaviors of the Chinese rice wine.

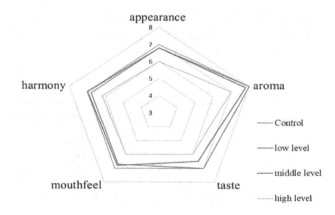

Figure 4. Average radar map of the Chinese rice wine including different level of biogenic amine-reduced *Lactobacillus plantarum* based on sensory scores.

3.7. Purification and Identification of the Amine Oxidases

To gain a deeper insight into the amine-degrading activity exhibited by *L. plantarum* CAU 3823, LC-MS/MS experiments were designed to show whether the amine oxidases existed in the strain. Cell-free extracts were obtained at a protein concentration of 5.5 mg/mL (Table 4). The MAO activity was 36.9 U/mg and the DAO activity was 128 U/L at 37 °C, pH = 7 in the cell-free extracts (Table 4). Three fractions were collected from a Sephadex G-100 column (Supplement Figure S1), and little DAO activity was detected in all three fractions, but only fraction 1 showed a good MAO activity of 19.8 U/mg compared to fraction 2 of 2.4 U/mg, and no amine oxidase activity was determined in fraction 3, which might be due to the low protein concentration.

To further investigate the amine degradation ability, the BA degradation rate (%) was calculated by incubating the three fractions with the eight BAs at pH 4.0, 33 °C for 2 h (Table 5). The BAs contents in fraction 1 significantly declined with the degradation rate of 41.9% for TYR, 41.1% for HIS, 40.3% for PUT, 44.3% for PHE, 41.1% for CAD, 41% for SPD, 43.5% for SPM and 47.9% for TRY. However, there were slight or little changes observed in the BA contents in the Fractions 2 and 3.

The fraction 1 was further identified by using LC-MSMS. Ten proteins including 9 amine oxidase proteins were identified in fraction 1, and hereinto, 8 amine oxidase proteins were monoamine oxidases, including 4 amine oxidase [flavin-containing] A (accession: P58027, P21396, Q5NU32 and A0A011QTL0), 2 amine oxidase [flavin-containing] B (accession: Q5RE98 and A0QU10), 1 monoamine oxidase [flavin-containing] (accession: A0A375EQX7) and 1 monoamine oxidase (accession: U2EF11) (Supplement Table S1). The MWs of the amine oxidases were closer and range from 46 to 60 kDa.

Table 4. The protein concentration, monoamine oxidase activity and diamine oxidase activity of the cell-free extracts (37 °C, pH = 7).

	Protein Concentration (mg/mL)	Monoamine Oxidase Activity (U/mg)	Diamine Oxidase Activity (×10^{-4} U/mg)
Cell-free extracts	5.5	36.9 ± 5.2	1.3 ± 0.1
Fraction 1	3.1	19.8 ± 2.6	ND
Fraction 2	1.6	2.4 ± 1.2	ND
Fraction 3	0.5	ND	ND

ND = Not determined.

Table 5. Degradation percentages (%) of the eight biogenic amines in the three fractions by Sephadex separation incubation with 50 mg/L of the eight biogenic amines at pH 4.0, 33 °C for 2 h.

	Tryptamine	Phenylethylamine	Putrescine	Cadaverine	Histamine	Tyramine	Spermidine	Spermine
Fraction 1	47.9	44.3	40.3	41.1	41.1	41.9	41	43.5
Fraction 2	ND	0.3	0.7	ND	1.2	3.8	ND	ND
Fraction 3	ND	ND	ND	ND	ND	ND	ND	ND

ND = Not determined.

3.8. Amine Oxidases Assays

As shown in Figure 5, the purified amine oxidases mixture (fraction 1) retained its activity in a wide temperature range from 15 to 80 °C and was shown to maintain the 50% MAO activity after on heat treatment at 80 °C for 2 h. The optimal temperature for the amine oxidase activity was 28 °C and the MAO activity was 36.9 U/mg (Figure 5A). The MAO activities increased from 22.3 U/mg to 35.9 U/mg accompanied by the pH value from 3.0 to 5.0 while the amine oxidases were incubated at 30 °C for 2 h (Figure 5B). All the ions could inhibit the MAO activity, as 73%, 31%, 58%, 64% and 79% activity retained when adding 0.2 mol/L Zn^{2+}, Cu^{2+}, Fe^{2+}, Ca^{2+} and Mg^{2+}, respectively (Figure 5C).

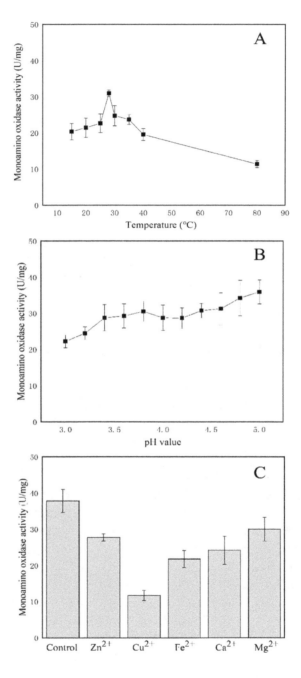

Figure 5. The monoamine oxidase (MAO) activity in the amine oxidase mixture at different temperatures (**A**) at pH 4.0 for 2 h; different pH (**B**) at 30 °C for 2 h and different metal ions (**C**) at 30 °C for 2 h (pH 4.0).

4. Discussion

Biogenic amines are considered as potential health risks since high amounts of them can lead to a series of health problems. The intake of foods with high level of BAs could induce the release of adrenaline and noradrenaline, provoking gastric acid secretion, increased cardiac output, migraine, tachycardia, increased blood sugar levels, and higher blood pressure [23]. Several researches supported the view that the BAs were formed in winemaking mainly by lactic acid bacteria carrying specific metabolic pathways that convert precursor amino acids into BAs [24]. In contrast, there is a lack of studies concerning BAs degradation by food sourced micro-organisms in wine, especially in Chinese rice wine.

In this paper, a *L. plantarum* CAU 3823, isolated from Chinese rice wine, can degrade more than 40% of the BAs, especially the five major BAs of TYR, PUT, HIS, PHE and CAD in Chinese rice wine. A similar research in grade wine showed that only one strain, *L. casei* IFI-CA 52, showed a strong ability to degrade the BAs (54% HIS, 55% TRY and 65% PUT) isolated from wine/ grape cell cultures of 85 strains [18]. However, the histamine-degrading ability of *L. casei* IFI-CA 52 was only 17% when addition of 12% ethanol, suggesting that the ability of *L. casei* IFI-CA 52 to reduce amine concentrations in wines would be rare. Regrettably, the ability of this strain to degrade other BAs was not analyzed. Moreover, a pilot scale fermentation, rather than addition of ethanol, would be a better choice to simulate accurately the complicated wine matrix.

In our experiment, pilot scale fermentation tests had proved that *L. plantarum* CAU 3823 was competent to be used as an extra starter in CRW industrial producing. Chinese koji was added at the beginning of brewing, which could bring in lots of bacteria, thus the BAs accumulated significantly at the beginning [13]. The BAs concentration showed a notably increase in the common CRW (the control group) from the starter addition stage to 10-days post-fermentation, indicating the proliferation of bacteria [13]. The concentration of BAs decreased at the end of post-fermentation, which might be due to the bacteria growth inhibition as the total acid increased during fermentation. According to our results, *L. plantarum* CAU 3823 could degrade the BAs in the CRW brewing process, and the formation of biogenic amines was further degraded by increasing the dose of strain. In this study, HIS, TYR, PUT and CAD were degraded significantly during the pilot scale fermentation, especially TYR, which indicated *L. plantarum* CAU 3823 could provide a more safety traditional fermented beverage for consumers.

Identification of functional microorganisms in CRW to reduce the formation of BAs has received more interest. Liu, Yu [13] utilized an in vivo screening process based on the next-generation sequencing technology to find BA-decreasing microorganism in CRW, and three *Lactobacillus* strains were detected that would not form biogenic amines, but only *L. plantarum* JN01 could grow under 15% ethanol, and the wine could form an unpleasant rancidity taste and more than 8 g/L total acid when the *L. plantarum* JN01 was more than 0.01 gDCW/t. Indeed, high level of functional bacteria could bring about unsatisfactory changes in CRW. A similar trend found in the current study showed that the total acid increased, and alcohol content decreased when 10^7 cfu/mL (high level) of *L. plantarum* CAU 3823 was added into the fermentation mash. Although the sensory scores were also decreased, the whole CRW was within the acceptable range for consumers at high level of the strain. Therefore, *L. plantarum* CAU 3823 could be the best choice to date to decrease BAs in CRW.

As a traditional alcoholic beverage, total sugar, alcoholic degree, pH value, total acid, amino acid nitrogen and non-sugar solid would play important roles in the flavor, taste and nutrition of Chinese rice wine [22]. Although high level (10^7 cfu/mL) of *L. plantarum* CAU 3823 could degrade the BAs maximally, undesirable influence on the acceptability was also noteworthy. Low level (10^5 cfu/mL) and middle level (10^6 cfu/mL) of *L. plantarum* CAU 3823 could eliminate the negative effect on the qualities of the wine, and what's more important, similar sensory characteristics were obtained in CRW. Thus, to degrade the content of BAs in CRW to the highest extent, middle level (10^6 cfu/mL) of the *L. plantarum* could be chosen in the CRW fermentation process.

Non-sugar solids, a major nutrition indicator to evaluate the quality grade of CRW, are mainly composed of dextrin, glycerin, non-volatile acid, protein and hydrolysates [25]. Interestingly, the content of non-sugar solids was increased remarkably when *L. plantarum* CAU 3823 was used, especially at middle level (10^6 cfu/mL), which provided a novel insight that the *L. plantarum* CAU 3823 could produce more non-sugar solids in CRW and thus have potential nutritional values.

BA can be converted into products via oxidation by microorganisms which can be used as a carbon and/or energy source or as a nitrogen source [26]. Limited studies attributed these transformations to amine oxidase activity derived from microorganisms. Yagodina [27] reported that flavoprotein oxidases existing in some microorganisms could catalyze the oxidation of BAs. Sekiguchi [28] found a histamine oxidase in the actinobacteria *Arthrobacter crystallopoietes* KAIT-B-007 isolated from soil. In this study, the amine oxidases from *L. plantarum* CAU 3823 were purified and characterized. Nine amine oxidase proteins, a mixture from *L. plantarum* CAU 3823, contributed the most of amine-degrading ability of *L. plantarum* CAU 3823. Eight MAOs were identified and thus confirmed a good monoamine oxidase activity shown in fraction 1. Amine oxidases can be divided into two subfamilies based on the cofactor they contain. MAO (EC 1.4.3.4) are a family of enzymes containing flavin that catalyze the oxidation of monoamines, employing oxygen to clip off their amine group [29]. The amine oxidases containing copper as cofactor (EC 1.4.3.6) are homodimers, which contain three subclass, namely, diamine oxidase, primary-amine oxidase and diamine oxidase [30]. Amine oxidase [flavin-containing] A and B can catalyze the oxidative deamination of biogenic amines [31]. Amine oxidase [flavin-containing] B that in humans was encoded by the MAOB gene could preferentially degrade PHE [32], which confirmed 44.3% PHE degradation in fraction 1. An "aromatic cage" has been found to play a steric role in substrate binding and in flavin accessibility and helps to increase the substrate amine nucleophilicity [33], which might enhance BA degradation. It is noted that no diamine oxidase was identified although cell-free extracts showed diamine oxidase activity.

To provide a seemingly feasible solution to degrade the BAs in foodstuffs, the biochemical character assays of the amine oxidases mixture from *L. plantarum* CAU 3823 were designed. The enzymes were very thermostable, as the activity remained stable at 80 °C, and were fully stable over the pH range of 3–5. Similar results were reported that a putrescine oxidase from *Rhodococcus erythropolis* NCIMB 11540 could be stable at 50 °C for 2 h [34] and a thermostable histamine oxidase was found in *Arthrobacter crystallopoietes* KAIT-B-007 [29]. These results indicated that the amine oxidases could be stable to use in fermented food processing.

5. Conclusions

In this paper, *Lactobacillus plantarum* CAU 3823 was a *L. plantarum* originating from Chinese rice wine which could effectively degrade the BAs. Middle level (10^6 cfu/mL) of *L. plantarum* could be an optimal choice to decrease the BAs maximally while keeping the CRW character and quality in the pilot scale fermentation. Nine amine oxidase proteins were identified from *L. plantarum* using Sephadex separation followed by LC-MS/MS analysis. The enzymes were very thermostable and fully stable at pH 3–5. All the ions can inhibit the amine oxidase to an extent. *L. plantarum* seemed to be an interesting species displaying BAS degradation, both in culture media conditions and in CRW fermentation, suggesting its suitability as a commercial malolactic starter. This paper provided an efficient method to decrease the biogenic amine contents in the traditional fermented food made by multiple microbes like wine, rice wine, sausages, vinegar, cheese, kimchi and so on.

Author Contributions: T.N. and Y.M. conceived and designed the experiments; T.N. and Y.G. performed the experiments; T.N. and Y.G. analyzed the data; T.N. and X.L. wrote the paper.

References

1. Kim, J.Y.; Kim, D.; Park, P.; Kang, H.-I.; Ryu, E.K.; Kim, S.M. Effects of storage temperature and time on the biogenic amine content and microflora in Korean turbid rice wine, Makgeolli. *Food Chem.* **2011**, *128*, 87–92. [CrossRef] [PubMed]
2. Guo, X.; Guan, X.; Wang, Y.; Li, L.; Wu, D.; Chen, Y.; Pei, H.; Xiao, D. Reduction of biogenic amines production by eliminating the PEP4 gene in Saccharomyces cerevisiae during fermentation of Chinese rice wine. *Food Chem.* **2015**, *178*, 208–211. [CrossRef] [PubMed]
3. Alvarez, M.A.; Moreno-Arribas, M.V. The problem of biogenic amines in fermented foods and the use of potential biogenic amine-degrading microorganisms as a solution. *Trends Food Sci. Technol.* **2014**, *39*, 146–155. [CrossRef]
4. Xia, X.; Zhang, Q.; Zhang, B.; Zhang, W.; Wang, W. Insights into the Biogenic Amine Metabolic Landscape during Industrial Semidry Chinese Rice Wine Fermentation. *J. Agric. Food Chem.* **2016**, *64*, 7385–7393. [CrossRef] [PubMed]
5. Ancín-Azpilicueta, C.; González-Marco, A.; Jiménez-Moreno, N. Current Knowledge about the Presence of Amines in Wine. *Crit. Rev. Food Sci. Nutr.* **2008**, *48*, 257–275. [CrossRef] [PubMed]
6. Guo, Y.Y.; Yang, Y.P.; Peng, Q.; Han, Y. Biogenic amines in wine: A review. *Int. J. Food Sci. Technol.* **2015**, *50*, 1523–1532. [CrossRef]
7. Liu, S.P.; Yu, J.X.; Wei, X.L.; Ji, Z.W.; Zhou, Z.L.; Meng, X.Y.; Mao, J. Sequencing-based screening of functional microorganism to decrease the formation of biogenic amines in Chinese rice wine. *Food Control* **2016**, *64*, 98–104. [CrossRef]
8. Torlois, S.; Joyeux, A.; Moreno-Arribas, V.; Lonvaud-Funel, A.; Bertrand, A. Isolation, properties and behaviour of tyramine-producing lactic acid bacteria from wine. *J. Appl. Microbiol.* **2000**, *88*, 584–593.
9. Moreno-Arribas, M.; Polo, M.; Jorganes, F.; Muñoz, R. Screening of biogenic amine production by lactic acid bacteria isolated from grape must and wine. *Int. J. Food Microbiol.* **2003**, *84*, 117–123. [CrossRef]
10. Zhang, F.; Xue, J.; Wang, D.; Wang, Y.; Zou, H.; Zhu, B. Dynamic changes of the content of biogenic amines in Chinese rice wine during the brewing process. *J. Inst. Brew.* **2013**, *119*, 294–302. [CrossRef]
11. Chen, S.; Xu, Y. The Influence of Yeast Strains on the Volatile Flavour Compounds of Chinese Rice Wine. *J. Inst. Brew.* **2010**, *116*, 190–196. [CrossRef]
12. Yongmei, L.; Xin, L.; Xiaohong, C.; Mei, J.; Chao, L.; Mingsheng, D. A survey of biogenic amines in Chinese rice wines. *Food Chem.* **2007**, *100*, 1424–1428. [CrossRef]
13. Callejon, S.; Sendra, R.; Ferrer, S.; Pardo, I. Identification of a novel enzymatic activity from lactic acid bacteria able to degrade biogenic amines in wine. *Appl. Microbiol. Biotechnol.* **2014**, *98*, 185–198. [CrossRef] [PubMed]
14. Capozzi, V.; Russo, P.; Ladero, V.; Fernandez, M.; Fiocco, D.; Alvarez, M.A.; Grieco, F.; Spano, G. Biogenic Amines Degradation by Lactobacillus plantarum: Toward a Potential Application in Wine. *Front. Microbiol.* **2012**, *3*, 122. [CrossRef] [PubMed]
15. Leuschner, R.; Hammes, W. Tyramine degradation by micrococci during ripening of fermented sausage. *Meat Sci.* **1998**, *49*, 289–296. [CrossRef]
16. Dapkevicius, M.L.; Nout, M.; Rombouts, F.M.; Houben, J.H.; Wymenga, W. Biogenic amine formation and degradation by potential fish silage starter microorganisms. *Int. J. Food Microbiol.* **2000**, *57*, 107–114. [CrossRef]
17. Landete, J.; Ferrer, S.; Pardo, I. Biogenic amine production by lactic acid bacteria, acetic bacteria and yeast isolated from wine. *Food Control* **2007**, *18*, 1569–1574. [CrossRef]
18. García-Ruiz, A.; González-Rompinelli, E.M.; Bartolomé, B.; Moreno-Arribas, M.V. Potential of wine-associated lactic acid bacteria to degrade biogenic amines. *Int. J. Food Microbiol.* **2011**, *148*, 115–120. [CrossRef]
19. Cui, Y.; Qu, X.; Li, H.; He, S.; Liang, H.; Zhang, H.; Ma, Y. Isolation of halophilic lactic acid bacteria from traditional Chinese fermented soybean paste and assessment of the isolates for industrial potential. *Eur. Food Res. Technol.* **2012**, *234*, 797–806. [CrossRef]
20. Yu, H.; Ying, Y.; Fu, X.; Lu, H. Quality Determination of Chinese Rice Wine Based on Fourier Transform near Infrared Spectroscopy. *J. Near Infrared Spectrosc.* **2006**, *14*, 37–44. [CrossRef]
21. Shen, F.; Ying, Y.; Li, B.; Zheng, Y.; Hu, J. Prediction of sugars and acids in Chinese rice wine by mid-infrared spectroscopy. *Food Res. Int.* **2011**, *44*, 1521–1527. [CrossRef]

22. Callejon, S.; Sendra, R.; Ferrer, S.; Pardo, I. Ability of Kocuria varians LTH 1540 To Degrade Putrescine: Identification and Characterization of a Novel Amine Oxidase. *J. Agric. Food Chem.* **2015**, *63*, 4170–4178. [CrossRef] [PubMed]

23. Caston, J.; Eaton, C.; Gheorghiu, B.; Ware, L. Tyramine induced hypertensive episodes and panic attacks in hereditary deficient monoamine oxidase patients. *J. S. C. Med. Assoc. 1975* **2002**, *98*, 187.

24. Beneduce, L.; Romano, A.; Capozzi, V.; Lucas, P.; Barnavon, L.; Bach, B.; Vuchot, P.; Grieco, F.; Spano, G. Biogenic amine in wines. *Ann. Microbiol.* **2010**, *60*, 573–578. [CrossRef]

25. Ouyang, Q.; Zhao, J.; Chen, Q. Measurement of non-sugar solids content in Chinese rice wine using near infrared spectroscopy combined with an efficient characteristic variables selection algorithm. *Spectrochim. Acta Part A Mol. Biomol. Spectrosc.* **2015**, *151*, 280–285. [CrossRef] [PubMed]

26. Levering, P.R.; Van Dijken, J.P.; Veenhuis, M.; Harder, W.; Dijken, J.P. Arthrobacter P1, a fast growing versatile methylotroph with amine oxidase as a key enzyme in the metabolism of methylated amines. *Arch. Microbiol.* **1981**, *129*, 72–80. [CrossRef] [PubMed]

27. Yagodina, O.V.; Nikol'Skaya, E.B.; Khovanskikh, A.E.; Kormilitsyn, B.N. Amine Oxidases of Microorganisms. *Zhurnal Evoliutsionnoǐ Biokhimii I Fiziol.* **2002**, *29*, 864–869.

28. Sekiguchi, Y.; Makita, H.; Yamamura, A.; Matsumoto, K. A thermostable histamine oxidase from Arthrobacter crystallopoietes KAIT-B-007. *J. Biosci. Bioeng.* **2004**, *97*, 104–110. [CrossRef]

29. Tipton, K.F.; Boyce, S.; O'Sullivan, J.; Davey, G.P.; Healy, J. Monoamine Oxidases: Certainties and Uncertainties. *Curr. Med. Chem.* **2004**, *11*, 1965–1982. [CrossRef]

30. Cona, A.; Rea, G.; Angelini, R.; Federico, R.; Tavladoraki, P. Functions of amine oxidases in plant development and defence. *Trends Plant Sci.* **2006**, *11*, 80–88. [CrossRef]

31. Grimsby, J.; Chen, K.; Wang, L.J.; Lan, N.C.; Shih, J.C. Human monoamine oxidase A and B genes exhibit identical exon-intron organization. *Proc. Natl. Acad. Sci. USA* **1991**, *88*, 3637–3641. [CrossRef] [PubMed]

32. Yang, H.Y.; Neff, N.H. Beta-phenylethylamine: A specific substrate for type B monoamine oxidase of brain. *J. Pharmacol. Exp. Ther.* **1973**, *187*, 365–371.

33. Li, M.; Binda, C.; Mattevi, A.; Edmondson, D.E. Functional Role of the "Aromatic Cage" in Human Monoamine Oxidase B: Structures and Catalytic Properties of Tyr435 Mutant Proteins. *Biochemistry* **2006**, *45*, 4775–4784. [CrossRef] [PubMed]

34. Van Hellemond, E.W.; Van Dijk, M.; Heuts, D.P.H.M.; Janssen, D.B.; Fraaije, M.W. Discovery and characterization of a putrescine oxidase from Rhodococcus erythropolis NCIMB 11540. *Appl. Microbiol. Biotechnol.* **2008**, *78*, 455–463. [CrossRef] [PubMed]

10

Extensive Chemometric Investigations of Distinctive Patterns and Levels of Biogenic Amines in Fermented Foods: Human Health Implications

Martin Grootveld [1,*], Benita C. Percival [1] and Jie Zhang [2]

[1] Leicester School of Pharmacy, De Montfort University, The Gateway, Leicester LE1 9BH, UK; p11279990@alumni365.dmu.ac.uk

[2] Green Pasture Products, 416 E. Fremont Street, O'Neill, NE 68763, USA; gpplab@greenpasture.org

* Correspondence: mgrootveld@dmu.ac.uk

Abstract: Although biogenic amines (BAs) present in fermented foods exert important health-promoting and physiological function support roles, their excessive ingestion can give rise to deleterious toxicological effects. Therefore, here we have screened the BA contents and supporting food quality indices of a series of fermented food products using a multianalyte-chemometrics strategy. A liquid chromatographic triple quadrupole mass spectrometric (LC-MS/MS) technique was utilized for the simultaneous multicomponent analysis of 8 different BAs, and titratable acidity, pH, total lipid content, and thiobarbituric acid-reactive substances (TBARS) values were also determined. Rigorous univariate and multivariate (MV) chemometric data analysis strategies were employed to evaluate results acquired. Almost all foods analyzed had individual and total BA contents that were within recommended limits. The chemometrics methods applied were useful for recognizing characteristic patterns of BA analytes and food quality measures between some fermented food classes, and for assessing their inter-relationships and potential metabolic sources. MV analysis of constant sum-normalized BA profile data demonstrated characteristic signatures for cheese (cadaverine only), fermented cod liver oil (2-phenylethylamine, tyramine, and tryptamine), and wine/vinegar products (putrescine, spermidine, and spermine). In conclusion, this LC-MS/MS-linked chemometrics approach was valuable for (1) contrasting and distinguishing BA catabolite signatures between differing fermented foods, and (2) exploring and evaluating the health benefits and/or possible adverse public health risks of such products.

Keywords: biogenic amines (BAs); fermented foods; chemometrics; multivariate (MV) statistical analysis; liquid chromatographic triple quadrupole mass spectrometric (LC-MS/MS) analysis; public health; lipid peroxidation; antioxidants

1. Introduction

Biogenic amines (BAs) may be biosynthesized and degraded via normal metabolic activities in animals, plants, and micro-organisms. As such, these amines occur in a wide variety of foods, such as fish, meat, and cheese products, and especially in fermented foods such as wines, and yoghurts, etc. [1–3]. BA formation in foods usually occurs via the decarboxylation of amino acids [3], of which there are rich sources in these matrices; for example, amino acids are present at very high levels in grapes, and comprise ca. 30–40% of the total nitrogen content of wines [1–3].

Metabolic pathways available in lactic acid bacteria, which have the ability to grow and thrive in foods and beverages, generate significant levels of BAs. Routes available for this are the enzymatic production of putrescine from ornithine (catalyzed by ornithine decarboxylase) and/or from arginine via agmatine, a scheme involving prior conversion of the amino acid substrate to agmatine with arginine decarboxylase, followed by transformation of agmatine to N-carbamoylputrescine via the action of agmatine imino-hydroxylase, and then on to putrescine (a second route for its generation involves the conversion of arginine to ornithine and then to this product via the above ornithine decarboxylase-catalyzed route); putrescine to spermine, a process involving the enzyme spermine synthase, and then spermine to spermidine via the actions of spermidine synthase; cadaverine from lysine with lysine carboxylase and a pyridoxal phosphate co-factor; 2-phenylethylamine from phenylalanine catalyzed by aromatic amino acid carboxylases, including tyrosine decarboxylase; tyramine from tyrosine via tyrosine decarboxylase action; histamine from histidine with histidine decarboxylase; tryptamine from tryptophan with trypotophan decarboxylase, another pyridoxal phosphate-dependent enzyme; and trimethylamine from trimethylamine-N-oxide with a trimethylamine-N-oxide reductase (enzymes involved in the conversion of amino acids to BAs are classified as decarboxylase deaminases) [4,5]. BAs may also be biosynthesized from the amination and transamination of aldehydes and ketones [5], and this may be of some relevance to their detection in marine oil products which have been allowed to autoxidize. Indeed, a range of aldehyde species arise from the fragmentation of conjugated hydroperoxydienes, which are lipid oxidation products resulting from the peroxidation of polyunsaturated fatty acids (PUFAs) [6].

Overall, microbial sources of BAs include yeasts, as well as gram-positive and -negative bacteria [7]. The physiological activity of BA synthesis in prokaryotic cells predominantly appears to be associated with bacterial defense mechanisms employed to combat environmental acidity [8–10]. Hence, amino acid decarboxylation in this manner enhances survival under harsh acidic stress states [9] via proton consumption, and amine and CO_2 excretion required to facilitate restorations of internal pH values [11].

As with their biosynthesis, the catabolism of BAs is extensively outlined and reviewed in [5]. In view of their potentially toxic nature, fortunately humans have detoxification enzyme systems which catabolically oxidize BAs in vivo. These enzymes principally comprise monoamine and diamine oxidases (MAOs and DAOs respectively). MAOs are flavoproteins acting by the oxidative deamination of BAs to their corresponding aldehydes, along with hydrogen peroxide (H_2O_2) and ammonia. Two different forms of MAO have been identified in humans [5]. DAOs are responsible for histamine catabolism, as is histamine-N-methyltransferase, the latter catalyzing a ring methylation process [5].

Evidence available indicates that BAs may confer a series of human health benefits, which involve their interactions with a wide variety of intracellular macromolecules such as proteins, DNA, and RNA. Indeed, monoamines are typically precursors of neuromodulators and neurotransmitters [12]. Moreover, evidence is accumulating that the polyamines spermine and spermidine are important for sexual function and fertility [13], and polyamines in general are associated with cell growth and differentiation, including protein biosynthesis [14]. Indeed, the generation of BAs in eukaryotic cells is essential, since they are required for the critical biosynthesis of hormones, alkaloids, proteins, and nucleic acids [15]. One further plausible health benefit offered by both monoamine and polyamine forms of BAs is their antioxidant potential [16], and recent studies have shown that they function efficiently in this context, and protect against adverse unsaturated fatty acid peroxidation reactions when present in or supplemented to culinary oils, and other foods rich in PUFAs [6] (details regarding the nature and mechanisms of these antioxidant actions are provided in Section S1 of the Supplementary Materials).

Notwithstanding, the availability of these amines in the diet has not been without its problems. Indeed, adverse toxicological events may be stimulated by the ingestion of foods which are known to provide high concentrations of these agents, and one notable example is the provocation of deleterious hypertensive events in patients receiving therapies with monoamine oxidase inhibitor (MAOI) drug treatments [17]. A further problem is the depression of histamine oxidation, a process which arises from the ingestion of putrescine and agmatine, which serve as potentiators of this process; this promotes histamine toxicity episodes in humans [18]. Moreover, it has been reported that BAs such as putrescine and agmatine give rise to their corresponding carcinogenic nitrosoamines from reactions with nitrite anion, dietary or in vivo [19].

Human sensitivity to BAs is contingent on the availability and activities of detoxifying enzymes featured in BA metabolism, i.e., specific ones such as histamine methyltransferase, and those less specific such as mono- and diamine oxidases. However, since these enzymes are inhibited by different classes of drugs, including neuromuscular blocking agents such as alcuronium, antidepressants [20], and ethanol [21], the accumulation of BAs by the consumption of selected foods and beverages can, at least in principal, give rise to clinical disorders, including the extremely hazardous serotonin syndrome [22]. Further details regarding the adverse health effects associated with the excessive intake of BAs are delineated in Section S2 of the Supplementary Materials.

Current consumer demands for safer and healthier foods has prompted a high level of research investigations focused on BAs, although it should be noted that further studies are required to expand this area. High levels of BAs can build up in fermented foods, including fish, fish sauce, and cheese products. Their biosynthesis and accumulation therein are critically dependent on the availability of bacteria with decarboxylase-deaminase enzyme activities, environmental conditions that are unrestrictive towards their growth and propagation, and the efficient functioning of BA-generating enzymes, together with the presence of sufficient amounts of the relevant amino acid substrates required.

Hence, supporting analytical methodologies for the identification and measurement of BAs are of much importance to the food industry, and also from a public health perspective. Such methods should ideally offer high levels of reliability in order to monitor the potential health benefits offered by fermented food products, and also to circumvent any toxicological risks to consumers arising from their excessive production therein; realistic estimates of their human consumption are also major factors for consideration. To date, BA determinations in foods have represented a major challenge for analytical chemists in view of their non-chromophoric nature, their natural occurrence in complex multicomponent food and biological matrices, and high polarities, factors which are further complicated by a requirement for high analytical sensitivity, potential interferences, and, where relevant, chromatographic separation/resolution issues arising from the presence of many structurally-related agents in samples requiring such analysis [23]. Methods previously available for this purpose, and those for the screening of BA-producing bacteria, are outlined in Section S3 of the Supplementary Materials.

Notwithstanding, in principle, the simultaneous and direct multicomponent determination of BAs by the LC-MS/MS method described here, or a newly-developed strategy focused on largely non-invasive high-resolution proton (^{1}H) nuclear magnetic resonance (NMR) analysis [6], serve as valuable assets which, in combination with MV chemometrics strategies, may be employed for the recognition of patterns of these bacterial catabolites which are characteristic of differential bacterial sources of these agents.

Multivariate (MV) data analysis of multicomponent analytical datasets serves as an extremely powerful means of probing and tracking metabolic signatures that are characteristic of differential groups or classifications of samples, and when applied to explore the biochemical basis of human disease etiology, this technique is commonly known as metabolomics [24]. Indeed, to date this combination of multianalyte-MV analysis has been copiously utilized in many biomedical and clinical investigations, mainly for the identification of diagnostic or prognostic monitoring biomarkers for human diseases. However, when applied in a non-biomedical context, the technique can best be described as chemometrics, a technology which also commonly employs many of the MV data analysis strategies used in metabolomics experiments.

In view of the rich sources of BAs in fermented food products, in this study we determined the contents of a total of 8 different BAs in a series of commercially-available fermented fish, fish sauce/paste, vegetable sauce, cheese, wine/vinegar, and cod liver oil (FCLO) products. For this purpose, we employed both univariate and MV chemometrics analysis techniques in order to recognize differential patterns of these catabolites, which may be representative or characteristic of their food, bacterial, metabolic pathway, and/or food processing technology sources. Such analytical information also serves to furnish us with valuable information regarding the provision of these important nutrients in the human diet, and to evaluate the toxicological/adverse health risks presented by the ingestion of fermented foods containing portentously excessive levels of these agents. Currently, a total BA content of *ca.* 1000 ppm is linked to toxicity, and in recommended manufacturing practices, 100 ppm histamine, or a total BA content of 200 ppm, are considered acceptable levels which do not give rise to any associated adverse health effects [25].

These studies were supported by the consideration of further food quality determinations on these fermented food products, which consisted of pH values, titratable acidities (TAs), and total lipid contents, along with an adapted method for determining lipid peroxidation status (thiobarbituric acid-reactive substances (TBARS)).

With the exception of a small number of studies focused on BAs detectable in selected wine products, e.g., [26], to the best of our knowledge this is the first time that MV chemometrics techniques have been applied to explore potentially valuable "between-food classification: distinctions between the concentrations and patterns of BAs in a series of different food products, albeit fermented ones. Therefore, the aims of this investigation are to explore the abilities and reliabilities of LC-MS/MS-based chemometrics analysis techniques to: (1) evaluate the possible public health benefits and/or risks of BAs arising from the human consumption of fermented foods; and (2) effectively compare and distinguish between differing patterns of BA molecules in different classes of fermented food products.

2. Materials and Methods

2.1. Fermented Food Products

Fermented food products (cheese, fish, fish sauce/paste, vegetable sauce, and wine/vinegar classifications) were randomly selected and purchased from a variety of US retail outlets based in the state of Nebraska. These comprised $n = 4$ fish samples, $n = 9$ fish sauce/paste samples, $n = 4$ vegetable sauce samples, $n = 5$ cheeses, and $n = 4$ wine/vinegar samples (Table 1). Details of the fermentation processes employed by the manufacturers involved were unavailable. Prior to analysis, all samples were stored in a darkened freezer at a temperature of −20 °C for a maximal duration of 72 h.

Table 1. Details of fermented food products investigated for each classification.

Fermented Food Classification	Products Investigated
Cheeses	Full-fat pasteurized cow's milk soft cheese (washed with brandy); full-fat French cow's milk soft-ripened cheese; semi-soft washed rind Limberger cheese; full-fat pasteurized cow's milk soft cheese; French cow's milk soft cheese.
Fish	Pickled mud fish; pickled gourami fish; dried gourami fish; salted crab.
Fish Sauce/Paste	Loc fish sauce; scad fish sauce; anchovy fish sauce; Vietnamese fish sauce (×2); Thai fish sauce; standard U.S. fish sauce; shrimp paste (×2).
Vegetable Sauce	Bean curd; chili bean sauce; kimchi sauce; spicy tofu sauce.
Wine/Vinegar	Balsamic vinegar (×2); red wine vinegar; Casella wine.

Fermented cod liver oil (FCLO) was a natural product that was manufactured and kindly donated by Green Pastures LLC, 416 E. Fremont O'Neill, NE 68763, USA for this study. Separate batches ($n = 10$) of this FCLO product were randomly selected by independent visitors to its manufacturing site throughout a 6-month period, as noted in [6].

FCLO products were prepared from the fermentation of Pacific cod livers. Livers were frozen (−20 °C) within 40 min following their harvest from the Pacific Ocean, and then transported to a preparation facility whilst remaining in the frozen state. Fermented CLO was produced from these cod liver sources using a novel and proprietary fermentation technology. Briefly, cod livers were loaded into a fermentation tank, and both salt and the fermentation starter agent were added to induce the process. The tank was completely sealed during the fermentation and, following periods of 28–84 days, the raw FCLO product accumulated and was then isolated from the tank. Following fermentation, products were centrifuged, filtered to remove particulates, and then packed.

On arrival at the laboratory, FCLO product sample batches were de-identified through their transfer to coded but unlabeled universal storage containers. Each sample was subsequently stored in a darkened freezer at −80 °C until ready for analysis (predominantly within 24 h of their arrival).

2.2. Analysis of BAs in Fermented Food Product Samples

A liquid chromatographic triple quadrupole mass spectrometric (LC-MS/MS) technique was employed for the simultaneous analysis of up to 11 BAs in fermented food products using an adaption of the LC-MS/MS method reported in [27]. A Shimadzu 8045 LC-MS/MS facility was used for this purpose, the MS/MS detection system for the monitoring and molecular characterization of eluting BA analytes. Primarily, pre-set accurately weighed masses of food samples were shaken with a 20.0 mL volume of 70% (v/v) methanol/30% (v/v) water for 20 min, which were then centrifuged at 7000 rpm at 4 °C for another 20 min period. The clear supernatant was subsequently transferred to 1.7 mL volume amber auto-sampler vials for LC-MS/MS analysis. For wine/vinegar and FCLO samples, fixed aliquots were filtered using a 0.45 μm filter paper prior to the above methanol/water extraction stage.

The LC facility comprised a pump, vacuum degasser, auto-sampler, and column compartment, and finally a secondary variable wavelength spectrophotometric detection system was used for these analyses. This system could operate up to 800 bar. The internal standard (IS) utilized was tetra-deuterated histamine (histamine-$\alpha,\alpha,\beta,\beta$−$d_4$, (2HCl)), which was purchased from C/D/N Isotopes Inc. (Pointe-Claire, Quebec, Canada). IS m/z values employed for quantification purposes were 116.1 and 99.0 for precursor and product ions, respectively (112.1 and 95.1 respectively for undeuterated histamine).

A 3-μm 50 × 2.1 mm Pinnacle® DB pentaflurophenyl (PFP) base with propyl spacer column was employed for optimal BA analysis. Mobile phase 1 contained water solutions of the ion-pair reagent trifluoroacetic acid (TFA) (either 0.05 or 0.10% (w/v)), and mobile phase 2 was acetonitrile containing equivalent TFA concentrations. BA analytes were monitored in positive ion mode for the MS/MS detection system. Reporting limit values for fermented food samples were 1 ppm for all BAs determined.

Authentic BA calibration standards were purchased from Sigma-Aldrich Chemical Co. (St. Louis, MO, USA) (histamine, H7125; cadaverine, 33220; putrescine, D13208; 2-phenylethylamine, P6513; spermidine, 85578; tyramine, T2879; tryptamine, 193747), and Alfa Aesar Inc. (Heysham, UK) (spermine, J63060). BA contents were determined from calibration curves developed with standard solutions of concentrations 0.5, 1.0, 10.0, 50.00, 100.0, 200.0, and 400.0 ppb for each BA.

2.3. Total Lipid Analysis

Total lipid (fat) analysis was performed according to the AOAC 922.06 method. Briefly, homogenized samples were treated with HCl, and then washed at least two-fold with both petroleum ether and diethyl ether; solutions arising therefrom were then placed in pre-weighed beaker containers. Subsequently, the lipid-containing ether solutions were evaporated, and the (w/w) % content of lipid was determined directly from the weight gain of the container.

2.4. Determination of Thiobarbituric Acid-Reactive Substances (TBARS) Values

Primarily, accurately-weighed samples were digested with perchloric acid ($HClO_4$) and subsequently the resulting clear filtered supernatant solution was reacted with thiobarbituric acid (TBA) for a period of 15–18 h at 27.5 °C according to the method outlined in [28]. The absorbance value at a wavelength of 532 nm was then determined, and TBA-reactive substance (TBARS) values were reported as mg/kg (ppm) units following their quantification from a calibration curve developed with MDA standards.

2.5. Titratable Acidity (TA) and pH Value Determinations

Titratable acidity values were determined using the AOAC 947.05 method [29], and pH measurements were made using a modified FO PROC 31 protocol which is based on the USDA PHM method. The latter approach is based on the formation of a homogenized food/water slurry which was allowed to stand prior to pH determination with a probe.

2.6. Experimental Design and Statistical Analysis

2.6.1. Univariate Statistical Analysis

The experimental design for univariate analysis of the individual BA, TA, pH, and further variable dataset involved an analysis-of-variance (ANOVA) model, which incorporated 1 prime factor and 2 sources of variation: (1) that "between-fermented food classifications", a qualitative fixed effect (FF_i); and (2) experimental error (e_{ij}). The mathematical model for this experimental design is shown in equation 1, in which y_{ij} represents the (univariate) BA or alternative analyte dependent variable values observed, and μ their overall population mean values in the absence of any significant, influential sources of variation.

$$y_{ij} = \mu + FF_i + e_{ij} \tag{1}$$

ANOVA was conducted with *XLSTAT2016* and *2020* software. Datasets were autoscaled (i.e., the mean value of each parameter monitored was subtracted from each entry, and the residual then divided by food class standard deviation, which was computed with an ($n - 1$) divisor) prior to analysis. In view of heterogeneities between the intra-sample variances of fermented food classifications, i.e., heteroscedasticities, the robust Welch test was employed to determine statistical significance of differences observed between the mean BA and other food quality variable values for each fermented food group. *post-hoc* ANOVA evaluations of the statistical significance of differences between the mean values of individual fermented food groups were performed using the Bonferroni test.

A similar ANOVA-based experimental design was applied to additional design models selected to determine the statistical significance and food class specificities of BA analytes only. For these purposes, the 8 BA dataset, which included those determined in the $n = 10$ batches of the FCLO product, was either constant sum (CS)-normalized or not, and then generalized logarithmically (glog)-transformed, and finally autoscaled prior to analysis. The CS normalization data preparation task was applied in order to evaluate the significance of fermented food classification-dependent BA profile patterns. The non-CS-normalized dataset also included total BA level as a further possible explanatory variable. *MetaboAnalyst 4.0* (University of Alberta and National Research Council, National Institute for Nanotechnology (NINT), Edmonton, AB, Canada) was utilized for the analysis of these data. Probability values obtained from *post-hoc* ANOVA comparisons of individual BA levels between fermented food classes were false discovery rate (FDR)-corrected.

Tests for the heteroscedasticity of ANOVA model residuals (Levene's test) were performed using *XLSTAT2020* (Addinsoft, Paris, France).

2.6.2. Multivariate Chemometrics and Algorithmic Computational Intelligence (CI) Analyses

Principal component analysis (PCA), partial least squares-discriminatory analysis (PLS-DA), correlation, and agglomerative hierarchical clustering (AHC) analyses of the combined BA dataset were performed using *XLSTAT2016* and *2020* and *MetaboAnalyst 4.0* [30] software module options. The dataset was generalized glog-transformed, and autoscaled prior *to MetaboAnalyst 4.0* analysis, but only autoscaled for *XLSTAT2016* and *2020* analyses. All these MV analysis strategies were primarily performed on non-CS-normalized data. For the PCA and PLS-DA analyses, limits for significant explanatory variable loadings vectors/coefficients were set at ≤ -0.40 or ≥ 0.40. Validation of PLS-DA models was performed by determining component number-dependent Q^2 values (predominantly for two classification comparisons), and permutation testing with 2000 permutations. The significance of variable contributions to these models was determined by the computation of variable importance parameter (VIP) values (values >0.90 were considered significant).

Additional PCA analysis was performed in order to explore associations or independencies of individual BAs and other active variables considered, e.g., pH and TA values, total lipid contents, etc. For this purpose, a maximal 5 PC limit was applied, and PCA was then conducted on autoscaled data using varimax rotation and Kaiser normalization. The loadings of each analytical variable on successive orthogonal PCs was then sequentially evaluated. Similarly, this form of PCA was employed to investigate possible inter-relationships and orthogonalities between BA variables analyzed in FCLO batches sampled from the same manufacturing source specified above.

A further PCA model involved its application to the 8 BA dataset alone, which was either CS-normalized or not, glog-transformed, and autoscaled prior to analysis. As noted above, the CS-normalization data preparation step was utilized in order to evaluate the significance of any differential patterns or distributions of BA analytes which may be characteristic of fermented food classifications. This analysis was performed using *MetaboAnalyst 4.0*.

The random forest (RF) machine-learning algorithm approach was also utilized for classification and discriminatory variable selection purposes (*MetaboAnalyst 4.0* Random Forest module), with 1000 trees (*ntree*) and 4 predictors selected at each node (mtry) subsequent to tuning. The dataset was randomly split into training and test sets containing approximately two-thirds and one-third of entries respectively. The training set was employed to construct the RFs model, and an out-of-the-bag (OOB) error value was determined to evaluate the classification performance of this. Again, this analysis was performed on the glog-transformed and autoscaled dataset, either with or without prior CS-normalization as specified in the manuscript.

Missing data, specifically total lipid and (TBARS):(total lipid) ratios for 2 × fish sauce/paste, 1 × vegetable sauce, 1 × wine/vinegar, and 1 × cheese samples, were estimated by the support vector machine (SVM) impute technique [31] (*MetaboAnalyst 4.0*), or supplementation with the explanatory variable column mean values, along with a corresponding reduction in degrees of freedom available for parametric univariate statistical testing (*XLSTAT2016 or 2020*).

3. Results and Discussion

3.1. BA Levels and Food Quality Indices in Fermented Food Products, and Univariate Analysis of These Analytical Data

Mean ± SEM values for the individual and total BA contents of the FF products investigated are provided in Table 2. The major contributors towards the relatively high BA levels observed in fermented cheese samples were cadaverine (mean 60% of total) and tyramine (mean 21.5% of total). Although three of the cheese products analyzed had total BA concentrations of 30–63 ppm, two of them were found to be as high as 666 and 780 ppm, which were markedly above the recommended 200 ppm content limit. The ANOVA Welch test demonstrated that there were highly significant differences between these total BA values (Table 3), as expected ($p = 2.84 \times 10^{-4}$); such differences were largely explicable by those observed between the cheese and wine/vinegar product classifications investigated.

Hence, characteristic "markers" of fermented cheese samples appeared to be cadaverine and tyramine, which had contents markedly elevated over those of the other fermented food products evaluated, although there were very high intra-fermented food classification variances for these estimates.

Table 2. Biogenic amines (BA) contents and quality indices of fermented foods investigated. Mean ± SEM BA levels, and titratable acidity (TA), pH, total lipid, thiobarbituric acid-reactive substances (TBARS) and (TBARS):(total lipid) ratio values, for five classes of fermented food products (cheese, fish, fish sauce/paste, vegetable sauce, and wine/vinegar) purchased at a range of U.S. retail outlets (bracketed numbers represent the number of different products analyzed for each classification).

BA Variable/ppm	Cheese (5)	Fish (4)	Fish Sauce (9)	Vegetable Sauce (4)	Wine/Vinegar (4)
Cadaverine	191.6 ± 99.8	30.7 ± 5.6	45.2 ± 8.2	30.6 ± 14.0	0.7 ± 0.7
Histamine	5.7 ± 1.6	10.6 ± 2.9	20.0 ± 5.8	17.7 ± 9.5	1.9 ± 1.9
2-Phenylethylamine	11.1 ± 6.8	13.1 ± 8.0	8.3 ± 4.2	5.00 ± 5.00	nd
Putrescine	21.2 ± 14.9	14.9 ± 7.2	18.9 ± 5.1	18.4 ± 7.7	3.3 ± 0.15
Spermidine	9.6 ± 3.2	10.9 ± 2.5	15.0 ± 1.9	24.5 ± 10.1	4.4 ± 1.5
Spermine	3.5 ± 2.2	12.6 ± 4.5	18.7 ± 3.0	11.4 ± 4.3	1.5 ± 1.5
Tryptamine	7.0 ± 5.8	2.0 ± 0.7	5.6 ± 1.7	3.4 ± 1.0	nd
Tyramine	69.4 ± 42.5	8.9 ± 3.7	17.2 ± 4.0	36.4 ± 20.5	0.6 ± 0.4
Total BAs	322.2 ± 166.0	103.8 ± 12.7	155.9 ± 18.7	147.9 ± 56.2	12.4 ± 5.5
Titratable Acidity (g acid/100 g)	1.3 ± 1.1	0.6 ± 0.2	0.6 ± 0.1	0.7 ± 0.2	3.6 ± 1.2
pH	6.09 ± 1.38	6.33 ± 0.62	5.47 ± 0.28	5.18 ± 0.53	2.99 ± 0.20
Total Lipid (% w/w)	23.3 ± 2.0	6.9 ± 2.7	4.2 ± 1.4	5.1 ± 2.8	1.1 ± 0.2
TBARS Value (ppm)	0.07 ± 0.05	0.35 ± 0.14	0.47 ± 0.27	0.09 ± 0.03	0.83 ± 0.51
10^2.(TBARS):(Total Lipid) Ratio (ppm(% w/w)$^{-1}$)	0.5 ± 0.4	10.6 ± 6.85	33.1 ± 22.3	5.6 ± 3.55	98.3 ± 45.7

nd: not determined.

Table 3. Statistical significance and nature of differences between the mean BA contents and other food quality indices for fermented food products. Both robust Welch and Bonferroni-corrected *post-hoc* ANOVA test significance (*p*) values are provided. Abbreviations: ns, not statistically significant. * These values were close to statistical significance, but did not attain a *p* value of ≤0.05 with the robust Welch test.

BA/Index	Welch Test (WT) *p* Value	*Post-hoc* Significant Differences (All *p* < 0.05: Bonferroni Test)
Cadaverine (ppm)	0.0016	Cheese > Wine/Vinegar; Cheese > Fish Sauce/Paste; Cheese > Fish; Cheese > Vegetable Sauce
Histamine (ppm)	0.087 *	All ns
2-Phenylethylamine (ppm)	ns	All ns
Putrescine (ppm)	0.068 *	All ns
Spermidine (ppm)	0.029	Vegetable Sauce > Wine/Vinegar; Vegetable Sauce > Cheese; Vegetable Sauce > Fish
Spermine (ppm)	0.010	Fish Sauce/Paste > Wine/Vinegar; Fish Sauce/Paste > Cheese
Tryptamine (ppm)	ns	All ns
Tyramine (ppm)	0.021	Cheese > Wine/Vinegar
Total BAs	2.84 × 10^{-4}	Cheese >> Wine/Vinegar

Table 3. *Cont.*

BA/Index	Welch Test (WT) p Value	*Post-hoc* Significant Differences (All $p < 0.05$: Bonferroni Test)
Titratable acidity (g acid/100 g)	0.024	Wine/Vinegar > Fish; Wine/Vinegar > Fish Sauce/Paste; Wine/Vinegar > Vegetable Sauce; Wine/Vinegar > Cheese
pH	6.91×10^{-4}	Wine/Vinegar < Fish; Wine/Vinegar < Fish Sauce/Paste; Wine/Vinegar < Vegetable Sauce Wine/Vinegar < Cheese
Total lipid (% *w/w*)	1.09×10^{-3}	Cheese > Wine/Vinegar; Cheese > Fish Sauce/Paste; Cheese > Fish; Cheese > Vegetable Sauce
TBARS value (ppm)	ns	ns
10^2.(TBARS):(Total Lipid) ratio (ppm(% *w/w*)$^{-1}$)	ns	Wine/Vinegar > Fish; Wine/Vinegar > Vegetable Sauce; Wine/Vinegar > Cheese

ns: not statistically significant.

Univariate statistical analysis performed by ANOVA (robust Welch test derivative), and also *post-hoc* Bonferroni test values, demonstrated that the mean values of each food classification examined were significantly or highly significantly different for 7 and 9 of the marker index variables respectively (p values ranging from <0.0003 to 0.04 for the former test, Table 3). Figure 1 shows a heatmap of the mean BA contents, and further variables included in this analysis; this clearly displays significantly higher tyramine, cadaverine, putrescine, and tryptamine levels in the fermented cheese products; higher histamine concentrations in the fish sauces/pastes explored, as expected (although vegetable sauces also had quite high levels of this BA); and also greater spermine contents in the fish paste/sauce products (*ca.* 1.5-fold greater than the mean value found for the fish classification, the next highest concentration). The vegetable sauce products had the highest mean spermidine levels, whereas the fermented fish group contained the largest amounts of 2-phenylethylamine detectable.

As expected, mean TA values were significantly greater for the wine/vinegar products than they were for all the other fermented food classes investigated, and correspondingly the mean pH value for the former group was significantly lower than those of all the other fermented foods. Of course, the mean total lipid content of the cheese group (23.3%) was significantly greater than all other food classifications tested (p *ca.* 10^{-3}), although no significant differences were found for the secondary lipid peroxidation TBARS marker. However, an examination of the mean ratio of TBARS index to total lipid content revealed that this value was markedly greater for the wine/vinegar group than that of all other food product types (Bonferroni-corrected *post-hoc* ANOVA tests), and significantly so over that of the cheese samples analyzed, as might be expected in view of the very low fat contents of fermented wine/vinegar samples (for example, it varies from 0.15–0.44% (*w/v*) in Zhenjiang aromatic vinegar samples [32]), and potentially substantially inflated TBARS levels resulting from quite high levels of TBA-reactive acetaldehyde and acrolein, amongst other aldehydes, present in such fermented products [33–36]. Indeed, many other aldehydes are reactive towards the TBA reagent, and also form chromophoric products on reaction with it [28]. Estimates for acetaldehyde in vinegar products can be as high as 1.0 g/kg respectively [33], but such levels are highly variable, with much lower levels being found, e.g., 2.6 mg/L (*ca.* 60 μmol/L) [37].

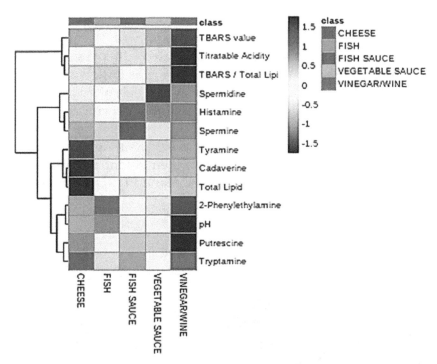

Figure 1. Heatmap diagram displaying the nature, extent and ANOVA-based significance of univariate differences between mean values of all 8 BA and further chemoanalytical food quality variables (near right-hand side *y*-axis) for the fermented cheese (red), fish (green), fish sauce/paste (dark blue), vegetable sauce (pale blue), and wine/vinegar (mauve) products. The complete dataset was glog-transformed and autoscaled prior to analysis, but not CS-normalized. Transformed analyte intensities are shown in the far right-hand side *y*-axis: deep blue and red colorations represent extremes of low and high contents respectively. The left-hand side of the plot shows results arising from an associated agglomerative hierarchical clustering (AHC) analysis of these variables, which reveals two major analyte clusterings, with three sub-clusterings for one of these. The top right-hand side major cluster comprises TBARS level, (TBARS):(total lipid) ratio and TA value, whereas the second contains all other analyte variables, including all BA contents. The first, second, and third sub-clusters within the bottom right-hand side major cluster feature spermine, spermidine, and histamine (the first two of these arising from the same putrescine and metabolically upstream ornithine and agmatine/arginine sources respectively); tyramine, cadaverine, and total lipid; and 2-phenylethylamine, putrescine, tryptamine, and pH respectively.

Acetaldehyde, a volatile flavor component of a variety of foods and beverages such as cheese, yoghurt, and wines [34], represents one of the most abundant carbonyl compounds detectable in wine, and typically accounts for *ca.* 90% of the total aldehydes present; its concentrations therein usually range from 10 to 200 mg/L (predominantly, it is generated as a yeast by-product during alcoholic fermentation processes [35], or from the chemical oxidation of ethanol [36]). However, very high levels of the unsaturated aldehyde acrolein are also present in red wine products [33]. Furthermore, a wide range of further aldehydes have been found to serve as major flavor constituents of traditional Chinese rose vinegar, and these include aliphatic *n*-alkanals such as heptanal, hexanal, nonanal, and dodecanal (ranging from 6–147 µg/kg), with larger amounts of benzaldehyde (851 µg/kg) [38].

Hence, overall these data clearly demonstrated that, in a univariate context, there were indeed significant differences between the mean contents of BAs and further parameters considered for the five classes of fermented food products studied.

Prior to the performance of MV statistical analysis of the dataset acquired, simple Pearson correlations were explored between all explanatory variables considered, and Figure 2 shows a correlation heatmap for these relationships. Clearly, there were moderate to strong positive correlations observed between all fermented food BAs present, the strongest observed between 2-phenylethylamine and tyramine (both aromatic BAs), tryptamine and spermine, and most notably, between cadaverine and histamine. Food pH values were found to have the strongest positive correlations with tyramine > putrescine > tryptamine, although spermidine was predominantly uncorrelated with this index. Moreover, as anticipated, TA was strongly negatively correlated with pH value > putrescine > tyramine ≈ histamine contents in that order. TBARS level, however, was largely independent of all BAs and their concentrations, with the exception of spermidine, which exhibited a weak positive relationship with this variable. Similarly, total lipid level was also mainly uncorrelated with all BA contents but was quite strongly anti-correlated with (TBARS):(total lipid) ratio and non-lipid-normalized TBARS value (both expected). The (TBARS):(total lipid) ratio was either strongly or moderately anti-correlated with all BA levels, and this may provide an indication of their potential antioxidant functions. In view of the complexity of these inter-relationships, the MV PCA and PLS-DA techniques were employed to explore them further.

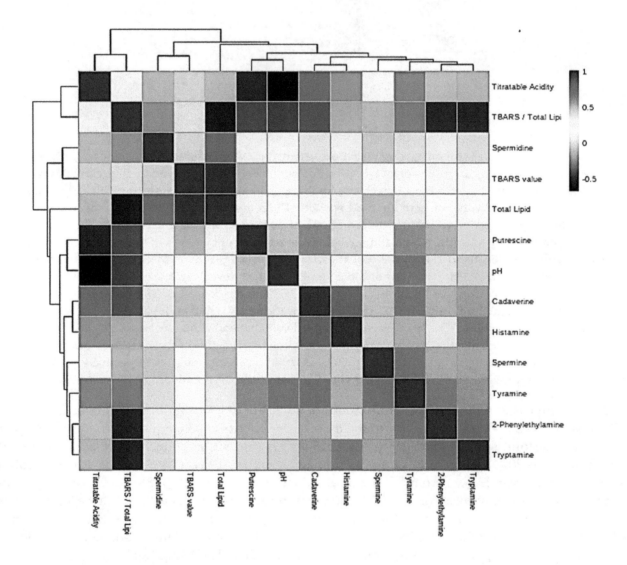

Figure 2. Correlation heatmap displaying positive and negative inter-relationships between BA concentrations, pH and TA values, total lipid contents, TBARS indices and (TBARS):(total lipid) ratios (TBARS/total lipid). The left-hand ordinate and top abscissa axes show AHC analysis based on these Pearson correlations (as a similarity criterion). From the top abscissa axis, of the two major clusterings revealed, that on the right-hand side contains all BA variable levels with the exception of spermidine, together with positively-correlated pH values, whereas the left-hand side one consists of all lipid- and lipid peroxidation-based variables, spermidine concentrations, and TA values.

3.2. Principal Component Analysis (PCA) of the Multivariate Fermented Food Dataset

PCA was primarily conducted in order to acquire an overview of the degree of distinctiveness between, i.e., clustering of, the fermented food classifications investigated, and also to identify any potential data outliers. An examination of two-dimensional (2D) scores plots from this analysis demonstrated that no significant outliers were detectable, and that PCs 1, 2, and 3 accounted for 41.5, 16.4, and 11.1% of the total variance respectively for the complete dataset which was glog-transformed and autoscaled. 2D and three-dimensional (3D) scores plots featuring these two most important PCs revealed that there was a reasonable level of distinction between the wine/vinegar and all other food product groups, and also between the cheese and fish classifications (Figure 3a); however, distinctions between the fish, fish sauce/paste, and vegetable sauce groups were not found, there being a significant degree of overlap between them. Notwithstanding, the sample sizes of the fermented fish and vegetable sauce groups involved were quite limited. A corresponding preliminary correlation circle diagram is shown in Figure 3b. Clear observations from this diagram are that (1) 2-phenylethylamine, tyramine, and cadaverine, and to a lesser extent, putrescine and tryptamine, are all correlated with PC1, and this observation indicates their communality in this model; (2) food pH values are also strongly correlated to PC1, and this indicates that higher values of this parameter may arise from the basicity of the above BAs (gas-phase primary amine basicity values increase with the length of its carbon chain substituents in view of their electron-donating positive charge-stabilizing effects—such values also increase with progression from primary to secondary to tertiary alkylamines [39]); (3) an at least partial correlation of histamine contents with PC2, which indicates distinction of this BA from those aligned with PC1; (4) an inverse correlation (anti-correlation) of total lipid level with the (TBARS):(total lipid) ratio index, as might be expected; and (5) a strong anti-correlation of TA value with BA levels, particularly tryptamine and putrescine, and this suggests that these amines serve to offer neutralization potential against acidic fermented food products. Also notable from this Figure are very strong correlations between the fermented food supplementary variable cheese and total lipid content, and between wine/vinegar and TA value, as indeed expected.

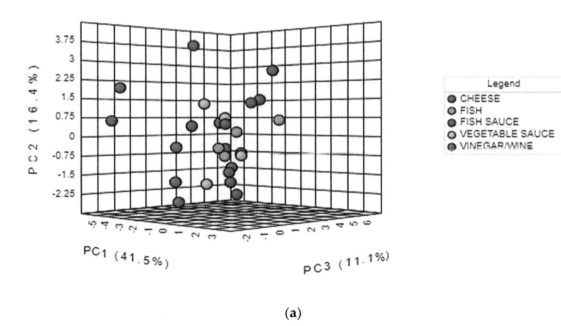

(a)

Figure 3. *Cont.*

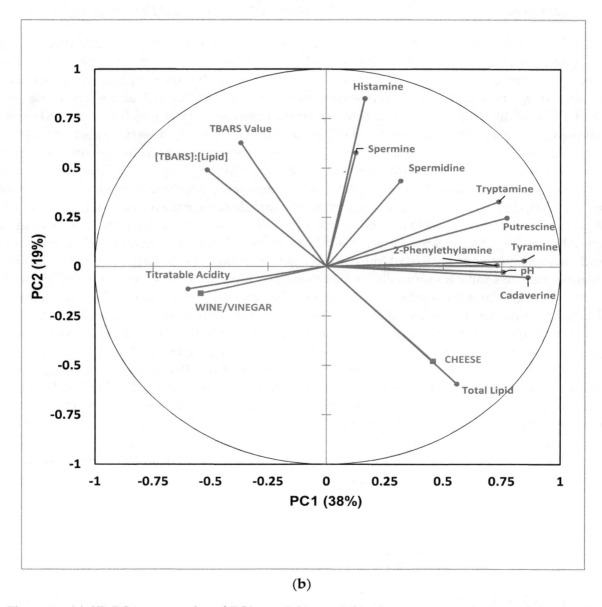

(b)

Figure 3. (a) 3D PCA scores plot of PC3 vs. PC2 vs. PC1, showing some degrees of distinction between different fermented food classes, i.e., those of cheese, fish, fish sauce/pastes, vegetable pastes, and wines/vinegars (particularly that between the wine/vinegar classification and all others). (b) Preliminary correlation circle diagram displaying correlations between all explanatory variables considered, and PCs 1 and 2 in a PCA model applied to the complete autoscaled (standardized) dataset. Active variables are depicted in red, whereas two of the supplementary variable classifications (cheese and wine/vinegar) are shown in blue. Variance contributions for PC1 and PC2 are indicated.

A more detailed analysis of these PCA loadings was made with the application of varimax rotation, Kaiser normalization, and a maximal number of 5 PCs considered. For this model, such variable loadings, and the percentage of total variance accounted for by each PC are available in Table 4. This analysis revealed that cadaverine, tryptamine, 2-phenylethylamine, and tyramine all strongly and positively loaded on PC1, spermidine and histamine strongly and positively loaded on PC3 (along with a more minor contribution from 2-phenylethylamine), and putrescine and spermidine loaded strongly and positively on PC5, albeit also with histamine to a much lesser extent. Interestingly, all aromatic BAs strongly loaded on PC1, as observed above (Figure 3b), whereas spermidine and its metabolic precursor putrescine both co-loaded onto the same PC (PC5).

Table 4. PCA loadings vectors for BAs and additional fermented food analyte parameters (including total lipid contents, and pH and TA values) for a 5 PC-limited model performed with varimax rotation and Kaiser normalization. Percentage variance contributions for PCs 1–5 and their (unrotated) analysis eigenvalues are also listed. Bold numbers are for a purpose specified in the Figure legends.

PC (Unrotated Eigenvalue):	PC1 (4.58)	PC2 (2.26)	PC3 (1.78)	PC4 (1.28)	PC5 (0.89)
% Variance Contribution	26.6	16.9	12.2	16.1	11.2
2-Phenylethylamine	**0.71**	−0.23	**0.43**	0.03	−0.06
Cadaverine	**0.90**	−0.07	−0.19	0.21	−0.04
Histamine	0.25	**0.45**	**0.63**	0.025	0.32
Putrescine	**0.57**	−0.04	−0.19	**0.42**	**0.50**
Spermidine	0.14	−0.19	0.20	0.06	**0.78**
Spermine	−0.08	−0.06	**0.85**	0.22	0.12
Tryptamine	**0.79**	0.07	0.11	0.13	0.30
Tyramine	**0.88**	−0.11	0.03	0.15	−0.03
Titratable acidity (TA)	−0.11	0.17	−0.30	**−0.86**	−0.09
pH	0.29	−0.10	0.04	**0.92**	−0.07
TBARS value	−0.08	**0.95**	0.08	−0.05	−0.05
Total lipid	**0.45**	−0.31	−0.21	0.37	**−0.61**
(TBARS):(Lipid) ratio	⁻0.14	**0.92**	⁻0.07	⁻0.23	⁻0.08

The TBARS secondary lipid oxidation index, along with its value normalized to total food lipid content, both loaded strongly and positively on PC2, as might be expected, although histamine also contributed somewhat towards this PC. Moreover, TA and pH values powerfully loaded on PC4 negatively and positively respectively, as would be expected from their anticipated negative correlation in fermented food products (putrescine also made a moderate positive contribution towards this component). Total lipid content was found to load significantly on PCs 1 and 5, positively and negatively so, respectively.

In a related study focused on PCA of both BAs and polyphenolics in Hungarian wines, Cosmos et al. [26] found that PC scores successfully clustered differential groups of these product classes, and that PC loadings vectors displayed significant patterns of BA and polyphenol levels. However, it should be noted that for this analysis, spermidine, and tyramine strongly loaded on PC1 (positively and negatively, respectively), agmatine and the sum total BA concentration loaded strongly and positively on PC2, spermine and cadaverine both strongly and negatively loaded on PC3, and that histamine loaded strongly and positively on PC4 alone. These associations between the BA analytes tested did not correspond to those found in the present study, although in the above MV analyses we elected not to include the total summed BA concentration value. Furthermore, our study also included the determinations of 2-phenylethylamine and putrescine, and not agmatine, but that reported in [26] monitored the latter BA but not 2-phenylethylamine and putrescine. However, as noted by the authors of [26], these PC loadings are only applicable to one region of Hungarian wine production, and their results will not be readily transferable to others, let alone other classes of fermented foods, especially in consideration of the often highly variable methods of fermentation, sources of fermentative micro-organisms, and conditions employed for these purposes. Notwithstanding, these researchers also concluded that in view of the loading patterns of BAs observed, it was unnecessary to measure all BA variables for quality assessments, and that only one per orthogonal PC was sufficient to provide acceptable levels of distinction between different sub-classes of such wines.

From this analysis, the unambiguously strong loadings vectors of the aromatic BAs 2-phenylethylamine and tyramine on PC1 provide evidence that they may indeed arise from the same biological and/or metabolic sources; however, this observation may also be rationalized by the natural production of tyrosine from phenylalanine, i.e., that involving the possible hydroxylation of the latter substrate to the former catalyzed by the enzyme phenylalanine hydroxylase (PAH) potentially available in fermentative lactobacilli employed for the production of fermented food products, followed by enzymatic transformation of the tyrosine product to tyramine by fermentative bacteria. To date, PAH is the only known aromatic amino acid hydroxylase found in bacteria [40].

The loadings of spermine and spermidine on different orthogonal PCs (PC3 and PC5, respectively) is not simply explicable, although the co-loading of spermidine's metabolic precursor putrescine on PC5 is consistent with them being featured in the same metabolic pathway. However, the co-loadings of BAs on differential PCs, particularly PC1, may reflect their engenderment from identical or related bacterial sources.

Notably, PC2 was dominated by powerful loading contributions from TBARS level and (TBARS):(total lipid) ratio (both positive), and PC4 by strong loadings from TA and pH values (negative and positive loadings vectors, respectively). These inter-relationships are, of course, expected, and are consistent with the data presented in Figure 3b. PC5 was retained in the model since it was the only one available which had a strong loading contribution from spermidine.

3.3. Distinction of Fermented Food Classifications Using PLS-DA

Similarly, PLS-DA of the dataset revealed an effective discrimination between the cheese and wine/vinegar classifications, although the fish, fish sauce/paste and vegetable sauce sample PC score datapoints were again unresolved; however, a visualized combination of these three fermented food classifications was at least partially resolved from the fermented cheese group (Figure 4). Permutation testing of the PLS-DA model confirmed its ability to distinguish between all the differing fermented food classifications evaluated ($p = 0.022$). For this model, key discriminatory variables were selected on the basis of their variable importance parameters (VIPs), and these were total lipid content (1.81) > cadaverine content (1.61) > (TBARS):(total lipid) ratio (1.36) > TA value (1.24) > histamine content (1.14) > 2-phenylethylamine content (0.78); data were glog-transformed and autoscaled prior to analysis. The top three discriminators largely arise from differential levels of lipids, cadaverine, and (TBARS):(total lipid) ratio between each of the fermented food groups, e.g., for the total lipids and cadaverine variables, the cheese content was significantly greater than that of all other fermented food groups, and for the above ratio, its value was significantly greater in the wine/vinegar group than it was in all other groups.

The quite strong distinctions observed between the cheese, wine/vinegar, and fish-fish sauce/paste-vegetable sauce composite products is readily explicable by significant or even substantial differences between the higher contents of cadaverine, tyramine, and, to a lesser extent, tryptamine in cheese, than those of the four other fermented food product classes. Further key discriminators are TA, pH, and total lipid contents, the latter of which is, of course, much higher in the cheese group.

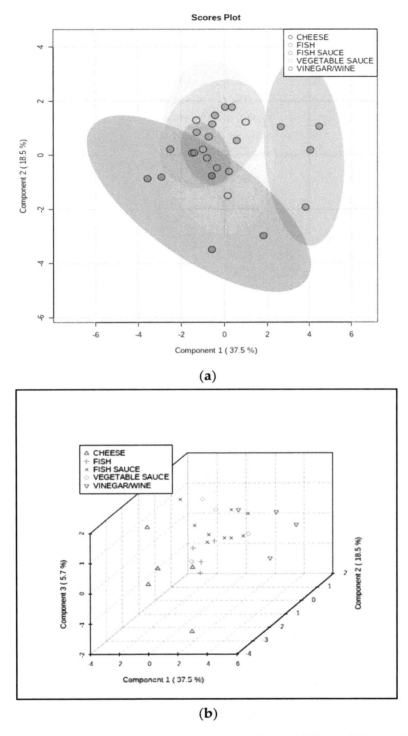

Figure 4. (a,b). 2D and 3D PLS-DA scores plots (PC2 vs. PC1, and PC3 vs. PC2 vs. PC1, respectively) revealing strong distinctions between the cheese, wine/vinegar, and a considered combination of fish, fish sauce/paste and vegetable sauce fermented food groups ((**a**) also shows 95% confidence ellipses for each fermented food classification). Little or no distinction between the latter three fermented food groups were discernable using this MV analysis approach.

3.4. RF Modelling of Fermented Food Classifications

Application of the RF CI classification technique was found to be only partially successful for the classification of the different fermented food groups investigated. Using the models described in Section 2.6.2, this approach correctly classified 4/4 wine vinegar, 6/9 fish sauce/paste, and 3/5 cheeses, but 0/4 for both fish and vegetable sauce products.

3.5. PCA of FCLO BAs

The FCLO product considered was primarily investigated separately since only BA contents, and not parameters such as pH and TA were available for it. Moreover, its total lipid content is, of course, not far removed from a value of 100%, and therefore it would be inappropriate to test this index in the above MV analysis models (similarly, total lipid level-normalized TBARS values would also be inappropriate to test in these systems). However, it was possible to explore inter-relationships between FCLO BA concentrations and/or their orthogonality status using a rigorous PCA approach featuring varimax rotation and Kaiser normaliszation in order to maximize success with the assignment of individual BA variables to PCs.

Table 5 lists the BA contents of $n = 10$ FCLO product batches. The total concentrations of BAs in these samples was higher than the recommended "limit" of 200 ppm in only two out of ten batches of the samples tested, albeit marginally so (only 14 and 20% higher). Similarly, bioactive histamine was completely undetectable in this product. As noted in [6], all BAs monitored were completely undetectable in three other natural, albeit unfermented, CLO products included for comparative purposes. All BAs tested were found to be reasonably soluble in FCLO lipidic matrices, and also in 1/3 (v/v) diluted solutions of this product in deuterochloroform (C^2HCl_3), presumably as the uncharged species with their amine functions deprotonated (solubility in these media is expected to increase with increasing amine function substituent chain length and hydrophobicity).

Table 5. (BA concentrations (ppm) of $n = 10$ separate batches of a FCLO product. Total BA and corresponding mean ± SEM values are also provided. Histamine and spermine were undetectable in all samples analyzed.

Biogenic Amine (ppm)	FCLO Batch										Mean ± SEM
	1	2	3	4	5	6	7	8	9	10	
2-PE	86	103	50	17	76	0	1.4	0	1.9	1.3	33.7 ± 13.0
Tyramine	70	88	43	8	32	0	1.8	0	1.1	0	24.4 ± 10.3
Tryptamine	35	24	26	3	8	0	0	0	1.7	1.5	9.9 ± 4.2
Cadaverine	23	11	25	0	7	0	0	0	0	0	6.6 ± 3.1
Putrescine	14	10	14	0	0	0	0	0	0	0	3.8 ± 2.0
Spermidine	0	4	0	0	0	0	0	0	0	0	0.4 ± 0.4
Total	228	240	158	28	123	0	3.2	0	4.7	2.8	78.8 ± 31.3

PCA performed on the FCLO BA dataset revealed that cadaverine, putrescine, and tryptamine all loaded strongly and positively on the first of two automatically-selected PCs (PC1), whereas the aromatic BAs 2-phenylethylamine and tyramine loaded strongly and positively on the second (PC2), along with spermidine (Table 6). These data displayed some consistency with PC loading values obtained on the full fermented food dataset (Table 4), which had 2-phenylethylamine and tyramine both strongly loading on one PC (PC1). However, such levels will, of course, be critically dependent on the microbial fermentation sources, parameters employed for fermented food production, and production conditions for these processes.

Table 6. PCA loadings vectors for FCLO BAs in a two PC-limited PCA model performed with varimax rotation and Kaiser normalization. Percentage variance contributions for these PCs and their (unrotated) analysis eigenvalues are also listed. Bold numbers are for a purpose specified in the Figure legends.

PC (unrotated Eigenvalue)	PC1 (4.03)	PC2 (1.58)
% Variance Contribution	52.8	40.6
2-PE	0.35	**0.84**
Tyramine	**0.54**	**0.84**
Tryptamine	**0.93**	0.33
Cadaverine	**0.99**	−0.01
Putrescine	**0.95**	0.23
Spermidine	−0.11	**0.93**

3.6. MV Chemometric Analysis of BA Data Only: Recognition of Fermented Food Class-Distinctive BA Patterns Using CS-Normalization

Additionally, we conducted univariate and MV analyses of datasets which were restricted to the BA profiles only, but also included the $n = 10$ FCLO samples reported above. Additionally, these analyses were performed with and without application of constant sum (CS) normalization. The CS-normalized data format was employed in order to facilitate the recognition of fermented food class-specific BA patterns. For the non-CS-normalized format, the total BA content value was also included as an explanatory variable, as indeed it was in [26].

Firstly, ANOVA performed on the CS-normalized, glog-transformed, and autoscaled dataset found very highly significant, albeit FDR-corrected p values for three of the sum-proportionate mean BA concentration differences observed between the fermented food classifications explored in this manner. Notably, these differences were observed for cadaverine, 2-phenylethylamine, and tryptamine (Table 7), and *post-hoc* testing revealed that for cadaverine, the cheese products had significantly greater proportionate levels than three others, and for both 2-phenylethylamine and tryptamine, FCLO had significantly higher ones than all other products examined. These differences in CS-normalized values are readily visualizable in the form of an ANOVA-based heatmap (Figure 5a), which revealed characteristic BA signatures for three of the fermented food product classifications. Clearly, the cheese, FCLO, and wine/vinegar sampling groups have high proportionate levels of cadaverine, 2-phenylethylamine/tyramine/tryptamine (all aromatic BAs), and metabolic pathway-associated putrescine/spermidine/spermine, respectively. However, when evaluated in this univariate system, "between-fermented food class" mean differences observed for putrescine, spermine, spermidine, histamine, and tyramine were not found to be statistically significant.

Secondly, both PCA and PLS-DA models were employed, and these approaches were successful in providing evidence for the MV distinctiveness of the FCLO, cheese, and wine/vinegar groups; however, as noted for the analyses conducted on the combined BA/further food quality parameter dataset, unfortunately no distinctions were observed between the fermented fish, fish sauce/paste, and vegetable sauce products (Figure 5b,c).

For the CS-normalized dataset (without total BA concentrations as an additional variable), PLS-DA variable importance parameter (VIP) values were in the order spermidine (1.48) > putrescine (1.34) > spermine (1.20) > histamine (1.06) > 2-phenylethylamine (0.94), whereas those for the non-CS-normalized dataset were spermidine (1.56) > spermine (1.35) > 2-phenylethylamine (1.32) > putrescine (0.84) (total BA level was a very poor predictor variable for the latter). As expected, there were significant differences between the sequential orders of these values when prior CS-normalization was implemented.

Table 7. Univariate statistical significance and nature of differences observed between the mean CS-normalized, glog-transformed, and autoscaled BA contents of fermented food samples (cheese, FCLO, fish, fish sauce/paste, vegetable sauce, and wine/vinegar products) in a completely randomized, one-way ANOVA model. The significance of FDR-corrected *post-hoc* ANOVA tests are also provided (significant differences are ranked in order of their decreasing statistical significance, i.e., increasing p value). The "between-fermented food class" source of variation was not statistically significant for putrescine, spermidine, spermine, histamine, or tyramine when tested in this model.

BA	FDR-Corrected p Value	Significant post-hoc *ANOVA* Differences
Cadaverine	1.49×10^{-5}	Cheese > FCLO; Cheese > Vegetable Sauce; Cheese > Wine/Vinegar; Fish > FCLO; Fish Sauce > FCLO; Vegetable Sauce > FCLO; Fish > Wine/Vinegar; Fish Sauce > Wine/Vinegar; Vegetable Sauce > Wine/Vinegar.
2-Phenylethylamine	8.25×10^{-4}	FCLO > Cheese; FCLO > Fish; FCLO > Fish Sauce; FCLO > Vegetable Sauce; FCLO > Wine/Vinegar; Fish > Vegetable Sauce.
Tryptamine	2.93×10^{-2}	FCLO > Cheese; FCLO > Fish; FCLO > Fish Sauce; FCLO > Vegetable Sauce; FCLO > Wine/Vinegar.

Moreover, for the PLS-DA model adopted without CS-normalization, histamine, spermidine, and spermine contents all loaded significantly on component 1 (loading vector coefficients 0.48, 0.57, and 0.47 respectively); 2-phenylethylamine, cadaverine, tyramine, and total BA levels on component 2 (loadings vector coefficients 0.42, −0.61, −0.57, and −0.61 respectively); 2-phenylethylamine and tryptamine levels on PC3 (loadings vector coefficients 0.50 and 0.57 respectively); and putrescine and spermine on PC4 (loadings vector coefficients 0.75 and −0.73 respectively). For this dataset, a four-component model was found to be most effective (permutation p value 0.0055)

Importantly, it should be noted that one now common issue in chemometrics/metabolomics experiments is the occurrence of a univariately-insignificant variable which remains multivariately-significant. Such observations are readily rationalized, firstly by the complementation (i.e., correlation) between explanatory variables, i.e., separately they do not, but when combined together as a MV composite (e.g., as a sufficiently-loading PC variable), they do serve to explain "between-classification" differences detected; secondly, consistency effects arising from the "masking" of potential univariately-significant differences by high levels of biological source sampling and/or measurement variation may be responsible (such variation may be averaged out via the conversion of datapoints to orthogonal component scores as in the PCA and PLS-DA models applied here); and thirdly, relatively small sample sizes for each classification involved (fermented foods in this case)—unfortunately, strategies applied to correct for FDRs promote the risk of statistical type II errors (i.e., false negatives) [24].

The PLS-DA evaluation was then extended and performed for pairwise comparisons of the differing fermented food classifications (CS-normalized dataset only). Firstly, as expected, Q^2 values for the fish vs. fish sauce/paste, fish sauce/paste vs. vegetable sauce, and fish vs. vegetable sauce comparisons were all moderately negative, and p values for associated permutation tests were all >0.10. However, these values for the wine/vinegar vs. FCLO, and FCLO vs. cheese two classification model comparisons revealed that Q^2 (permutation p values) indices for these comparisons were 0.71 (0.059) and 0.72 (0.090), but only 0.38 (0.16) for the wine/vinegar vs. cheese one (values were based on models containing two, five, and one components respectively). Hence, these results provide some evidence for the success of this strategy in distinguishing between the FCLO product, and both the cheese and wine/vinegar ones, although permutation test p values obtained for these models were a little higher than the 0.05 significance level, i.e., they were close to statistical significance.

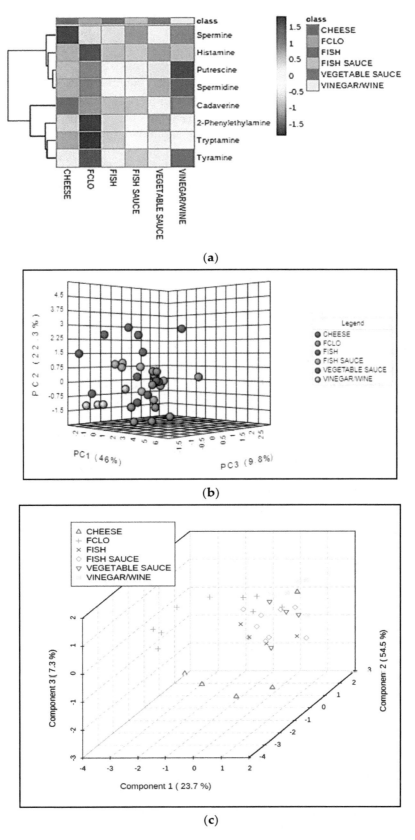

(a)

(b)

(c)

Figure 5. (**a**) Heatmap diagram displaying the most univariately-significant differences between mean values of eight BA explanatory variables (near right-hand side *y*-axis) for the fermented cheese (red), FCLO (green), fish (dark blue), fish sauce/paste (pale blue), vegetable sauce (purple), and wine/vinegar (yellow) products. The complete BA dataset was CS-normalized, glog-transformed, and autoscaled prior to analysis. AHC analysis shown on the left-hand side ordinate axis demonstrated two major analyte clusterings, the upper one consisting of putrescine, spermidine, and spermine pathway biomolecules (and histamine), whereas the lower one features all aromatic BAs, along with cadaverine. (**b**) 3D PCA PC3 vs. PC2 vs. PC1 scores plot for the same CS-normalized dataset shown in (**a**), showing reasonable or strong distinctions between the cheese, wine/vinegar and FCLO fermented food classes. (**c**) 3D PLS-DA PC3 vs. PC2 vs. PC1 scores plot for the corresponding non-CS-normalized dataset, which also incorporated total BA content as a potential explanatory variable (again, effective distinctions between the cheese, FCLO, and wine/vinegar classes were notable).

We then elected to statistically combine the fish, fish sauce/paste, and vegetable sauce groups, and repeated the PLS-DA modelling in order to compare the sauce/fish composite, cheese, FCLO, and wine/vinegar groups using the CS-normalized dataset. This analysis exhibited a quite high level of classification success (Figure 6a); Q^2 for this comparative four-classification analysis was 0.44, and a PLS-DA permutation test confirmed its significance ($p = 0.031$). The loadings of each BA variable on PLS-DA components 1 and 2 is shown in Figure 6b, and this demonstrates three groups of these predictors: the first with highly positive component 1 and highly negative component 2 loadings (all aromatic BAs, i.e., 2-phenylethylamine, tyramine, and tryptamine); the second with low to intermediate positive component 1 but highly positive component 2 loadings (metabolically-related putrescine, spermidine, and spermine, together with histamine); and the third with highly negative loadings on component 1, but negligible loadings on component 2 (cadaverine only). These grouped BA loadings vectors were very consistent with other observations made from the MV analysis of these data as a full six fermented food classification dataset. Specifically, they are completely reflective of the patterns of BA "markers" found in fermented FCLO, wine/vinegar, and cheese products respectively (Figure 5a).

Finally, RF analysis of this revised dataset showed that this approach had an at least reasonable level of classification success, with all (10/10) FCLO and 88% (15/17) of the fish/sauce combination samples being correctly classified; notwithstanding, only 60 and 50% of the cheese and wine/vinegar fermented food products, respectively, were.

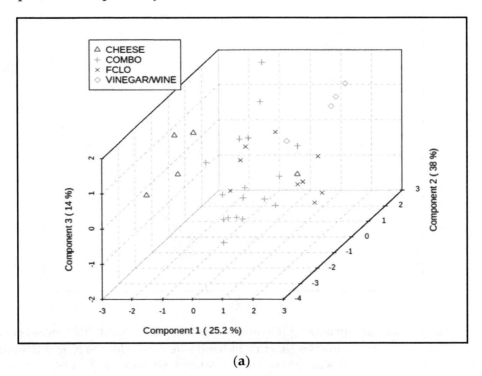

(**a**)

Figure 6. *Cont.*

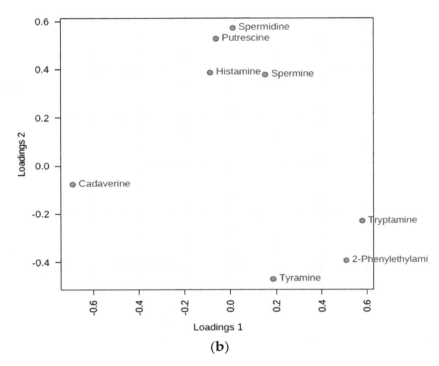

(b)

Figure 6. PLS-DA evaluation of revised dataset with combined fish, fish sauce/paste, and vegetable sauce classifications (abbreviated COMBO); CS-normalization was applied to the dataset prior to analysis. (**a**) 3D PLS-DA component 3 vs. component 2 vs. component 1 scores plot revealing some clustering of the fermented food classifications (i.e., cheese, wine/vinegar, FCLO, and COMBO). (**b**) Corresponding component 2 vs. component 1 loadings plot for this PLS-DA analysis.

3.7. Scientific Significance and Human Health Implications of Results Acquired

Results acquired from the combined applications of univariate and MV chemometrics techniques in this study clearly demonstrated that the latter strategy was valuable for distinguishing between fermented wine/vinegar products and cheeses, and the discrimination between both of these food classes from either fish, fish sauce/paste, or vegetable sauce products (or a statistical combination of them) was possible on the basis of their BA, total lipid, pH, and TA values; nevertheless, such techniques were not readily able to distinguish between the latter three fermented food classes. However, a rigorously-constrained univariate analysis method selected to overcome complications arising from intra-food classification heteroscedasticities and FDRs was able to successfully distinguish between the vegetable sauce and fish groups through significantly higher and lower levels of spermidine and 2-phenylethylamine, respectively, present in the former class. Moreover, experimental results indicated that cadaverine, tyramine, putrescine, and tryptamine concentrations may all contribute significantly towards food pH values in view of their strong positive correlations with this parameter found, together with corresponding negative ones with TA values (Figure 3b).

Moreover, BA-targeted univariate and multivariate analyses of CS-normalized data was found to be valuable for providing useful discriminatory information, which highlighted the characteristic patterns of BA biomolecules, which may be valuable for further investigations of the particular nature and/or geographic origins of fermented foods, and the mechanisms involved in their formation. Indeed, the present study found that such patterns comprised cadaverine only for cheese samples, three aromatic BAs (2-phenylethylamine, tyramine, and tryptamine), for FCLOs (sourced from fermented cod livers), and those from the sequential metabolic pathway which transforms the amino acid substrates ornithine or arginine to spermine (i.e., putrescine, spermidine, and spermine itself) for wine/vinegar products. Such idiosyncratic, fermented food product-dependent signatures for CS-normalized fermented food BA concentrations may serve to provide valuable information regarding the fermentative bacterial sources, routes involved in fermentation, and product manufacturing conditions employed for them.

For the putrescine → spermidine → spermine metabolic pathway, which was identified as representing a wine/vinegar-specific one from analysis of the CS-normalized dataset, and which accounted for >70% of total BAs in this fermented food class (Table 2), both positive or negative correlations could arise between a BA catabolite and its immediate upstream precursor, but not necessarily between the terminal spermine metabolite and that upstream of its spermidine substrate (i.e., putrescine).

With regard to toxic concentrations and health risk recommendations available in [25], it should be noted that all mean histamine levels determined in the fermented food samples tested here lie markedly below the recommended 100 ppm limit for it (with no single product exceeding this value—the highest level observed was 57 ppm in one of the fish sauce products assessed). Furthermore, with the exception of the cheese products evaluated, the mean total BA values all food groups were <200 ppm, the wine/vinegar classification substantially so (Table 2). However, although three of the cheese products tested had total BA contents of <200 ppm, two of them had levels ranging from 600–800 ppm, and therefore their dietary consumption may present a health risk for susceptible individuals.

Mean BA concentrations for the FCLO product examined ranged from 0 (histamine) to only 34 ppm (2-phenylethylamine), with the highest levels observed for the most predominant species, 2-phenylethylamine and tyramine, being 103 and 88 ppm. Since the United States of America's recommended dietary intake of health-friendly, highly unsaturated omega-3 (O-3) fatty acids (FAs) is a maximum of 1.0 g/day [41], and the oil explored here contains a mean of 29% (w/w) total O-3 FAs (predominantly the sum of eicosapentaenoic and docosahexaenoic acids) [6], then daily consumption of $100/29\% \times 1.0$ g = 3.45 g of this FCLO product would provide estimated absolute maximal daily intake levels of 3.45×103 μg = 355 μg, and 3.45×88 μg = 304 μg of 2-phenylethylamine and tyramine, respectively. Based on the 10 samples of this product analyzed, estimated mean daily intakes of these BAs will be 111 and 95 μg only. Therefore, it appears that daily consumption of this product at the recommended U.S.-recommended dosage levels will certainly not provide any health risks to consumers, even if they are susceptible to the adverse effects experienced by their excessive intake (e.g., migraines induced by 2-phenylethylamine).

As noted above, one potentially important health benefit offered by the ingestion of dietary BAs is their novel antioxidant properties, both for the prevention of food spoilage during storage or transport episodes, but also in vivo following their ingestion. Indeed, our laboratory recently explored the powerful antioxidant capacities of BA-containing natural FCLO products, and their resistivities to thermally-mediated oxidative damage to unsaturated FAs therein, particularly O-3 PUFAs [6]. These marine oil products, which arise from the pre-fermentation of cod livers (Section 2), were indeed found to display a very high level of antioxidant activity, and PUFAs therein were also more resistant towards thermally-mediated peroxidation than other natural cod liver oil products evaluated. Resonances assignable to aromatic BAs, specifically those arising from 2-phenylethylamine and tyramine, were directly observable in the ^{1}H NMR profiles of ca. 1/3 (v/v) diluted solutions of these products with C^2HCl_3. Additionally, corresponding spectra acquired on both ^2H_2O and $C^2H_3O^2H$ extracts of these oils confirmed the presence of both these BAs, together with a series of others, both aromatic and aliphatic. In the present study, mean concentrations of 2-phenylethylamine and tyramine detectable in these products were found to be 34 and 24 ppm, respectively. Antioxidant actions of the phenolic BA antioxidant tyramine found in this FCLO product may be explicable by its chain-breaking antioxidant effects, and this may offer contributions towards the potent resistance of PUFAs, particularly O-3 FAs, present therein. However, in view of the absence of a phenolic function in 2-phenylethylamine, its antioxidant potential is likely to involve an alternative radical-scavenging mechanism, presumably that involving O_2-consuming carbon-centered pentadienyl radical species, as found in [22,23].

The TBARS method employed here to determine the lipid peroxidation status of fermented foods, which involved an extended low temperature equilibration process [28], successfully avoids the artefactual generation of TBA-reactive aldehydes, including malondialdehyde (MDA), during commonly-employed alternative protocols for this assay system, which generally involve a short (ca. 10–15 min) heating stage at 95–98 °C in order to develop the monitored pink/red chromophore rapidly. However, from an analysis of TBARS and (TBARS):(total lipid) ratios determined on the preliminary FCLO-excluded fermented food samples, there appears to be only little evidence for the ability of BAs to offer any protection against lipid peroxidation in such products. Although Table 3 shows that the above ratio is significantly greater in the wine/vinegar group, this observation is perceived to be derived from their very low lipid contents, and the presence of a range of non-MDA TBA-reactive aldehydes present therein, including acetaldehyde and acrolein, for example, although these are also lipid oxidation products. Moreover, despite taking steps to avoid the artefact-generating heating stage of this assay, this test still remains poorly specific in view of the reactions of a variety of non-aldehydic substrates to react with it to form interfering chromophores, which also absorb at a monitoring wavelength of 532 nm. Nevertheless, TBARS level appeared to be positively correlated with fermented food spermidine concentration (Figure 2), and both this lipid peroxidation index and its lipid-normalized value appeared to be positively correlated with fermented food histamine content (loading on PC2, Table 4). However, in view of the many complications associated with this TBARS lipid peroxidation index, which offers only a very limited and still often erroneous viewpoint on the highly complex lipid peroxidation process [42], such observations cannot be rationally considered at this stage. As expected, the lipid-normalized TBARS value was negatively correlated with total lipid content (Figure 3b). The latter variable also appeared to be negatively correlated with histamine and putrescine levels (loading on PC5, Table 4). Unfortunately, results from unspecific TBARS assays are still widely employed as important quality indices throughout the food industry.

One quite surprising observation made in the current study was the detection of lipids, albeit at low levels, in wine and vinegar samples. Notwithstanding, as noted above, FAs have been detected in Zhenjiang aromatic vinegar products at similar contents to those found here [32]. Furthermore, Yunoki et al. [43] explored the FA constituents of some commercially-available red wine products, and found that lipid constituent concentrations varied from 27 to 96 mg/100 mL for $n = 6$ domestic (Japanese) wines, and 31 to 56 mg/100 mL for $n = 6$ foreign products, and that a total of 12 different FAs were detectable, mainly saturated ones. Although the extraction method described in the latter report was a 2:1 chloroform:methanol (Folch) one that targets non-polar triacylglycerols (TAGs) and more polar phospholipids, it is likely that the FAs detectable in the wine/vinegar products explored here, and also those present in Zhenjiang aromatic vinegars [32], are present as free non-glycerol-esterified species and their corresponding anions, and this would account for their higher levels detectable in these studies than those reported in [43]. Indeed, fermentation processes readily induce the hydrolysis of TAGs to free FAs, together with mono- and diacylglycerol adducts, and free glycerol [44]; such FAs will be expected to contribute towards the food pH values determined here. Similarly, Phan et al. [45] found a broad spectrum of lipidic species, specifically TAGs, polar lipids, free FAs, sterols, and cholesterol esters present in pinot noir wines.

The official AOAC gravimetric method for lipid determination employed in the current study involves an acid hydrolysis step involving HCl in any case, followed by extraction with mixed ethers, i.e., both diethyl and petroleum ethers. Hence, the HCl added will be sufficient to hydrolyze any residual TAGs present to free FAs and glycerol, and also fully protonate the former so that they are extractable as such into ether solvents. Indeed, it has been demonstrated that such free FAs are readily soluble and extractable into these ether solvent systems [46,47]. Hence, the passage of lipidic species from grapes and/or micro-organisms to finalized bottled wine and vinegar products has been confirmed in further investigations.

Interestingly, ^{1}H NMR analysis of [2]H$_2$O extracts of the FCLO product investigated found proportionately high concentrations of free FAs and free glycerol therein (data not shown). These FAs were mainly present as PUFAs, as would be expected from the overall lipid composition of this product which contains high levels of omega-3 FAs as TAG species prior to fermentation induction. This observation is fully consistent with the ability of lactobacilli-mediated fermentation processes to partially hydrolyze TAGs in such a product. High levels of the short-chain organic acids propionic and acetic acids (as their propionate and acetate anions in neutral solution media), both lactobacilli fermentation catabolites, the former arising from the metabolic reduction of lactate [48], were also detectable in these extracts. These results will be reported in detail elsewhere.

4. Limitations of the Study

One important limitation of this study is the limited sample sizes of some of the fermented food sampling classes incorporated into our primary experimental design. This was largely a consequence of only small numbers of differing fermented food products being available for purchase locally, for example vegetable sauce and fish products. However, it should be noted that the cheese and wine/vinegar classifications had BA contents and patterns which markedly contrasted with those of the other fermented food groups evaluated. These differences, along with those for other food quality markers observed (Table 3, Figures 1 and 3–5), were found to be very highly statistically significant, even with these limited sample sizes. Hence, this did not present a major constraining issue. Moreover, the performance of additional MV analyses on a revised model including a combined fish, fish sauce/paste, and vegetable sauce classification (on the basis of only a limited level of significant differences between them) with $n = 17$ overall served to overcome this problem (Figure 6), and this incentive did not distract from the main objectives and focus of the investigation in view of their predominant MV similarities in BA contents. However, univariate analysis found that the mean spermidine concentration was significantly higher in fermented vegetable sauces than it was in corresponding fish products (Table 3), and vice-versa for mean 2-phenylethylamine levels (Table 7). Further evidentiary support was provided by data analysis strategies applied, which were highly rigorous, and included the preliminary tracking of sample outliers. Furthermore, rigorous Welch tests were implemented for the ANOVA models employed, and either Bonferroni or FDR corrections were applied for *post-hoc* "between-fermented food classification" tests in order to circumvent potential problems with false positives (type I errors).

Another limitation of the current study was the unavailability of differing manufacturing sources of FCLO products, and therefore unlike other fermented food products assessed here, statistical evaluations involved an investigation of 10 separate, randomly-selected batches of a single product, both separately (Table 5) and jointly with all other classes involved in the primary statistical analysis conducted (Table 7, and Figures 5 and 6). However, the very wide between-batch variance of all FCLO samples explored facilitated this approach.

Finally, one further limitation is the poor specificity and interpretability of the TBARS method employed for the quality assessment of fermented food products here, specifically for assessments of their degrees of lipid peroxidation. However, one major precautionary step was taken in this study to minimize problems and potential interferences in this assay system, and this involved the avoidance of an aldehydic artefact-forming heating stage. Future investigations of the lipid oxidation status of fermented foods should therefore employ more reliable and specific methodologies such as those involving high-resolution ^{1}H NMR analysis for the direct, simultaneous, multicomponent analysis of a series of both primary and secondary lipid oxidation products, e.g., conjugated hydroperoxydienes and their aldehydic fragmentation products, respectively. This protocol may be applied directly to solution-state products, or indirectly to either aqueous or lipid/deuterochloroform extracts of fermented food products.

5. Conclusions

This study demonstrated that almost all fermented foods tested had total BA levels which lay below the maximum recommended values for them. A composite application of univariate and MV chemometrics techniques clearly demonstrated that the MV approach applied was valuable for discriminating between fermented wine/vinegar products and cheeses, and the distinction between these two fermented food classes and a combination of fish, fish sauce/paste, and vegetable sauce products. Further MV analysis performed on CS-normalized BA profiles revealed distinctive patterns for cheese (cadaverine only), FCLOs (the aromatic BAs 2-phenylethylamine, tyramine, and tryptamine), and wine/vinegar products (pathway-associated putrescine, spermidine, and spermine). Such distinctive signatures for fermented food BA contents may offer useful information regarding the nature of, and regulatory conditions employed for, fermentation processes utilized during their commercial production.

The simultaneous untargeted analysis of eight or more BAs using the LC-MS/MS analysis strategy employed here offers major advantages which are unachievable by alternative, more targeted techniques with the ability to determine only single or very small numbers of chemometrically-important analytes. Notably, the diagnostic potential of a series of n (for example, five or more) BA content analyte variables in a MV chemometrics investigation offers major advantages over the analytical acquisition of only a single possible marker. Indeed, food sample patterns of BAs and related food quality indices, which are characteristic of a particular fermented food product classification, will be expected to provide a much higher level of statistical power, reliability, and confidence concerning the accurate distinction between these classifications, and their accurate and selective assignment to one of them, than that discernable from a single BA analyte level only. Secondly, the patterns of BAs and associated food quality criteria determined, together with their correlations to particular factors or components (predominantly linear, but occasionally quadratic or higher combinations of predictor BA and supporting variables), may potentially serve to supply extensive information regarding the sources of such BAs, bacterial, commercial, or otherwise.

Author Contributions: J.Z. was responsible for the manufacture of FCLO samples and the random distribution of these samples from different batches for analysis; he was also responsible for surveys of the availabilities, and purchases of all fermented food products from US retail outlets, together with their distribution for analysis. M.G. and B.C.P. monitored and validated all chemical analysis methods for fermented food products, involving those for BAs, TA and pH values, total lipid contents, and TBARS levels. M.G. was responsible for study experimental design, and also performed the univariate and MV chemometrics analyses of analytical datasets acquired, with assistance from B.C.P. M.G. also prepared, drafted, and finalized the manuscript for submission purposes. J.Z., B.C.P., and M.G. reviewed and edited manuscript drafts, and also contributed towards the interpretation of experimental results obtained. M.G. also fully supervised the complete study. All authors have read and agreed to the published version of the manuscript.

Acknowledgments: All authors are very grateful to the Weston A. Price Foundation (DC, USA) for part-funding the study, and to Midwest Laboratories (13611 B Street, Omaha, NE 68144-3693, USA) for performing the laboratory analysis of BAs. We are also grateful to Dave Wetzel of Green Pastures Products Inc. (NE, USA) for valuable discussions.

References

1. Vidal-Carou, M.C.; Ambatle-Espunyes, A.; Ulla-Ulla, M.C.; Marine Â-Font, A. Histamine and tyramine in Spanish wines: Their formation during the winemaking process. *Am. J. Enolog. Viticul.* **1990**, *41*, 160–167.
2. Izquierdo-Pulido, M.; Marine Â-Font, A.; Vidal-Carou, M.C. Biogenic amine formation during malting and brewing. *J. Food Sci.* **1994**, *59*, 1104–1107. [CrossRef]
3. Perpetuini, G.; Tittarelli, F.; Battistelli, N.; Arfelli, G.; Suzzi, G.; Tofalo, R. Biogenic amines in global beverages. In *Biogenic Amines in Food: Analysis, Occurrence and Toxicity*; Saad, B., Tofalo, R., Eds.; The Royal Society of Chemistry: Cambridge, UK, 2020; pp. 133–156.
4. Halaasz, A.; Barath, A.; Simon-Sakardi, L.; Holzapel, W. Biogenic amines and their production by microorganisms in food. *Trends Food Sci. Technol.* **1994**, *5*, 42–49. [CrossRef]
5. Tittarelli, F.; Perpetuini, G.D.; Gianvito, P.; Tofalo, R. Biogenic amines producing and degrading bacteria: A snapshot from raw ewes' cheese. *LWT Food Sci. Technol.* **2019**, *101*, 1–9. [CrossRef]
6. Percival, B.C.; Wann, A.; Zbasnik, R.; Schlegel, V.; Edgar, M.; Zhang, J.; Ampem, G.; Wilson, P.; Le-Gresley, A.; Naughton, D.; et al. Evaluations of the peroxidative susceptibilities of cod liver oils by a ^{1}H NMR analysis strategy: Peroxidative resistivity of a natural collagenous and biogenic amine-rich fermented product. *Nutrients* **2020**, *12*, 3075. [CrossRef]

7. Alvarez, M.A.; Moreno-Arribas, M.V. The problem of biogenic amines in fermented foods and the use of potential biogenic amine-degrading microorganisms as a solution. *Trends Food Sci. Technol.* **2014**, *39*, 146–155. [CrossRef]

8. Vandekerckove, P. Amines in dry fermented sausage: A research note. *J. Food Sci.* **1977**, *42*, 283–285. [CrossRef]

9. Rhee, J.E.; Rhee, J.H.; Ryu, P.Y.; Choi, S.H. Identification of the cadBA operon from *Vibrio vulnificus* and its influence on survival to acid stress. *FEMS Microbiol. Lett.* **2002**, *208*, 245–251. [CrossRef] [PubMed]

10. Lee, Y.H.; Kim, B.H.; Kim, J.H.; Yoon, W.S.; Bang, S.H.; Park, Y.K. CadC has a global translational effect during acid adaptation in Salmonella enterica serovar typhimurium. *J. Bacteriol.* **2007**, *189*, 2417–2425. [CrossRef]

11. Van de Guchte, M.; Serror, P.; Chervaux, C.; Smokvina, T.; Ehrlich, S.D.; Maguin, E. Stress responses in lactic acid bacteria. *Antonie van Leeuwenhoek* **2002**, *82*, 187–216. [CrossRef]

12. D'Aniello, E.; Periklis, P.; Evgeniya, A.; Salvatore, D.A.; Arnone, M.I. Comparative neurobiology of biogenic amines in animal models in deuterostomes. *Front. Ecol. Evolut.* **2020**, *8*, 322.

13. Bendera, R.; Wilson, L.S. The regulatory effect of biogenic polyamines spermine and spermidine in men and women. *Open J. Endocrin. Metab. Dis.* **2019**, *9*, 35–48. [CrossRef]

14. Tabor, C.; Tabor, H. Polyamines. *Ann. Rev. Biochem.* **1984**, *53*, 749–790. [CrossRef] [PubMed]

15. Premont, R.T.; Gainetdinov, R.R.; Caron, M.G. Following the trace of elusive amines. *Proc. Natl. Acad. Sci. USA* **2001**, *98*, 9474–9475. [CrossRef]

16. Santos, M.H.S. Biogenic amines: Their importance in foods. *Int. J. Food Microbiol.* **1996**, *29*, 213–231. [CrossRef]

17. Arena, M.E.; de Nadra, M.C.M. Biogenic amine production by Lactobacillus. *J. Appl. Microbiol.* **2001**, *90*, 158–162. [CrossRef]

18. Taylor, S.L. Histamine food poisoning: Toxicology and clinical aspects. *Crit. Rev. Toxicol.* **1986**, *17*, 91–128. [CrossRef]

19. Hotchkiss, J.H.; Scanlan, R.A.; Libbey, L.M. Formation of bis (hydroxyalkyl)-N-nitrosamines as products of the nitrosation of spermidine. *J. Agric. Food Chem.* **1977**, *25*, 1183–1189. [CrossRef]

20. Livingston, M.G.; Livingston, H.M. Monoamine oxidase inhibitors. An update on drug interactions. *Drug Saf.* **1996**, *14*, 219–227. [CrossRef]

21. Prell, G.D.; Mazurkiewicz-Kwilecki, I.M. The effects of ethanol, acetaldehyde, morphine and naloxone on histamine methyltransferase activity. *Prog. Neuro-Psychopharmacol.* **1981**, *5*, 581–584. [CrossRef]

22. Lonvaud-Funel, A. BAs in wines: Role of lactic acid bacteria. *FEMS Microbiol. Lett.* **2001**, *199*, 9–13. [CrossRef] [PubMed]

23. Jastrzębska, A.; Piasta, A.; Kowalska, S.; Krzemiński, M.; Szłyk, E. A new derivatization reagent for determination of biogenic amines in wines. *J. Food Comp. Anal.* **2016**, *48*, 111–119. [CrossRef]

24. Grootveld, M. *Metabolic Profiling: Disease and Xenobiotics*; Issues in Toxicology Series; Royal Society of Chemistry: Cambridge, UK, 2014; ISBN 1849731632.

25. Nout, M.J.R. Food Technologies: Fermentation. *Encycl. Food Saf.* **2014**, *3*, 168–177. [CrossRef]

26. Cosmos, E.; Heberger, K.; Simon-Sarkadi, L. Principal component analysis of biogenic amines and polyphenols in Hungarian wines. *J. Agric. Food Chem.* **2002**, *50*, 3768–3774.

27. Quilliam, M.A.; Blay, P.; Hardstaff, W.; Wittrig, R.E.; Bartlett, V.; Bazavan, D.; Schreiber, A.; Ellis, R.; Fernandes, D.; Khalaf, F.; et al. LC-MS/MS analysis of biogenic amines in foods and beverages. In Proceedings of the 123rd AOAC Annual Meeting and Exposition, Philadelphia, PA, USA, 13–16 September 2009.

28. Witte, V.C.; Krause, G.F.; Bailey, M.E. A new extraction method for determining 2-thiobarbituric acid value of pork and beef during storage. *J. Food Sci.* **1970**, *35*, 582–585. [CrossRef]

29. FAO. Appendix XXV, Replies of the 6th Session of the CCMMP to Questions Referred by the 23rd Session of the CCMAS. Available online: http://www.fao.org/3/j2366e/j2366e25.htm (accessed on 27 September 2020).

30. Chong, J.; Soufan, O.; Li, C.; Caraus, I.; Li, S.; Bourque, G.; Wishart, D.S.; Xia, J. MetaboAnalyst 4.0: Towards more transparent and integrative metabolomics analysis. *Nucleic Acids Res.* **2018**, *46*, W486–W494. [CrossRef]

31. Pelckmans, K.; De Brabanter, J.; Suykens, J.A.K.; De Moor, B. Handling missing values in support vector machine classifiers. *Neural Netw.* **2005**, *18*, 684–692. [CrossRef]

32. Zhao, C.; Xia, T.; Du, P.; Duan, W.; Zhang, B.; Zhang, J.; Zhu, S.; Zheng, Y.; Wang, M.; Yu, Y. Chemical

composition and antioxidant characteristic of traditional and industrial Zhenjiang aromatic vinegars during the aging process. *Molecules* **2018**, *23*, 2949. [CrossRef]

33. Grootveld, M.; Percival, B.C.; Leenders, J.; Wilson, P. Potential adverse public health effects afforded by the ingestion of dietary lipid oxidation product toxins: Significance of fried food sources. *Nutrients* **2020**, *12*, 974. [CrossRef]

34. Fugelsang, K.C. *Wine Microbiology*; Chapman & Hall: New York, NY, USA, 1997.

35. Romano, P.; Suzzi, G.; Turbanti, L.; Polsinelli, M. Acetaldehyde production in *Saccharomyces cerevisiae* wine yeasts. *FEMS Microbiol. Lett.* **1994**, *118*, 213–218. [CrossRef]

36. Liu, S.-Q.; Pilone, G.J. An overwiew of formation and roles of acetaldehyde in winemaking with emphasis on microbiological implications. *Int. J. Food Sci. Technol.* **2000**, *35*, 49–61. [CrossRef]

37. Uebelacker, M.; Lachenmeier, D.W. Quantitative determination of acetaldehyde in foods using automated digestion with simulated gastric fluid followed by headspace gas chromatography. *J. Anal. Meth. Chem.* **2011**. [CrossRef] [PubMed]

38. Zhao, G.; Kuang, G.; Li, J.; Hadiatullah, H.; Chen, H.; Wang, X.; Yao, Y.; Pan, Z.-H.; Wang, Y. Characterization of aldehydes and hydroxy acids as the main contribution to the traditional Chinese rose vinegar by flavor and taste analyses. *Food Res. Int.* **2020**, *129*, 108879. [CrossRef] [PubMed]

39. Vijisha, K.R.; Muraleedharan, K. The pKa values of amine based solvents for CO_2 capture and its temperature dependence—An analysis by density functional theory. *Int. J. Greenh. Gas. Cont.* **2017**, *58*, 62–70. [CrossRef]

40. Flydal, M.I.; Martinez, A. Phenylalanine hydroxylase: Function, structure, and regulation. *IUBMB Life* **2013**, *65*, 341–349. [CrossRef] [PubMed]

41. Harris, W.S. Fish oil supplementation: Evidence for health benefits. *Cleveland Clin. J. Med.* **2004**, *71*, 208–221. [CrossRef]

42. Janero, D.R. Malondialdehyde and thiobarbituric acid-reactivity as diagnostic indices of lipid peroxidation and peroxidative tissue injury. *Free Rad. Biol. Med.* **1990**, *9*, 515–540. [CrossRef]

43. Yunoki, K.; Tanji, M.; Murakami, Y.; Yasui, Y.; Hirose, S.; Ohnishi, M. Fatty acid compositions of commercial red wines. *Biosci. Biotechnol. Biochem.* **2004**, *68*, 2623–2626. [CrossRef]

44. Anihouvi, V.B.; Kindossi, J.M.; Hounhouigan, J.D. Processing and quality characteristics of some major fermented fish products from Africa: A critical review. *Int. Res. J. Biol. Sci.* **2012**, *1*, 72–84.

45. Phan, Q.; Tomasino, E.; Osborne, J. Influences of yeast product addition and fermentation temperature on changes in lipid compositions of pinot noir wines (abstract). In Proceedings of the 3rd Edition of International Conference on Agriculture and Food Chemistry, Rome, Italy, 23–24 July 2018.

46. Markley, K.S.; Sando, C.E.; Hendricks, S.B. Petroleum ether-soluble and ether-soluble constituents of grape pomace. *J. Biol. Chem.* **1938**, *123*, 641–654.

47. Sun, R.C.; Tomkinson, J. Comparative study of organic solvent and water-soluble lipophilic extractives from wheat straw I: Yield and chemical composition. *J. Wood Sci.* **2003**, *49*, 47–52. [CrossRef]

48. Zhang, C.; Brandt, M.J.; Schwab, C.; Gänzle, M.G. Propionic acid production by cofermentation of *Lactobacillus buchneri* and *Lactobacillus diolivorans* in sourdough. *Food Microbiol.* **2010**, *27*, 390–395. [CrossRef] [PubMed]

11

Quality Assessment of Fresh Meat from Several Species Based on Free Amino Acid and Biogenic Amine Contents during Chilled Storage

Mehdi Triki [1],*, Ana M. Herrero [2], Francisco Jiménez-Colmenero [2] and Claudia Ruiz-Capillas [2],*

[1]   Ministry of Public Health, P.O. Box 42, Doha, Qatar
[2]   Department of Products, Institute of Food Science, Technology and Nutrition, ICTAN-CSIC, Ciudad Universitaria, 28040 Madrid, Spain; ana.herrero@ictan.csic.es (A.M.H.); fjimenez@ictan.csic.es (F.J.-C.)
*   Correspondence: mrmehditriki@gmail.com (M.T.); claudia@ictan.csic.es (C.R.-C.)

Abstract: This paper studies the changes that occur in free amino acid and biogenic amine contents of raw meats (beef, pork, lamb, chicken and turkey) during storage (2 °C, 10 days). The meat cuts samples were harvested from a retail outlet (without getting information on the animals involved) as the following: Beef leg (four muscles), pork leg (five muscles), lamb leg (seven muscles), turkey leg (four muscles), and chicken breast (one muscle). Meat composition varied according to meat types. In general, pH, microbiology counts, biogenic amine (BA), and free amino acid (FAA) contents were also affected by meat types and storage time ($p < 0.05$). Chicken and turkey presented the highest levels ($p < 0.05$) of FAAs. Total free amino acids (TFAA) were higher ($p < 0.05$) in white meats than in red ones. The behavior pattern, of the total free amino acids precursors (TFAAP) of Bas, was saw-toothed, mainly in chicken and turkey meat during storage, which limits their use as quality indexes. Spermidine and spermine contents were initially different among the meats. Putrescine was the most prevalent BA ($p < 0.05$) irrespective of species. In general, chicken and turkey contained the highest ($p < 0.05$) levels of BAs, and TFAAP of BAs. In terms of the biogenic amine index (BAI), the quality of chicken was the worst while beef meat was the only sample whose quality remained acceptable through the study. This BAI seems to be more suitable as a quality index for white meat freshness than for red meat, especially for beef.

Keywords: meat species; free amino acid; biogenic amines; quality index

1. Introduction

Meat and meat products constitute an important protein group of foods, that can be consumed directly or as products after undergoing different processes. Consumers nowadays are asking for safe and high quality meat products. This quality is influenced by various factors, and complex interactions between the biological traits of the live animal, including mainly the biological processes that occur postmortem as muscles conversion to meat, processing and storage phases, etc. [1,2]. These meat and meat products, especially when they are fresh, undergo spoilage even during refrigerated storage [3]. This deterioration is associated with major proteolysis and microbial growth. Due to proteolysis, peptides, dipeptides, and free amino acids (FAA) are formed and used by microorganisms for their growth. Following these processes, different compounds, including biogenic amines (BA), are formed by amino acid decarboxylase action from microbial origin (Figure 1). The biogenic amine content depends on a number of interrelated factors such as the raw material (meat composition, pH, handling and hygienic conditions, etc.), additives (salt, sugar, nitrites), etc. These factors affect free amino acid availability, microbiological aspects (bacterial species and strain, bacterial growth, etc.), technical processing of the meat or meat products (e.g., steaks, roasts and hams, and ground, restructured, comminuted, fresh, cooked, smoked, and fermented meats, etc.), and storage conditions (time/temperature, packaging, temperature abuse, etc.). The combined action of all these factors will determine the final biogenic amine profile and concentrations by directly or indirectly determining substrate and enzyme presence and activity.

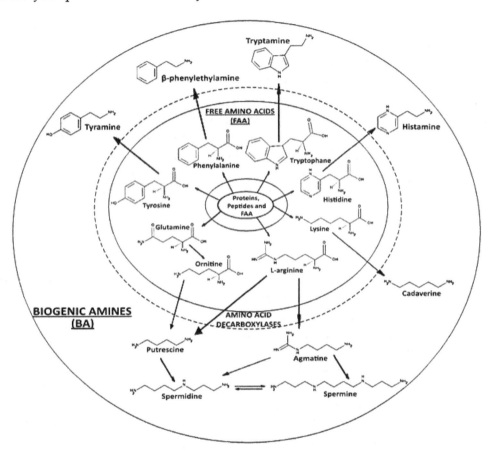

Figure 1. Biogenic amine formation from free amino acids.

Therefore, biogenic amines are of particular concern in food hygiene. Indeed, they have been used as quality indices, mainly in fish and meat, under different processing and storage conditions, whether considered individually or in combined forms [4–9]. In this regard, Hernández-Jover et al. [10] suggested a biogenic amine index (BAI) as a sum of tyramine, histamine, putrescine and cadaverine, with four-scale classification intended for cooked meat products which were based on the sum of total BA concentration in the BAI. Tyramine is also widely used individually as a quality indicator for vacuum-packed beef and cooked ham; this is also the case for spermidine and spermine [11]. Other authors as Vinci & Antonelli [12] proposed the use of cadaverine and tyramine concentrations to assess beef and chicken deterioration during storage. Moreover, some BAs were studied for their potential toxicity for consumers, especially tyramine, histamine, cadaverine and putrescine [5,13,14]. Based on these considerations, regulations have been introduced to limit BA intake levels in various kinds of food [15,16].

Consequently, the determination of biogenic amines and free amino acid fractions can provide useful information for the industry regarding freshness or spoilage and sanitary quality of fresh muscle that could be consumed directly or used as raw material for meat products preparation. High concentrations of certain amines in food may be interpreted as a consequence of poor quality of the raw materials used, contamination, or inappropriate conditions during food processing and storage.

In this regard, numerous studies were carried out in order to understand how FAA and biogenic amine formation is associated with microorganisms development in various meat products, mainly in fermented ones [5,7,8,17,18], and rarely in fresh meat [12]. On the other hand, "already-formed BAs" in the raw meat materials cannot be destroyed by thermal action during meat product processing or cooking, and this can lead to higher amine levels at the end of the pool [5,9]. However, some authors reported that BA formation not only depends on the conservation method used (refrigeration, protective atmosphere, etc.) or the type of processing (fermented, cooked, fresh, etc.), but it also depends on the type of raw material or animal species studied. Delgado-Pando et al. [19] and Triki et al. [20,21] observed differences in BA levels in reformulated (pork) frankfurters and fresh (beef) "merguez" sausages. Thus, FAA and BA production and microbial growth in fresh meat is of great interest for understanding and controlling the influence of raw meat as a factor in the final quality of fresh meat products (hamburger, fresh sausages, etc.). This approach might improve the safety as well as hygiene aspects of raw meat whether when they are used in the preparation for other meat products or employed for direct consumption or during the chilled storage of such foodstuffs. The aim of this study is, then, to assess the changes that take place in free amino acid and biogenic amine contents during chilled storage of fresh meats from some of the most frequently consumed species (beef, pork, lamb, chicken and turkey), which are used in the preparation of meat products as raw material.

2. Material and Methods

2.1. Fresh Meat Samples

Approximately 4 kg of commercial cuts of each type of fresh lean muscle meat from five species were purchased from a local supermarket. Leg cuts were taken from the four species: Beef (*Rectus femoris* M., *Semitendinosus* M., *Flexor digitorum longus* M., *Gastrocnemius* M.), pork (*Biceps femoris* M., *Semimembranosus* M., *Semitendinosus* M., *Gracilis* M., *Adductor* M.), turkey (*Flexor perforans* M., *Gastrocnemius pars external* M., *Gastrocnemius pars internal* M., *Fiburalis longus* M.), and lamb (*Quadriceps femoris* M., *Biceps femoris* M., *Semimembranosus* M., *Gluteus medius* M., *Gastrocnemius* M., *Adductor* M., *Semitendinosus* M.). Breast cuts were taken from chicken (*Pectoralis Major*). Two hundred to two hundred and fifty grams of each type of meat cut were representative of the pieces. Then meat cuts were placed on expanded polystyrene (EPS) trays (Type 89 white SPT-Linpac Packaging Pravia, S.A., Pravia, Spain) and covered with oxygen-permeable cling film (LINPAC Plastics, Pontivy, France) in aerobic conditions. From the 15 trays of each meat type that were kept in chilled storage (0 to 4 °C), three were taken periodically for further analysis. First was taken the sample for the microbiological analysis and then the sample was homogenized for the other analysis (protein content, pH, FAA and BA). Samples were assessed at 0, 3, 6, and 10 days of chilled storage.

2.2. Protein Content and pH Determination

Protein content was measured in quadruplicate with a LECO FP-2000 Nitrogen Analyzer (Leco Corporation, St. Joseph, MI, USA). For pH determination, 10 g homogenate samples in 100 mL of distilled water were prepared using a pH meter (827 pH Lab Methrom, Herisau, Switzerland). Both analyses followed the methodology used by Triki et al. [20]. Three measurements were performed per sample.

2.3. Microbiological Analysis

Ten grams of each representative sample were taken and placed in a sterile plastic bag with 90 mL of peptone water (0.1%) with 0.85% NaCl. After 2 min in a stomacher blender (Stomacher Colworth 400, Seward, UK), appropriate decimal dilutions were pour-plated (1 mL) on the following media: Plate Count Agar (PCA) (Merck, Darmstadt, Germany) for the total viable count (TVC) (30 °C for 72 h); De Man, Rogosa, Sharpe Agar (MRS) (Merck, Darmstadt, Germany) for lactic acid bacteria (LAB) (30 °C for 3–5 days); and Violet Red Bile Glucose Agar (VRBG) (Merck, Darmstadt, Germany) for

Enterobacteriaceae (37 °C for 24 h). All microbial counts were converted to logarithms of colony-forming units per gram (Log cfu/g), following the methodology used by Triki et al. [20].

2.4. Determination of Free Amino Acids (FAA)

Free amino acids extracts were prepared with 5 g of fresh meat samples from each species. They were homogenized with 10 mL of perchloric acid 6% (*w/v*) (to extract the FAA and precipitate the proteins and peptides) in an Ultraturrax homogenizer (IKA-Werke, Janke, & Kunkel, Staufen, Germany) then centrifuged at 27,000× g (Sorvall RTB6000B, DuPont, Wilmington, DE, USA) for 10 min at 4 °C. 2 mL of KOH 1M were added to the centrifugation tube and the whole was centrifuged again. Afterwards, the supernatant was filtered through a Millipore filter (45 μm) (Millipore, Ireland) and put into vials until use [6].

Free amino acids (FAA) were determined by cation-exchange chromatography, using a Biochron 20 automatic amino acid analyser (Amersham Pharmacia LKB, Biotech Biocom, Uppsala, Sweden) with an Ultropac high-resolution cation-exchange resin column (9 ± 0.5 μm particle size, Pharmacia, Biotech) 200 × 4.6 mm. Amino acids were determined and measured using a ninhydrin derivative reagent at 570 nm, while proline was measured at 440 nm. It should be noted that the derivatization used did not allow the determination of tryptophan. Results are means of at least three determinations.

Total FAA precursors (TFAAP) of BA was calculated by summing the levels of the following FAA: Tyrosine + Phenylalanine + Histidine + Lysine + Arginine

Total FAA (TFAA) was calculated by summing the levels of the following FAAs: Aspartic acid + Threonine+ Serine + Glutamic acid + Glycine + β alanine + Cysteine + Valine + Methionine + Isoleucine + Leucine + TFAAP

2.5. Determination of Biogenic Amines (BA)

Tyramine, phenylethylamine, histamine, putrescine, cadaverine, tryptamine, agmatine, spermidine, and spermine were determined using an acid based extraction prepared with trichloroacetic acid (7.5%) in fresh meat samples from each specie. They were analyzed in a HPLC model 1022 with a Pickering PCX 3100 post-column system (Pickering Laboratories, Mountain View, Ca, USA) following the methodology of Triki et al. [22] by ion-exchange chromatography. Briefly, 15 g of/from each sample were mixed with 30 mL of 7.5% trichloroacetic acid in an omnimixer (Omni Internacional, Waterbury, CT, USA) (20,000 rpm, 3 min) and centrifuged at 5000 g for 15 min at 4 °C in a desktop centrifuge (Sorvall RTB6000B, DuPont, Wilmington, DE, USA) (for proteins and peptides precipitation and BA extraction in the supernatant). The supernatants were filtered through a Whatman No. 1 filter, passed back through a 0.22 μm Nylon filter (Millipore, Ireland), and then placed in opaque vials in the auto-sampler of the HPLC. The results are averages of at least 3 determinations.

Biogenic amine index (BAI) was calculated by summing tyramine, histamine, putrescine and cadaverine levels in the different meat types according to Hernández-Jover et al. [10]. When one of the BA involved was not detected (ND), its value was considered as being 0. BAI < 5 mg/kg means good meat quality; between 5–20 mg/kg means acceptable meat quality; between 20–50 mg/kg means poor meat quality; and BAI > 50 mg/kg means spoiled meat.

2.6. Statistical Analysis

A One-way ANOVA analysis of variance was performed in order to evaluate the statistical significance ($p < 0.05$) of the meat type effect. Analysis of the main effect of each independent variable and any interaction between them were carried out with two-way ANOVA, which was performed as a function of meat type and storage days, using SPSS Statistics general linear model (GLM) procedure (v.14, SPSS Inc.; Chicago, IL, USA). The types of meat and storage time were assigned as fixed effects and the replication (samples were taken from meat types of different animals) was considered as a random effect. Least squares differences were used for comparison between the mean values among meat types and Tukey's HSD (honestly significant difference) test was used to identify significant differences ($p < 0.05$) between sample type and storage time. The error terms used throughout this study are standard deviations (SD). For the presented tables throughout the study: Different superscript letters in the same row indicate significant difference ($p < 0.05$) between storage days for the same meat type. Different superscript numbers in the same column indicate significant difference ($p < 0.05$) between meat types for the same storage day.

3. Results and Discussion

3.1. Protein Content and pH

Protein contents of meat (pork, 21.30 ± 0.16; lamb 19.32 ± 0.39; beef, 21.78 ± 0.25; turkey 20.02 ± 0.21; chicken 25.55 ± 0.36) were within the normal ranges for each species and type of meat cut [23]. Differences between protein amounts were due to the nature of species and cuts. As reported previously, protein is an important precursor of various compounds involved in BA formation.

The initial pH levels were significantly higher in pork (6.70) followed by turkey, chicken, lamb and beef (Table 1). This variance could be due to post mortem metabolism difference between species. As a matter of fact, both intrinsic (species, animal age, type of muscle and position of the muscle, concentration of glycogen etc.) and extrinsic (pre-slaughter stress, slaughter conditions, post-slaughter handling and temperature) factors can affect the extent of post-mortem glycolysis, and consequently the ultimate pH [1,24].

Table 1. pH values of the fresh different meat types during chilled storage at 2 °C.

Meat Type	Chilling Storage at 2 °C			
	Day 0	Day 3	Day 6	Day 10
Pork leg (5 muscles)	6.70 ± 0.02 [c5]	5.89 ± 0.02 [a2]	6.26 ± 0.03 [b3]	6.32 ± 0.01 [b3]
Lamb leg (7 muscles)	5.90 ± 0.01 [b2]	6.05 ± 0.01 [c3]	5.77 ± 0.01 [a2]	6.01 ± 0.03 [c1]
Turkey leg (4 muscles)	6.54 ± 0.01 [a4]	6.89 ± 0.01 [b5]	6.63 ± 0.02 [a4]	7.34 ± 0.01 [c5]
Chicken breast (1 muscle)	6.39 ± 0.26 [b3]	6.30 ± 0.02 [a, b4]	6.26 ± 0.01 [a3]	6.67 ± 0.00 [c4]
Beef leg (4 muscles)	5.71 ± 0.01 [a1]	5.71 ± 0.10 [a1]	5.52 ± 0.02 [a1]	6.20 ± 0.01 [b2]

Each value is the mean of three replicates per meat sample and storage day $\pm$ standard deviation (SD). Different superscript letters in the same row indicate significant difference ($p < 0.05$) between storage days for the same meat type. Different superscript numbers in the same column indicate significant difference ($p < 0.05$) between meat types for the same storage day.

Initial pH levels for fresh beef and lamb meats were considered normal in comparison with reported levels in the literature [25,26]. However, they were high in pork, turkey and chicken compared with other studies [27–29]. These initial values were related to the kind of the cut for each species and its quality according to the rigor's resolution. The pH increased in all meat types during storage, except for pork, and at the end of the storage they were between 6.01 and 7.34 (Table 1). These results agree with previous reports in different types of meats, also during refrigerated storage [30]. The increases were associated with the production of nitrogenized basic compounds, mainly aminic, which are the main results of microbial spoilage and are conditioned by the type of packaging. On the other hand, some studies reported lower pH values particularly for pork, turkey and chicken [31].

3.2. Microbiology

Microbial growth (TVC, LAB and *Enterobaceteriaceae*) was affected by meat type and storage time ($p < 0.05$) (Table 2). At the beginning of the experiment, the highest level of TVC was observed in lamb (5.64 Log cfu/g) and the lowest in chicken (4.13 Log cfu/g). Similar behavior was observed in LAB counts. Initial levels of *Enterobacteriaceae* were the lowest in pork (2.95 Log cfu/g). Generally, similar amounts were also reported by other authors for meat and meat products based on the different meat types [17,20,21,32,33].

Table 2. Microbial counts Log (cfu/g) of the fresh different meat types during chilled storage at 2 °C.

Microorganisms	Meat Type	Chilling Storage at 2 °C			
		Day 0	Day 3	Day 6	Day 10
Total Viable Count (TVC)	Pork leg (5 muscles)	5.53 ± 0.12 [a4]	7.79 ± 0.05 [b3]	9.74 ± 0.01 [c4]	10.04 ± 0.01 [d1, 2]
	Lamb leg (7 muscles)	5.64 ± 0.25 [a4]	7.54 ± 0.40 [b2]	9.99 ± 0.17 [c5]	10.04 ± 0.02 [c2]
	Turkey leg (4 muscles)	4.98 ± 0.01 [a3]	6.45 ± 0.13 [b1]	8.96 ± 0.05 [c1]	9.91 ± 0.09 [d1]
	Chicken breast (1 muscle)	4.13 ± 0.17 [a1]	7.79 ± 0.03 [b3]	9.55 ± 0.02 [c3]	10.04 ± 0.05 [d1, 2]
	Beef leg (4 muscles)	4.74 ± 0.10 [a2]	6.55 ± 0.09 [b1]	9.25 ± 0.04 [c2]	10.22 ± 0.03 [d2]
Lactic acid bacteria (LAB)	Pork leg (5 muscles)	3.76 ± 0.07 [a3]	5.64 ± 0.01 [b3]	7.57 ± 0.03 [c5]	8.34 ± 0.08 [d2]
	Lamb leg (7 muscles)	4.24 ± 0.01 [a4]	5.70 ± 0.05 [b3]	6.05 ± 0.11 [c3]	8.26 ± 0.09 [d2]
	Turkey leg (4 muscles)	4.65 ± 0.03 [a5]	5.61 ± 0.10 [b3]	7.40 ± 0.02 [c4]	8.08 ± 0.13 [d1]
	Chicken breast (1 muscle)	2.96 ± 0.02 [a1]	5.38 ± 0.06 [b2]	6.26 ± 0.02 [b1]	8.99 ± 0.09 [c1]
	Beef leg (4 muscles)	3.20 ± 0.14 [a2]	4.34 ± 0.06 [b1]	5.89 ± 0.01 [c2]	8.04 ± 0.06 [d1]
Enterobacteriaceae	Pork leg (5 muscles)	2.95 ± 0.00 [a1]	4.95 ± 0.04 [b1]	6.11 ± 0.05 [c1]	7.53 ± 0.10 [d1]
	Lamb leg (7 muscles)	4.71 ± 0.00 [a4]	6.55 ± 0.05 [b3, 4]	6.37 ± 0.02 [b3]	7.99 ± 0.04 [c2]
	Turkey leg (4 muscles)	4.24 ± 0.02 [a3]	5.88 ± 0.06 [b2]	6.79 ± 0.02 [c4]	7.68 ± 0.03 [d4]
	Chicken breast (1 muscle)	4.07 ± 0.76 [a2, 3]	6.73 ± 0.03 [b4]	6.72 ± 0.04 [b2]	7.27 ± 0.09 [c1]
	Beef leg (4 muscles)	3.92 ± 0.03 [a2]	6.42 ± 0.12 [b3]	6.62 ± 0.01 [c2]	7.66 ± 0.05 [d3]

Each value is the mean of three replicates per meat sample and storage day ± standard deviation (SD). For every type of microorganism: Different superscript letters in the same row indicate significant difference ($p < 0.05$) between storage days for the same meat type, and different superscript numbers in the same column indicate significant difference ($p < 0.05$) between meat types for the same storage day.

Microbial growth (TVC, LAB, *Enterobacteriaceae*) increased by 1 and 2 logarithmic units during refrigerated storage (Table 2). After three days, all meat samples registered 6 Log cfu/g of TVC. These levels reached more than 8.9 Log cfu/g on day six and up to one more Log cfu/g unit until day 10. Other authors reported similar microbial behavior during chilled storage of fresh meat [17,20,21]. TVC values are commonly associated with meat spoilage when it reaches levels higher than 6 Log cfu/g [34]. The high levels reported during the experiment were in relation of the high pH levels of the samples (Table 1).

Microorganisms levels in raw meat are influenced by many factors which directly affect meat quality such as animal stress susceptibility, pre- and post-slaughter handling, processing, transport, packaging, storage, composition, etc. The experiment could have been finalized at microorganism levels that limit meat consumption, but samples were analyzed for a longer period in order to better understand BA and FAA formation and the relationship between microbial counts, FAA, and BA levels.

3.3. Free Amino Acids (FAA)

During storage, free amino acid levels and their behavior varied considerably depending on the type of meat (Table 3 and Figure 2).

Initially, the most abundant FAA in all meat samples was β-alanine, with levels of 18.70–37.64 mg/100 g, followed by the glutamic acid (4.42–28.95 mg/100 g) (Table 3). High levels of glycine, threonine and aspartic acid were also detected at the beginning of the storage. The lowest amounts were observed in cysteine, while histidine was not detected in any sample. Serine was highly present in turkey and chicken (37.63 and 27.67 mg/100 g, respectively) but it is not present in pork as much as in the aforementioned meat types (5.18 mg/100 g). However, it was detected neither in lamb nor beef until the sixth and tenth days of storage respectively (Table 3).

In general, chicken and turkey presented the highest levels of total free amino acids (TFAA) and TFAAP of Bas, while the lowest levels were observed in lamb which are followed by beef and pork (Figure 2a,b). These differences seem to be related to the type of meat (poultry or mammals). Indeed, white meats (chicken and turkey) registered approximately twice the amounts of TFAAs (234.30 and 201.33 mg/100 g respectively) as lamb (86.56 mg/100 g) and three times more than pork (77.50 mg/100 g) or beef (67.5 mg/100 g) at the beginning of storage (Figure 2a). These results are consistent with reports by the USDA [35] and other authors such as Leggio et al. [18], who reported levels between 47.5 and 93.07 mg/kg of TFAA of industrial "sopressata" pork sausage. Even though higher levels have been reported in chicken [36], low concentrations were found by Cowieson et al. [37] and Rabie et al. [7] in chicken, and in other meats (horse, beef, and turkey).

These differences could be due to many variables possibly influencing the formation and destruction of FAAs in the various meat types. These include factors such as the intrinsic properties of the product (biological factors relating to species, breed, sex, etc.; physiological aspects—genetic background, stress responses, etc.; and production practice—feeding, finishing weight, age at slaughter, etc.), handling, and processing conditions, etc. [5,38].

Table 3. Free amino acids (FAA) concentration (mg/100 g) of the fresh different meat types during chilled storage at 2 °C.

FAA	Meat Type	Chilling Storage at 2 °C			
		Day 0	Day 3	Day 6	Day 10
Aspartic acid	Pork leg (5 muscles)	2.96 ± 0.04 [a2]	6.82 ± 0.86 [c3]	9.46 ± 2.59 [d2]	4.87 ± 0.02 [b1]
	Lamb leg (7 muscles)	1.72 ± 0.00 [a1]	1.80 ± 0.31 [a1]	4.00 ± 0.12 [b1]	5.98 ± 0.04 [c2]
	Turkey leg (4 muscles)	7.95 ± 0.41 [a3]	13.61 ± 0.32 [b4]	20.68 ± 0.02 [d3]	15.92 ± 0.39 [c3]
	Chicken breast (1 muscle)	15.38 ± 0.83 [b3]	23.87 ± 0.16 [c5]	29.18 ± 0.29 [d4]	4.57 ± 0.11 [a1]
	Beef leg (4 muscles)	2.37 ± 1.00 [a1, 2]	5.15 ± 0.18 [b2]	3.99 ± 0.00 [b1]	4.96 ± 0.11 [b1]
Threonine	Pork leg (5 muscles)	2.21 ± 0.04 [a2]	4.23 ± 0.23 [b3]	11.23 ± 2.71 [c3]	12.30 ± 0.04 [c2]
	Lamb leg (7 muscles)	2.67 ± 0.04 [a b2]	2.10 ± 0.02 [a1]	4.08 ± 0.05 [b c2]	4.97 ± 0.19 [c1]
	Turkey leg (4 muscles)	8.30 ± 0.80 [a3]	11.06 ± 0.48 [b4]	12.40 ± 0.18 [b3]	24.41 ± 0.14 [c3]
	Chicken breast (1 muscle)	13.71 ± 0.59 [a4]	21.17 ± 0.34 [b5]	30.82 ± 0.23 [c4]	12.51 ± 0.26 [a2]
	Beef leg (4 muscles)	1.40 ± 0.04 [a1]	2.58 ± 0.07 [b2]	2.75 ± 0.07 [b1]	5.10 ± 0.10 [c1]
Serine	Pork leg (5 muscles)	5.18 ± 0.45 [a1]	9.12 ± 0.74 [b1]	10.51 ± 2.41 [c2]	8.45 ± 0.06 [b3]
	Lamb leg (7 muscles)	ND	ND	5.84 ± 0.12 [b1]	4.36 ± 0.00 [a1]
	Turkey leg (4 muscles)	37.63 ± 1.04 [c3]	41.59 ± 0.32 [d2]	31.69 ± 0.17 [b3]	15.26 ± 0.19 [a4]
	Chicken breast (1 muscle)	27.67 ± 1.42 [b2]	41.41 ± 0.52 [d2]	36.99 ± 0.73 [c4]	8.85 ± 0.28 [a3]
	Beef leg (4 muscles)	ND	ND	ND	7.40 ± 0.04 [a2]
Glutamic acid	Pork leg (5 muscles)	8.09 ± 0.96 [a2]	16.78 ± 0.94 [b3]	48.67 ± 1.93 [d3]	28.97 ± 0.19 [c1]
	Lamb leg (7 muscles)	5.94 ± 0.14 [a1]	11.05 ± 0.26 [b2]	19.09 ± 0.39 [c2]	27.66 ± 1.73 [d1]
	Turkey leg (4 muscles)	28.95 ± 1.66 [a4]	50.35 ± 0.09 [b5]	55.64 ± 0.04 [b4]	87.58 ± 0.19 [c3]
	Chicken breast (1 muscle)	21.07 ± 0.88 [a3]	36.66 ± 0.54 [c4]	51.58 ± 0.73 [d3]	30.52 ± 0.66 [b1]
	Beef leg (4 muscles)	4.42 ± 0.38 [a1]	8.81 ± 0.08 [b1]	15.06 ± 0.30 [c1]	65.41 ± 0.14 [d2]
Glycine	Pork leg (5 muscles)	6.17 ± 0.37 [a2]	12.65 ± 0.77 [b c2]	13.71 ± 3.14 [c2]	11.23 ± 0.06 [b2]
	Lamb leg (7 muscles)	18.10 ± 0.02 [a4]	19.89 ± 0.06 [b3]	17.88 ± 0.33 [a3]	17.69 ± 0.42 [a3]
	Turkey leg (4 muscles)	18.24 ± 0.99 [a4]	20.88 ± 0.22 [b3,4]	18.62 ± 0.18 [a3]	28.28 ± 0.01 [c4]
	Chicken breast (1 muscle)	16.76 ± 0.75 [b3]	21.93 ± 0.44 [c4]	27.41 ± 0.45 [d4]	11.74 ± 0.21 [a2]
	Beef leg (4 muscles)	3.14 ± 0.07 [a1]	10.31 ± 0.08 [c1]	6.73 ± 0.18 [b1]	7.04 ± 0.03 [b1]
β-alanine	Pork leg (5 muscles)	18.70 ± 1.06 [a1]	29.68 ± 1.51 [b1]	38.51 ± 8.70 [c2]	27.69 ± 0.03 [b1]
	Lamb leg (7 muscles)	30.83 ± 0.00 [a3]	36.65 ± 0.09 [c2]	32.65 ± 0.68 [b1]	41.50 ± 0.87 [d2]
	Turkey leg (4 muscles)	37.64 ± 2.14 [a4]	42.34 ± 0.10 [b3]	41.90 ± 0.45 [b2]	48.33 ± 0.11 [c3]
	Chicken breast (1 muscle)	22.68 ± 0.86 [a2]	38.59 ± 0.67 [c2]	52.54 ± 0.87 [d3]	28.75 ± 0.72 [b1]
	Beef leg (4 muscles)	29.09 ± 0.20 [a3]	47.69 ± 0.34 [c4]	41.59 ± 0.77 [b2]	42.99 ± 0.30 [b2]
Cysteine	Pork leg (5 muscles)	0.53 ± 0.12 [a2]	0.92 ± 0.04 [b2]	1.61 ± 0.27 [c3]	1.71 ± 0.03 [c3]
	Lamb leg (7 muscles)	0.20 ± 0.04 [a1]	1.05 ± 0.09 [b2]	1.17 ± 0.00 [b2]	1.20 ± 0.04 [b1]
	Turkey leg (4 muscles)	0.33 ± 0.00 [a1]	0.39 ± 0.00 [a1]	0.42 ± 0.04 [a1]	2.01 ± 0.09 [b4]
	Chicken breast (1 muscle)	0.22 ± 0.00 [a1]	0.33 ± 0.00 [a1]	3.95 ± 0.04 [c4]	1.74 ± 0.09 [b3]
	Beef leg (4 muscles)	0.98 ± 0.04 [a3]	0.98 ± 0.12 [a2]	1.06 ± 0.00 [a2]	1.42 ± 0.03 [b2]
Valine	Pork leg (5 muscles)	5.77 ± 0.04 [a2]	8.17 ± 0.45 [b3]	13.16 ± 2.74 [c3]	16.25 ± 0.06 [d3]
	Lamb leg (7 muscles)	2.59 ± 0.08 [a1]	4.64 ± 0.08 [b1]	7.58 ± 0.14 [c1]	7.09 ± 0.11 [c1]
	Turkey leg (4 muscles)	8.29 ± 0.37 [a3]	11.28 ± 0.43 [b4]	11.49 ± 0.18 [b2]	23.19 ± 0.14 [c4]
	Chicken breast (1 muscle)	12.97 ± 0.47 [a4]	20.28 ± 0.34 [c5]	23.45 ± 0.18 [d4]	16.67 ± 0.61 [b3]
	Beef leg (4 muscles)	5.30 ± 0.08 [a2]	6.23 ± 0.56 [b2]	8.26 ± 0.16 [c1]	9.37 ± 0.04 [d2]
Methionine	Pork leg (5 muscles)	2.67 ± 0.29 [a2]	3.17 ± 0.10 [a2]	9.31 ± 1.94 [b3]	12.30 ± 0.04 [c3]
	Lamb leg (7 muscles)	1.10 ± 0.00 [a1]	1.44 ± 0.30 [a1]	3.97 ± 0.01 [b1]	4.12 ± 0.10 [b1]
	Turkey leg (4 muscles)	3.19 ± 0.07 [a2]	4.84 ± 0.34 [b3]	5.00 ± 0.04 [b2]	16.79 ± 0.01 [c4]
	Chicken breast (1 muscle)	7.10 ± 0.30 [a3]	11.58 ± 0.27 [b4]	14.63 ± 0.21 [c4]	12.34 ± 0.29 [b3]
	Beef leg (4 muscles)	1.31 ± 0.00 [a1]	2.20 ± 0.68 [a1, 2]	4.09 ± 0.05 [b1]	6.62 ± 0.08 [c2]
Isoleucine	Pork leg (5 muscles)	2.98 ± 0.12 [a3]	4.36 ± 0.25 [b2]	8.28 ± 1.74 [c2]	9.88 ± 0.08 [d3]
	Lamb leg (7 muscles)	1.73 ± 0.13 [a1]	2.40 ± 0.10 [b1]	3.92 ± 0.08 [c1]	4.05 ± 0.00 [c1]
	Turkey leg (4 muscles)	5.25 ± 0.21 [a4]	7.24 ± 0.18 [b3]	6.97 ± 0.08 [b2]	16.33 ± 0.00 [c4]
	Chicken breast (1 muscle)	8.16 ± 0.35 [a5]	13.98 ± 0.23 [c4]	15.75 ± 0.40 [d3]	10.08 ± 0.20 [b3]
	Beef leg (4 muscles)	2.21 ± 0.04 [a2]	3.02 ± 0.51 [a b1]	3.90 ± 0.05 [b1]	5.88 ± 0.07 [c2]
Leucine	Pork leg (5 muscles)	5.35 ± 0.08 [a2]	6.88 ± 0.38 [b3]	14.18 ± 2.92 [c3]	16.65 ± 0.06 [d3]
	Lamb leg (7 muscles)	3.09 ± 0.34 [a1]	4.20 ± 0.14 [b1]	7.05 ± 0.19 [c1]	7.22 ± 0.04 [c1]
	Turkey leg (4 muscles)	7.66 ± 0.40 [a3]	11.28 ± 0.10 [b4]	10.84 ± 0.07 [b2]	26.19 ± 0.06 [c4]
	Chicken breast (1 muscle)	14.29 ± 0.44 [a4]	24.10 ± 0.46 [c5]	26.76 ± 0.46 [d4]	16.86 ± 0.60 [b3]
	Beef leg (4 muscles)	3.79 ± 0.04 [a1]	5.65 ± 0.54 [b2]	6.99 ± 0.05 [c1]	9.99 ± 0.09 [d2]

(FAA no precursors of BA)

Table 3. *Cont.*

	FAA	Meat Type	Chilling Storage at 2 °C			
			Day 0	Day 3	Day 6	Day 10
FAA precursors of BA	Tyrosine	Pork leg (5 muscles)	3.82 ± 0.24 [a3]	5.02 ± 0.35 [b3]	7.90 ± 1.79 [c3]	8.94 ± 0.27 [c3]
		Lamb leg (7 muscles)	2.13 ± 0.06 [a2]	2.36 ± 0.04 [b1]	2.62 ± 0.17 [c1]	3.09 ± 0.47 [c1]
		Turkey leg (4 muscles)	6.21 ± 0.34 [a4]	8.75 ± 0.13 [b4]	8.20 ± 0.41 [b3]	9.24 ± 0.07 [a3]
		Chicken breast (1 muscle)	11.23 ± 0.54 [b5]	18.14 ± 0.43 [d5]	15.80 ± 0.20 [c4]	17.17 ± 0.27 [c4]
		Beef leg (4 muscles)	1.63 ± 0.18 [a1]	2.96 ± 0.17 [b2]	5.22 ± 0.18 [c2]	6.04 ± 0.56 [d2]
	Phenylalanine	Pork leg (5 muscles)	3.94 ± 0.33 [a3]	4.39 ± 0.37 [a3]	14.82 ± 3.31 [b3]	17.41 ± 0.22 [b3]
		Lamb leg (7 muscles)	2.36 ± 0.22 [a2]	2.50 ± 0.12 [a1]	4.86 ± 0.63 [b1]	5.82 ± 0.59 [b1]
		Turkey leg (4 muscles)	4.10 ± 0.46 [a3]	6.87 ± 0.28 [b4]	6.45 ± 0.15 [b2]	19.92 ± 0.05 [c4]
		Chicken breast (1 muscle)	8.24 ± 0.12 [a4]	13.70 ± 0.35 [b5]	15.74 ± 0.34 [c3]	18.41 ± 0.78 [d3]
		Beef leg (4 muscles)	1.72 ± 0.27 [a1]	3.54 ± 0.06 [b2]	4.72 ± 0.22 [c1]	13.18 ± 0.00 [d2]
	Histidine	Pork leg (5 muscles)	ND	3.26 ± 0.16 [a1]	3.85 ± 0.91 [a1]	5.00 ± 0.03 [b1]
		Lamb leg (7 muscles)	ND	ND	ND	ND
		Turkey leg (4 muscles)	ND	ND	ND	ND
		Chicken breast (1 muscle)	ND	ND	ND	5.62 ± 0.78 [a1]
		Beef leg (4 muscles)	ND	ND	ND	ND
	Lysine	Pork leg (5 muscles)	4.96 ± 0.29 [a1]	9.92 ± 056 [b1]	24.46 ± 5.63 [c3]	25.84 ± 0.04 [c3]
		Lamb leg (7 muscles)	7.69 ± 0.00 [a3]	9.64 ± 0.06 [b1]	12.69 ± 0.16 [c1]	12.76 ± 0.24 [c1]
		Turkey leg (4 muscles)	15.09 ± 0.92 [a4]	20.66 ± 0.17 [b3]	26.19 ± 0.26 [c3]	60.33 ± 0.04 [d5]
		Chicken breast (1 muscle)	32.38 ± 1.74 [b5]	37.50 ± 0.73 [c4]	69.68 ± 0.89 [d4]	27.13 ± 0.52 [a4]
		Beef leg (4 muscles)	5.91 ± 0.05 [a2]	12.09 ± 0.17 [b2]	13.58 ± 0.16 [c2]	18.56 ± 1.63 [d2]
	Arginine	Pork leg (5 muscles)	4.19 ± 0.40 [b1]	7.56 ± 0.58 [c1]	0.12 ± 0.01 [a2]	ND
		Lamb leg (7 muscles)	6.39 ± 0.28 [b2]	7.06 ± 0.12 [c1]	0.08 ± 0.01 [a1]	ND
		Turkey leg (4 muscles)	12.49 ± 0.89 [b3]	16.17 ± 0.13 [c2]	1.45 ± 0.23 [a3, 4]	0.72 ± 0.04 [d5]
		Chicken breast (1 muscle)	22.44 ± 1.21 [b4]	26.32 ± 0.62 [c3]	1.41 ± 0.06 [a3]	ND
		Beef leg (4 muscles)	4.27 ± 0.11 [b1]	6.74 ± 0.33 [c1]	1.57±0.06 [a4]	ND
TFAAP of BA		Pork leg (5 muscles)	16.91 ± 1.25 [a2]	30.15 ± 0.55 [b3]	51.15 ± 11.76 [c3]	57.18 ± 0.04 [c3]
		Lamb leg (7 muscles)	18.58 ± 0.56 [a2]	21.56 ± 0.02 [c1]	20.25 ± 0.85 [b1]	20.68 ± 0.12 [b1]
		Turkey leg (4 muscles)	37.89 ± 2.61 [a3]	52.46 ± 0.70 [c4]	42.29 ± 0.52 [b3]	90.21 ± 0.27 [d5]
		Chicken breast (1 muscle)	74.29 ± 3.60 [b4]	95.66 ± 2.12 [c5]	102.63 ± 1.48 [d4]	62.71 ± 1.37 [a4]
		Beef leg (4 muscles)	13.53 ± 0.61 [a1]	25.33 ± 0.27 [b2]	25.09 ± 0.62 [b2]	37.79 ± 1.07 [c2]
TFAA		Pork leg (5 muscles)	77.50 ± 4.35 [a2]	132.94 ± 6.63 [b3]	229.78 ± 42.86 [c3]	207.48 ± 0.23 [c3]
		Lamb leg (7 muscles)	86.56 ± 0.10 [a3]	106.78 ± 0.98 [b1]	127.48 ± 2.46 [c2]	147.51 ± 3.50 [d1]
		Turkey leg (4 muscles)	201.33 ± 10.70 [a4]	267.32 ± 3.10 [c4]	257.94 ± 1.62 [b3]	394.50 ± 0.36 [d5]
		Chicken breast (1 muscle)	234.30 ± 10.49 [b5]	349.56 ± 6.08 [c5]	415.66 ± 6.08 [d4]	217.34 ± 5.40 [a4]
		Beef leg (4 muscles)	67.54 ± 0.94 [a1]	117.96 ± 2.37 [b2]	119.51 ± 2.12 [b1]	203.98 ± 0.26 [c2]

ND: Not Detected. Each value is the mean of three replicates per meat sample and storage day ± standard deviation (SD). For every type of free amino acid: Different superscript letters in the same row indicate significant difference ($p < 0.05$) between storage days for the same meat type, and different superscript numbers in the same column indicate significant difference ($p < 0.05$) between meat types for the same storage day.

In general, the relative initial differences in TFAAs and TFAA precursors of BAs in chicken and turkey were maintained until the end of storage. Contents were at their highest ($p < 0.05$) in these species, although they were smaller in chicken than in turkey at the end of the experiment following a decrease after day six (Figure 2a,b). Glutamic acid was the most prevalent FAA at the end of the storage (reaching 87.58 mg/100 g in turkey and 65.41 mg/100 g in beef), followed by β-alanine (48.33 mg/100 g in turkey and 41.50 mg/100 g in lamb), and lysine (27.13 mg/100 g in chicken and 60.33 mg/100 g in turkey), which is a precursor of cadaverine. Serine registered a considerable decrease in all samples while valine, methionine and isoleucine increased from the initial levels (Table 3). The concentration of histidine, a precursor of histamine, was very low or beneath the threshold of detection throughout the study since it is not typically present in meat, but it is rather one of the characteristics of fish products [6]. On the other hand, Arginine was the most prevalent FAA precursor at the beginning of the storage and went undetected at its end. This decrease in arginine was due to agmatine and putrescine formation, which can also lead to spermidine and spermine production since the formation of these three amines is interrelated [13,39].

Some meat types (mainly chicken and turkey) presented a saw-toothed pattern for TFAA and TFAAP of BA over storage (Figure 2a,b). The saw-toothed pattern of the FAAs observed during the experiment is typical of the one reported in myosystems such as meat [35] and in various research studies on fish and seafood [40]. This pattern is related to both formation and destruction of FAAs [6,41], which are associated with meat proteolysis (breakdown of proteins into small peptides and free amino acids) during storage. This hydrolysis of the peptide bonds may be of endogenous (endogenous proteolytic enzymes as exopeptidases) or exogenous origin. The latter origins are associated with microbial activity and the transformation of FAAs into other compounds through chemical and metabolic reactions. Other authors [36] also reported significant increases of amino acid levels in chicken during refrigerated storage that were associated with proteolysis. The saw-toothed pattern could limit the use of FAAs as reliable quality indexes for fresh refrigerated meat, as reported elsewhere [6].

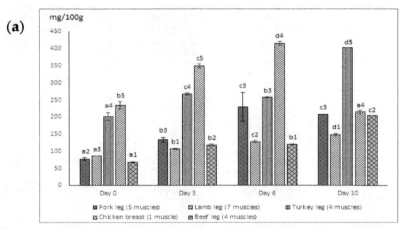

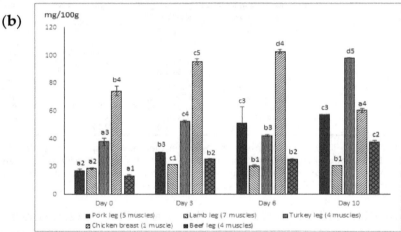

Figure 2. Total free amino acid (TFAA) (**a**) and Total free amino acid precursors (TFAAP) (**b**) of biogenic amines in mg/100 g of the different meat types during chilled storage at 2 °C. (Each value is the mean of three replicates per meat sample and storage day ± standard deviation (SD). Different letters indicate significant difference ($p < 0.05$) between storage days for the same meat type and different numbers indicate significant difference ($p < 0.05$) between meat types for the same storage day).

Nevertheless, the presence of certain amino acids such as glutamic acid, β-alanine, and phenylalanine were associated with typical flavors in meat and myosystems [42]. They are also very important as potential flavor and odor precursors through their interactions during heating, which contributes to the flavor and/or odor of cooked meat [43]. Indeed, glutamic acid and glycine are used as flavor enhancing additives in the food industry [44,45].

3.4. Biogenic Amines

Biogenic amine contents of the different meat types were affected ($p < 0.05$) by storage (Table 4). Except for the physiological amines spermidine and spermine (Spd and Spm), the initial levels of BAs were very low, and in some cases they were not detected. Spermidine and spermine presented the highest ($p < 0.05$) concentrations of BAs at the beginning of the experiment in all meat types, with spermine as the most abundant one (27.60–45.03 mg/kg). Levels of Spd and Spm were significantly lower in pork and beef as opposed to the rest of the meats, with the highest levels detected in chicken.

The reported amounts of these amines in the literature are wide-ranging. Similar values of Spd and Spm were reported in meat products formulated with pork and beef [5,7,12,20–22,39], in raw turkey [46], lamb, and sheep liver [47]. On the other hand, Rokka et al. [48] reported higher levels of Spm and Spd in chicken in comparison with our results, while other authors observed smaller amounts for Spm [12,32]. In addition, significantly lower levels of the physiological amines were observed in lamb and mutton as in our study [47].

Table 4. Biogenic amines (BA) concentration (mg/kg) of fresh different meat types during chilled storage at 2 °C.

BA	Meat Type	Chilling Storage at 2 °C			
		Day 0	Day 3	Day 6	Day 10
Tyramine	Pork leg (5 muscles)	0.67 ± 0.03 [a3]	1.10 ± 0.02 [b3]	11.20 ± 0.04 [c4]	16.58 ± 0.04 [d4]
	Lamb leg (7 muscles)	0.10 ± 0.00 [a1]	0.19 ± 0.01 [a1]	7.05 ± 0.25 [b3]	10.71 ± 0.01 [c3]
	Turkey leg (4 muscles)	ND	0.40 ± 0.04 [a2]	1.72 ± 0.02 [b2]	6.88 ± 0.14 [c2]
	Chicken breast (1 muscle)	ND	0.47 ± 0.07 [a2]	27.54 ± 0.86 [b5]	35.16 ± 0.36 [c5]
	Beef leg (4 muscles)	0.34 ± 0.04 [a2]	0.42 ± 0.02 [a2]	0.53 ± 0.03 [b1]	1.57 ± 0.07 [c1]
Histamine	Pork leg (5 muscles)	ND	ND	ND	ND
	Lamb leg (7 muscles)	ND	ND	ND	ND
	Turkey leg (4 muscles)	ND	ND	ND	ND
	Chicken breast (1 muscle)	0.53 ± 0.03 [a1]	1.23 ± 0.03 [b2]	1.73 ± 0.03 [c2]	2.11 ± 0.01 [d2]
	Beef leg (4 muscles)	ND	0.10 ± 0.00 [a1]	0.21 ± 0.01 [b1]	0.50 ± 0.02 [c1]
Phenylethylamine	Pork leg (5 muscles)	ND	0.77 ± 0.03 [a1]	1.28 ± 0.02 [b1]	1.66 ± 0.06 [c1]
	Lamb leg (7 muscles)	0.76 ± 0.00 [a3]	4.80 ± 0.08 [b3]	7.57 ± 0.27 [c3]	9.05 ± 0.15 [d3]
	Turkey leg (4 muscles)	0.21 ± 0.05 [a1]	15.07 ± 0.01 [d4]	12.85 ± 0.47 [c4]	11.33 ± 0.29 [b4]
	Chicken breast (1 muscle)	ND	16.87 ± 0.13 [c5]	17.99 ± 0.15 [b5]	12.81 ± 0.05 [a5]
	Beef leg (4 muscles)	0.47 ± 0.01 [a2]	2.33 ± 0.05 [b2]	2.47 ± 0.05 [b2]	2.62 ± 0.02 [c2]
Putrescine	Pork leg (5 muscles)	0.57 ± 0.09 [a1]	0.72 ± 0.02 [a1]	5.10 ± 0.48 [b2]	14.55 ± 0.09 [c3]
	Lamb leg (7 muscles)	1.19 ± 0.07 [a2]	3.22 ± 0.00 [b4]	6.40 ± 0.24 [c3]	10.11 ± 0.01 [d2]
	Turkey leg (4 muscles)	1.23 ± 0.03 [a2]	4.70 ± 0.08 [b5]	8.44 ± 0.44 [c5]	68.72 ± 0.02 [d5]
	Chicken breast (1 muscle)	1.23 ± 0.03 [a2]	1.83 ± 0.01 [b2]	7.47 ± 0.07 [c4]	51.99 ± 0.29 [d4]
	Beef leg (4 muscles)	1.34 ± 0.02 [a3]	2.07 ± 0.03 [b3]	3.99 ± 0.03 [c1]	7.40 ± 0.04 [d1]
Cadaverine	Pork leg (5 muscles)	ND	ND	1.11 ± 0.07 [a1]	16.16 ± 0.28 [b4]
	Lamb leg (7 muscles)	ND	ND	3.42 ± 0.02 [a2]	5.08 ± 0.12 [b1]
	Turkey leg (4 muscles)	ND	ND	1.27 ± 0.03 [a1]	13.25 ± 0.27 [b2]
	Chicken breast (1 muscle)	ND	ND	3.98 ± 0.10 [a3]	14.31 ± 0.11 [b3]
	Beef leg (4 muscles)	ND	ND	ND	ND
Tryptamine	Pork leg (5 muscles)	ND	ND	6.37 ± 0.07 [a1]	6.56 ± 0.10 [a2]
	Lamb leg (7 muscles)	ND	ND	ND	ND
	Turkey leg (4 muscles)	ND	ND	ND	ND
	Chicken breast (1 muscle)	3.82 ± 0.12 [b1]	15.78 ± 0.16 [d1]	7.47 ± 0.07 [c2]	0.37 ± 0.01 [a1]
	Beef leg (4 muscles)	ND	ND	ND	ND
Agmatine	Pork leg (5 muscles)	ND	ND	ND	ND
	Lamb leg (7 muscles)	ND	0.15 ± 0.01 [a1]	1.14 ± 0.02 [b1]	2.30 ± 0.08 [c1]
	Turkey leg (4 muscles)	ND	ND	ND	ND
	Chicken breast (1 muscle)	ND	ND	ND	ND
	Beef leg (4 muscles)	ND	ND	ND	ND
Spermidine	Pork leg (5 muscles)	2.63 ± 0.13 [a2]	2.70 ± 0.00 [a1]	3.18 ± 0.14 [b1]	3.88 ± 0.18 [c1]
	Lamb leg (7 muscles)	8.09 ± 0.13 [a4]	11.99 ± 0.13 [c4]	8.69 ± 0.21 [a4]	10.16 ± 0.12 [b5]
	Turkey leg (4 muscles)	7.33 ± 0.05 [a3]	18.27 ± 0.07 [d5]	12.14 ± 0.10 [c5]	9.67 ± 0.09 [b4]
	Chicken breast (1 muscle)	9.78 ± 0.06 [d5]	6.69 ± 0.13 [b3]	6.24 ± 0.20 [a3]	7.02 ± 1.10 [c3]
	Beef leg (4 muscles)	2.29 ± 0.05 [a1]	3.39 ± 0.05 [b2]	4.35 ± 0.01 [c2]	5.39 ± 0.05 [d2]
Spermine	Pork leg (5 muscles)	27.60 ± 0.96 [b1]	27.10 ± 0.12 [b1]	25.23 ± 1.17 [a1]	26.87 ± 0.31 [b2]
	Lamb leg (7 muscles)	31.36 ± 0.58 [a2]	40.85 ± 0.37 [c3]	31.60 ± 0.98 [a3]	36.42 ± 0.00 [b4]
	Turkey leg (4 muscles)	35.44 ± 1.28 [b3]	49.20 ± 0.40 [d4]	36.95 ± 0.53 [c4]	32.55 ± 0.09 [a3]
	Chicken breast (1 muscle)	45.03 ± 0.81 [b4]	53.60 ± 0.24 [d5]	47.26 ± 0.54 [c5]	41.92 ± 0.16 [a5]
	Beef leg (4 muscles)	30.86 ± 0.36 [b2]	33.02 ± 0.24 [c2]	29.54 ± 0.40 [b2]	25.08 ± 0.04 [a1]
BAI	Pork leg (5 muscles)	1.24	1.82	17.41	47.29
	Lamb leg (7 muscles)	1.29	3.41	16.87	25.90
	Turkey leg (4 muscles)	1.23	5.10	11.43	88.85
	Chicken breast (1 muscle)	1.76	3.53	40.72	103.57
	Beef leg (4 muscles)	1.68	2.59	4.73	9.47

Each value is the mean of three replicates per meat sample and storage day ± standard deviation (SD). ND: Not detected (average limit of detection being 0.065 mg/L). For every biogenic amine: Different superscript letters in the same row indicate significant difference ($p < 0.05$) between storage days for the same meat type, and different superscript numbers in the same column indicate significant difference ($p < 0.05$) between meat types for the same storage day.

During refrigerated storage, the levels of physiological amines fluctuated in turkey for Spd and in turkey and chicken for Spm following a saw-toothed pattern (Table 4). Irrespective of the meat type. Both amines increased, peaking after the third day (18.27 mg/kg of Spd in turkey and 53.60 mg/kg of Spm in chicken) which was followed by a decrease until the end of the storage. This pattern, which is reported in other meat products in chilled storage [7], could reflect the relationship between these FAAs and the evolution pattern of their precursor, arginine (Figure 1). Initial arginine levels were considerable (Table 3). In fact, these levels increased up to day six and then decreased. Finally, they disappeared in all the samples except in turkey, which registered very low levels at the end of the experiment.

Putrescine (Put) was also detected throughout the study in all samples. Initial levels of these physiological amines were low in pork (0.57 mg/kg) while the other meat types contained the double of that amount (Table 4). Put levels increased significantly in all meats during the experiment but not at the same rate for all samples. For instance, in lamb and turkey, levels increased considerably by day three but the highest Put levels were recorded in chicken and turkey at the end of the study (51.99 mg/kg and 68.72 mg/kg, respectively). These were the highest levels among all BAs, even higher than tyramine, which is the most prevalent BA in meat products. At the end of the storage, only in the case of pork and lamb, similar levels of putrescine and tyramine were observed (Table 4). These putrescine levels were mainly related to *Enterobacteriaceae* growth (Table 2), which registered the highest levels. In addition, put production is associated with a reduction in arginine, which was also observed in all samples (Table 3) as noted earlier. Put is identified as one of the toxic biogenic amines, together with cadaverine (Cad), since they favor intestinal absorption of HIS and Tyr and contribute to catabolism reduction, thus enhancing their toxicity [5,14].

Cadaverine (Cad) is another important amine that was not detected in the meat samples until day six of chilled storage, except for beef, where it was not detected throughout the study. Other authors [20] reported that cadaverine was undetectable in beef whereas the meats in which it was detected, concentrations rose quickly, except for lamb. However, considerable levels were observed in pork, chicken, and turkey (16.16, 14.31 and 13.25 mg/kg, respectively). In most cases, no clear relationship was observed regarding the levels of its FAA lysine precursor, except for chicken and turkey, in which there was some correlation (coefficient > 0.92). It is worth noticing that at the end of storage, lysine was the precursor with the highest levels, peaking in turkey and chicken at day six of the experiment (Table 3). High levels of the FAA precursors that did not correlate with high levels of their corresponding BAs could be due to the lower BA decarboxylation capacity of the microorganisms that grow in this type of meat. These amines are also associated with *Enterobacteriaceae* growth; but, in this study, there was no clear relationship. However, there are other factors that can also affect BA presence in meat or meat products such as processing, meat matrix nature, etc. [9,20–22,49].

Agmatine (Agm) was present only in lamb meat with very low levels (0.15–2.30 mg/kg) (Table 4) starting from the third day of storage. However, its precursor arginine was widely represented in all meats, especially chicken, throughout the study (Table 3). This inverse behavior in lamb could be due to the weakness or lack of aminogenetic capacity of the microbiota [8] and the fact that arginine can be transformed into Putrescine (Figure 1) depending on the microbial flora type present in the matrix. As explained before, put levels were indeed consistent with the arginine levels observed in this study.

In general, the other BAs' levels increased, from day three until the end of the storage. Levels of toxic amines such as tyramine (Tyr) and β-phenylethylamine (Phe) were very low (<0.8 mg/kg) at the beginning of the study and were not even detected in some meats such as in chicken and turkey (Tyr), and in pork and chicken (Phe). However, they were detected afterwards in all the studied kinds of meats during storage. These amounts increased during the experiment until day 10, but levels remained below 36 mg/kg in the case of Tyr, which reached its highest levels in chicken and lowest in beef (1.57 mg/kg), followed by turkey (6.88 mg/kg), lamb (10.71 mg/kg), and pork (16.58 mg/kg). Phe levels registered lower amounts in all species at the end of the storage. Chicken also contained the highest levels of Phe (12.81 mg/kg), followed by turkey, lamb, beef, and pork (Table 4) at the end of the experiment. These levels are very low in terms of toxicological limits, especially for Tyr (800 mg/kg) [50] while the limit for Phe is 30 mg/kg, which is twice the level found in this study [51]. These BAs are formed from the decarboxylation reactions of tyrosine and phenylalanine, respectively. The highest levels of tyrosine were observed in chicken, which can be associated with the high levels of tyramine in this species. However, no clear relationship was observed for the other species regarding tyramine and tyrosine even though the evolution of this FAA was clear throughout the storage and was species-dependent (Table 3). As a matter of fact, in some species, such as pork, lamb, beef or turkey, tyramine increased over storage, and these changes correlated with tyrosine production. On the other hand, the evolution in chicken followed a saw-toothed pattern, thus the possibility for establishing a correlation became more unlikely with Tyr production. Several authors reported similar evolutions of these FAAs and their relationship with BAs [6].

Phenylalanine levels (Table 3) were higher than tyrosine's although its corresponding biogenic amine (Phe) levels were significantly lower than tyramine in some species. Phenylalanine levels were significantly higher at the end of the storage in all meat types, particularly turkey, chicken, and pork (Table 3). Other authors also found little correlation between Phe and its FAA precursor, which seems to be closely related to the nature of the flora and its aminogenic capacity [8,52]. Lamb, for example, contained higher levels of Phe (9.05 mg/kg) at the end of the experiment than of phenylalanine (5.82 mg/100 g) suggesting that its flora has high Phenylalanine decarboxylation capacity. The rest of the meats presented the opposite pattern suggesting that their flora presented less phenylalanine decarboxylation capacity.

On the other hand, histamine (HIS), another toxicological amine, was not detected in the majority of meat types (pork, lamb and turkey), except in chicken and beef where levels were very low (2.11 and 0.50 mg/kg, respectively at the end of storage) (Table 4). This was consistent with histidine levels, the FAA precursor of HIS (Table 3), which were not initially detected in any type of meat, except for pork and chicken, with very low levels (<6 mg/100 g). Final HIS levels in the meat types samples were significantly lower than the legal limit of 50 mg/kg [15], meaning that there is no potential health risk after consumption of these meats in their fresh status. Moreover, HIS levels are related to those of its FAA precursor, which is poorly represented in the studied meats (Table 3). Some authors also reported very low to undetectable HIS levels in fresh beef sausages and dry fermented pork sausages [20,21,39]. These low amounts are consistent with the type of samples analyzed since HIS is not typically found in fresh meat and meat products [5].

Tryptamine (Trp) was detected only in chicken and pork. While its presence in chicken was observed from the beginning of the storage, in the case of pork it was only detected after six days.

In general, higher BA levels (Table 4) were observed in chicken, turkey and pork, and these results were associated with the total free amino acid BA precursors (TFAAP) (Table 3). Given the importance of BAs as quality indexes, the application of the biogenic amines index (BAI) in this study showed a clear increase over storage, which was related to the meat type (Table 4). At the outset, all the BAIs registered less than 2 mg/kg, with the highest levels in chicken. According to the index's classification, all meat types presented good quality, which was maintained until day six, afterwards there was a considerable increase in BAI levels, with the highest registered once more in chicken. The latter reached levels of 40.72 mg/kg, followed by pork (17.41 mg/kg), lamb (16.87 mg/kg), turkey (11.43 mg/kg), and beef (4.73 mg/kg). According to this index, chicken was considered of poor quality at day six while the other meats were still classified as acceptable. However, this classification is not related to microorganism levels on the same day of analysis (Table 2). As a matter of fact, a delay was observed in the formation of biogenic amines with respect to microbial growth. Such a delay was also reported by other authors [21,53] and constituted one of the factors that some authors used to set the limit of 6 Log/cfu for TVC as indicating unfitness for consumption.

At the end of storage, the highest BAI levels were observed in chicken (103.87 mg/kg) and turkey (88.85 mg/kg), which were spoiled, and the lowest BAI level was found in beef (9.47 mg/kg), which was still the only meat in the range of BAI acceptability (Table 4), while pork and lamb were considered as exhibiting poor quality.

Several authors demonstrated that beef is less spoilable than the other types of meats and that chicken is the first to undergo deterioration reactions [12,54]. BAI levels in turkey, beef, and pork presented a high correlation with TFAAP (0.97, 0.92, and 0.86, respectively) as well as with TFAA levels in lamb (0.97), beef (0.96, and turkey (0.95) throughout the storage (Table 4).

These results support the theory of rapid white meat spoilage compared with red meat [12,54], as shown in the FAA section. This behavior was also observed in tuna, where amine levels were generally higher in the white than in the red muscle [40]. In this study; the BAI levels of white meats (turkey and chicken) close to and above 50, at 10 days of chilling storage.

These BAI levels reflect the rate of deterioration of each type of meat and thus provide useful information when planning meat product processing and manufacture strategies. These BAI levels showed better results than some BA contents such as Tyramine or Spermidine and Spermine individually. However, BAI results did not show any clear relationship with microbial levels, which exceeded the permitted limits of microorganisms in all types of meats from day three of storage (Table 2). Therefore, in this case, the BAI index seems to be a more suitable indicator for white meat freshness than for red ones, especially beef, which was also reported by other authors [10].

4. Conclusions

The evolutions of free amino acids (FAA) and biogenic amines (BA) were clearly influenced by the meat type. The largest amounts were observed in chicken and turkey followed by the other meat types. Even though a clear relationship was observed for certain meat species between the biogenic amine index (BAI) and total free amino acids (TFAA), this index did not correlate with microbial growth. The relationship between overall TFAAs and BAIs was closer between an FAA precursor and its corresponding biogenic amine when considered individually. BAIs showed that only beef maintained acceptable quality throughout the study (<10 at day 10 of chilling storage), while chicken presented a quality (103.57) followed by turkey, lamb and pork. Overall, BA levels were higher in white meats than in red ones. During storage, some TFAAP of BA followed a saw-toothed pattern mainly in chicken and turkey meat. This limits its use as a quality index for fresh meat during chilled storage. Finally, the BAI index seems to be more suitable as a quality index for white meat freshness than for red meat, especially for beef.

Author Contributions: Conceived and designed the experiments; C.R.-C. and M.T.; performed the experiments: M.T., C.R-C and A.M.H. analyzed the data; M.T., C.R-C, A.M.H., F.J.-C.; wrote the paper: M.T., C.R-C, A.M.H., F.J.-C.

Acknowledgments: The authors wish to thank the AECID-MAE for Mehdi Triki's outstanding scholarship.

References

1.   Jiménez-Colmenero, F.; Herrero, A.; Cofrades, S.; Ruiz-Capillas, C. Meat: Eating Quality and Preservation. In *The Encyclopedia of Food and Health*; Caballero, B., Finglas, P., Toldrá, F., Eds.; Oxford Academic Press: Kidlington, UK, 2016; Volume 3, pp. 685–692.
2.   Shabbir, M.A.; Raza, A.; Anjum, F.M.; Khan, M.R.; Suleria, H.A.R. Effect of Thermal Treatment on Meat Proteins with Special Reference to Heterocyclic Aromatic Amines (HAAs). *Crit. Rev. Food Sci. Nutr.* **2015**, *55*, 82–93. [CrossRef] [PubMed]
3.   Jiménez Colmenero, F.; Herrero, A.M.; Ruiz-Capillas, C.; Cofrades, S. Meat and functional foods. In *Handbook of Meat and Meat Processing*; Hui, Y.H., Ed.; CRC Press. Taylor & Francis Group: Boca Raton, FL, USA, 2012; pp. 225–248.

4. Mietz, J.L.; Karmas, E. Polyamine and histamine content of rockfish, salmon, lobster, and shrimp as an indicator of decomposition. *J. Assoc. Off. Anal. Chem.* **1978**, *61*, 139–145.
5. Ruiz-Capillas, C.; Jiménez-Colmenero, F. Biogenic amines in meat and meat products. *Crit. Rev. Food Sci. Nutr.* **2004**, *44*, 489–499. [CrossRef] [PubMed]
6. Ruiz-Capillas, C.; Moral, A. Changes in free amino acids during chilled storage of hake (*Merluccius merluccius* L.) in controlled atmospheres and their use as a quality control index. *Eur. Food Res. Technol.* **2001**, *212*, 302–307. [CrossRef]
7. Rabie, M.A.; Peres, C.; Malcata, F.X. Evolution of amino acids and biogenic amines throughout storage in sausages made of horse, beef and turkey meats. *Meat Sci.* **2014**, *96*, 82–87. [CrossRef] [PubMed]
8. Latorre-Moratalla, M.L.; Bover-Cid, S.; Bosch-Fusté, J.; Veciana-Nogués, M.T.; Vidal-Carou, M.C. Amino acid availability as an influential factor on the biogenic amine formation in dry fermented sausages. *Food Control* **2014**, *36*, 76–81. [CrossRef]
9. Jairath, G.; Singh, P.K.; Dabur, R.S.; Rani, M.; Chaudhari, M. Biogenic amines in meat and meat products and its public health significance: A review. *J. Food Sci. Technol.* **2015**, *52*, 6835–6846. [CrossRef]
10. Hernández-Jover, T.; Izquierdo-Pulido, M.; Veciana-Nogués, M.T.; Vidal-Carou, M.C. Biogenic Amine Sources in Cooked Cured Shoulder Pork. *J. Agric. Food Chem.* **1996**, *44*, 3097–3101. [CrossRef]
11. Silva, C.M.G.; Glória, M.B.A. Bioactive amines in chicken breast and thigh after slaughter and during storage at 4 ± 1 °C and in chicken-based meat products. *Food Chem.* **2002**, *78*, 241–248. [CrossRef]
12. Vinci, G.; Antonelli, M.L. Biogenic amines: Quality index of freshness in red and white meat. *Food Control* **2002**, *13*, 519–524. [CrossRef]
13. Halász, A.; Baráth, Á.; Simon-Sarkadi, L.; Holzapfel, W. Biogenic amines and their production by microorganisms in food. *Trends Food Sci. Technol.* **1994**, *5*, 42–49. [CrossRef]
14. Önal, A. A review: Current analytical methods for the determination of biogenic amines in foods. *Food Chem.* **2007**, *103*, 1475–1486. [CrossRef]
15. European Food Safety Authority (EFSA). Scientific Opinion on risk based control of biogenic amine formation in fermented foods. *EFSA J.* **2011**, *9*, 2393. [CrossRef]
16. World Health Organization. Joint FAO/WHO Expert Meeting on the Public Health Risks of Histamine and other Biogenic Amines from Fish and Fishery Products: Meeting Report. Available online: http://www.fao.org/fileadmin/user_upload/agns/news_events/Histamine_Final_Report.pdf (accessed on 10 July 2018).
17. Cofrades, S.; López-López, I.; Ruiz-Capillas, C.; Triki, M.; Jiménez-Colmenero, F. Quality characteristics of low-salt restructured poultry with microbial transglutaminase and seaweed. *Meat Sci.* **2011**, *87*, 373–380. [CrossRef] [PubMed]
18. Leggio, A.; Belsito, E.L.; De Marco, R.; Di Gioia, M.L.; Liguori, A.; Siciliano, C.; Spinella, M. Dry fermented sausages of Southern Italy: A comparison of free amino acids and biogenic amines between industrial and homemade products. *J. Food Sci.* **2012**, *77*, S170–S175. [CrossRef] [PubMed]
19. Delgado-Pando, G.; Cofrades, S.; Ruiz-Capillas, C.; Solas, M.T.; Triki, M.; Jiménez-Colmenero, F. Low-fat frankfurters formulated with a healthier lipid combination as functional ingredient: Microstructure, lipid oxidation, nitrite content, microbiological changes and biogenic amine formation. *Meat Sci.* **2011**, *89*, 65–71. [CrossRef] [PubMed]
20. Triki, M.; Herrero, A.M.; Jiménez-Colmenero, F.; Ruiz-Capillas, C. Storage stability of low-fat sodium reduced fresh merguez sausage prepared with olive oil in konjac gel matrix. *Meat Sci.* **2013**, *94*, 438–446. [CrossRef] [PubMed]
21. Triki, M.; Herrero, A.M.; Jiménez-Colmenero, F.; Ruiz-Capillas, C. Effect of preformed konjac gels, with and without olive oil, on the technological attributes and storage stability of merguez sausage. *Meat Sci.* **2013**, *93*, 351–360. [CrossRef] [PubMed]
22. Triki, M.; Jiménez-Colmenero, F.; Herrero, A.M.; Ruiz-Capillas, C. Optimisation of a chromatographic procedure for determining biogenic amine concentrations in meat and meat products employing a cation-exchange column with a post-column system. *Food Chem.* **2012**, *130*, 1066–1073. [CrossRef]
23. ANSES Ciqual Table de Composition Nutritionnelle Des Aliments. Available online: https://ciqual.anses.fr/ (accessed on 10 July 2018).
24. Obanor, F.O. Biochemical Basis of the Effect of Pre-Slaughter Stress and Post-Slaughter Processing Conditions on Meat Tenderness. Master's Thesis, Lincoln University, Christchurch, New Zealand, 2002.

25. Page, J.K.; Wulf, D.M.; Schwotzer, T.R. A survey of beef muscle color and pH. *J. Anim. Sci.* **2001**, *79*, 678–687. [CrossRef] [PubMed]

26. Fleck, C.; Kozačinski, Ž.L.; Njari, B.; Marenčić, D.; Mršić, G.; Špiranec, K.; Špoljarić, D.; Čop, M.J.; Živković, M.; Popović, M. Technological properties and chemical composition of the meat of sheep fed with *Agaricus bisporus* supplement. *Vet. Arh.* **2015**, *85*, 591–600.

27. Van Laack, R.L.J.M.; Kauffman, R.G.; Sybesma, W.; Smulders, F.J.M.; Eikelenboom, G.; Pinheiro, J.C. Is color brightness (*L*-value) a reliable indicator of water holding capacity in porcine muscle? *Meat Sci.* **1994**, *38*, 193–201. [CrossRef]

28. Patterson, B.A.; Matarneh, S.K.; Stufft, K.M.; England, E.M.; Scheffler, T.L.; Preisser, R.H.; Shi, H.; Stewart, E.C.; Eilert, S.; Gerrard, D.E. Pectoralis major muscle of turkey displays divergent function as correlated with meat quality. *Poult. Sci.* **2017**, *96*, 1492–1503. [PubMed]

29. Cortez-Vega, W.R.; Pizato, S.; Prentice, C. Quality of raw chicken breast stores at 5 °C and packaged under different modified atmospheres. *J. Food Saf.* **2012**, *32*, 360–368. [CrossRef]

30. Yang, C.C.; Chen, T.C. Effects of Refrigerated Storage, pH Adjustment, and Marinade on Color of Raw and Microwave Cooked Chicken Meat. *Poult. Sci.* **1993**, *72*, 355–362. [CrossRef]

31. Debut, M.; Berri, C.; Baeza, E.; Sellier, N.; Arnould, C.; Guemene, D.; Jehl, N.; Boutten, B.; Jego, Y.; Beaumont, C. Variation of chicken technological meat quality in relation to genotype and preslaughter stress conditions. *Poult. Sci.* **2003**, *82*, 1829–1838. [CrossRef] [PubMed]

32. Lázaro, C.A.; Conte-Júnior, C.A.; Canto, A.C.; Monteiro, M.L.G.; Costa-Lima, B.; da Cruz, A.G.; Mársico, E.T.; Franco, R.M. Biogenic amines as bacterial quality indicators in different poultry meat species. *LWT Food Sci. Technol.* **2015**, *60*, 15–21. [CrossRef]

33. Laranjo, M.; Gomes, A.; Agulheiro-Santos, A.C.; Potes, M.E.; Cabrita, M.J.; Garcia, R.; Rocha, J.M.; Roseiro, L.C.; Fernandes, M.J.; Fraqueza, M.J.; et al. Impact of salt reduction on biogenic amines, fatty acids, microbiota, texture and sensory profile in traditional blood dry-cured sausages. *Food Chem.* **2017**, *218*, 129–136. [CrossRef] [PubMed]

34. Commission Regulation (EC) No 2073/2005 of 15 November 2005 on Microbiological Criteria for Foodstuffs. Available online: https://eur-lex.europa.eu/LexUriServ/LexUriServ.do?uri=CONSLEG: 2005R2073:20060101:EN:PDF (accessed on 14 August 2018).

35. USDA. USDA National Nutrient Database for Standard Reference. Available online: https://ndb.nal.usda. gov/ndb/search/list (accessed on 14 August 2018).

36. Niewiarowicz, A. Meat Anomalies in Broilers. *Poult. Int.* **1978**, *17*, 50–51.

37. Cowieson, A.J.; Acamovic, T.; Bedford, M.R. The effects of phytase and phytic acid on the loss of endogenous amino acids and minerals from broiler chickens. *Br. Poult. Sci.* **2004**, *45*, 101–108. [CrossRef] [PubMed]

38. Iida, F.; Miyazaki, Y.; Tsuyuki, R.; Kato, K.; Egusa, A.; Ogoshi, H.; Nishimura, T. Changes in taste compounds, breaking properties, and sensory attributes during dry aging of beef from Japanese black cattle. *Meat Sci.* **2016**, *112*, 46–51. [CrossRef] [PubMed]

39. Ruiz-Capillas, C.; Triki, M.; Herrero, A.M.; Jiménez-Colmenero, F. Biogenic amines in low- and reduced-fat dry fermented sausages formulated with konjac gel. *J. Agric. Food Chem.* **2012**, *60*, 9242–9248. [CrossRef] [PubMed]

40. Ruiz-Capillas, C.; Moral, A. Free amino acids and biogenic amines in red and white muscle of tuna stored in controlled atmospheres. *Amino Acids* **2004**, *26*, 125–132. [CrossRef] [PubMed]

41. Freiding, S.; Gutsche, K.A.; Ehrmann, M.A.; Vogel, R.F. Genetic screening of *Lactobacillus sakei* and *Lactobacillus curvatus* strains for their peptidolytic system and amino acid metabolism, and comparison of their volatilomes in a model system. *Syst. Appl. Microbiol.* **2011**, *34*, 311–320. [CrossRef] [PubMed]

42. Subramaniyan, S.A.; Kang, D.R.; Belal, S.A.; Cho, E.-S.-R.; Jung, J.-H.; Jung, Y.-C.; Choi, Y.-I.; Shim, K.-S. Meat Quality and Physicochemical Trait Assessments of Berkshire and Commercial 3-way Crossbred Pigs. *Korean J. Food Sci. Anim. Resour.* **2016**, *36*, 641–649. [CrossRef] [PubMed]

43. Dermiki, M.; Phanphensophon, N.; Mottram, D.S.; Methven, L. Contributions of non-volatile and volatile compounds to the umami taste and overall flavour of shiitake mushroom extracts and their application as flavour enhancers in cooked minced meat. *Food Chem.* **2013**, *141*, 77–83. [CrossRef] [PubMed]

44. Dashdorj, D.; Amna, T.; Hwang, I. Influence of specific taste-active components on meat flavor as affected by intrinsic and extrinsic factors: An overview. *Eur. Food Res. Technol.* **2015**, *241*, 157–171. [CrossRef]

45. Ruiz-Capillas, C.; Nollet, L.M.L. *Flow Injection Analysis of Food Additives*; CRC Press. Taylor & Francis Group: Boca Raton, FL, USA, 2016.

46. Fraqueza, M.J.; Alfaia, C.M.; Barreto, A.S. Biogenic amine formation in turkey meat under modified atmosphere packaging with extended shelf life: Index of freshness. *Poult. Sci.* **2012**, *91*, 1465–1472. [CrossRef] [PubMed]

47. Dadáková, E.; Pelikánová, T.; Kalač, P. Concentration of biologically active polyamines in meat and liver of sheep and lambs after slaughter and their changes in mutton during storage and cooking. *Meat Sci.* **2011**, *87*, 119–124. [CrossRef] [PubMed]

48. Rokka, M.; Eerola, S.; Smolander, M.; Alakomi, H.-L.; Ahvenainen, R. Monitoring of the quality of modified atmosphere packaged broiler chicken cuts stored in different temperature conditions: B. Biogenic amines as quality-indicating metabolites. *Food Control* **2004**, *15*, 601–607. [CrossRef]

49. Roig-Roig-Sagués, A.X.; Ruiz-Capillas, C.; Espinosa, D.; Hernández, M. The decarboxylating bacteria present in foodstuffs and the effect of emerging technologies on their formation. In *Biological Aspects of Biogenic Amines, Polyamines and Conjugates*; Dandrifosse, G., Ed.; Transworld Research Network: Kerala, India, 2009.

50. Ten Brink, B.; Damink, C.; Joosten, H.M.; Huis in't Veld, J.H. Occurrence and formation of biologically active amines in foods. *Int. J. Food Microbiol.* **1990**, *11*, 73–84. [CrossRef]

51. Gardini, F.; Martuscelli, M.; Caruso, M.C.; Galgano, F.; Crudele, M.A.; Favati, F.; Guerzoni, M.E.; Suzzi, G. Effects of pH, temperature and NaCl concentration on the growth kinetics, proteolytic activity and biogenic amine production of Enterococcus faecalis. *Int. J. Food Microbiol.* **2001**, *64*, 105–117. [CrossRef]

52. Curiel, J.A.; Ruiz-Capillas, C.; de Las Rivas, B.; Carrascosa, A.V.; Jiménez-Colmenero, F.; Muñoz, R. Production of biogenic amines by lactic acid bacteria and enterobacteria isolated from fresh pork sausages packaged in different atmospheres and kept under refrigeration. *Meat Sci.* **2011**, *88*, 368–373. [CrossRef] [PubMed]

53. Ruiz-Capillas, C.; Herrero, A.; Triki, M.; Jiménez-Colmenero, F. Biogenic Amine Formation in Reformulated Cooked Sausage Without Added Nitrite. *J. Nutr. Med. Diet Care* **2017**, *3*, 2–6. [CrossRef]

54. James, S.; James, C. Raw material selection: meat and poultry. In *Chilled Foods: A Comprehensive Guide*, 3rd ed.; Brown, M., Ed.; Woodhead Publishing: Cambridge, England, 2008; pp. 61–82.

12

Effect of Brine Concentrations on the Bacteriological and Chemical Quality and Histamine Content of Brined and Dried Milkfish

Chiu-Chu Hwang [1,*]**, Yi-Chen Lee** [2]**, Chung-Yung Huang** [2]**, Hsien-Feng Kung** [3]**,
Hung-Hui Cheng** [4] **and Yung-Hsiang Tsai** [2,*]

[1] Department of Hospitality Management, Yu Da University of Science and Technology, Miaoli 361027, Taiwan
[2] Department of Seafood Science, National Kaohsiung University of Science and Technology,
    Kaohsiung 811213, Taiwan; lionlee@nkust.edu.tw (Y.-C.L.); cyhuang@nkust.edu.tw (C.-Y.H.)
[3] Department of Pharmacy, Tajen University, Pingtung 907391, Taiwan; khfeng@mail.tajen.edu.tw
[4] Mariculture Research Center, Fisheries Research Institute, Council of Agriculture, Tainan 724028, Taiwan;
    cheng.hunghui@msa.hinet.net
* Correspondence: omics1@ydu.edu.tw (C.-C.H.); yhtsai01@seed.net.tw (Y.-H.T.)

Abstract: In this research, the occurrence of hygienic quality and histamine in commercial brined and dried milkfish products, and the effects of brine concentrations on the quality of brined and dried milkfish, were studied. Brined and dried milkfish products ($n = 20$) collected from four retail stores in Taiwan were tested to investigate their histamine-related quality. Among them, five tested samples (25%, 5/20) had histamine contents of more than 5 mg/100 g, the United States Food and Drug Administration guidelines for scombroid fish, while two (10%, 2/20) contained 69 and 301 mg/100 g of histamine, exceeding the 50 mg/100 g potential hazard level. In addition, the effects of brine concentrations (0%, 3%, 6%, 9%, and 15%) on the chemical and bacteriological quality of brined and dried milkfish during sun-drying were evaluated. The results showed that the aerobic plate count (APC), coliform, water activity, total volatile basic nitrogen (TVBN), and histamine content values of the brined and dried milkfish samples decreased with increased brine concentrations, whereas those of salt content and thiobarbituric acid (TBA) increased with increasing brine concentrations. The milkfish samples prepared with 6% NaCl brine had better quality with respect to lower APC, TVBN, TBA, and histamine levels.

Keywords: histamine; dried milkfish; hygienic quality; brine-salting

1. Introduction

Histamine is a biogenic amine in charge of histamine fish poisoning (HFP) or scombroid poisoning. Histamine fish poisoning is a food outbreak with allergy-like symptoms arising from ingesting mishandled scombroid fish that have high levels of histamine in their flesh [1]. Histamine is formed mainly through the decarboxylation of free histidine in fish muscles by histidine decarboxylases produced by a number of histamine-forming bacteria present in seafood [2]. HFP has occasionally been associated with the consumption of milkfish, marlin, mackerel, and tuna in Taiwan [2–6]. However, there is compelling evidence to implicate that other factors, such as other biogenic amines, can potentiate histamine toxicity, as spoiled fish containing histamine tends to be more toxic than the equivalent amount of pure histamine that is ingested orally [1,2]. Putrescine and cadaverine were shown to enhance histamine toxicity when present in spoiled fish by inhibiting the intestinal histamine metabolizing enzyme, including diamine oxidase [1,2].

Milkfish (*Chanos chanos*) is widely distributed throughout the Indo-Pacific region and is the second most important inland aquaculture fish in Taiwan [7,8]. This fish has been cultivated in Taiwan for more than 350 years. Taiwan's total milkfish production is approximately 50,000–60,000 tons each year [8]. Chiou et al. [9] demonstrated that histidine at 441 mg/100 g is the most prominent free amino acid (FAA) in the white muscles of milkfish and accounts for 80% of the total FAAs in the fish. Therefore, milkfish products have become most often associated with HFP in Taiwan, including dried milkfish [6], milkfish sticks [10], and milkfish surimi [11]. In addition, our research team determined that 78% of commercial dry-salting and dried milkfish products have histamine contents greater than the 5 mg/100 g recommended value of the United States Food and Drug Administration's (USFDA) guidelines, while 43.7% of the fish samples were found to exceed 50 mg/100 g of histamine [12].

In general, there are two major salting methods for milkfish preservation, namely, dry-salting and brine-salting. In Taiwan, the traditional processes of dry-salting and dried milkfish include scaling, back-cutting, degutting, and dry-salting with 3–12% NaCl (*w/w*) followed by sun-drying for 5–7 days [12]. However, the consumption of high salt levels from seafood can result in several chronic diseases, such as hypertension and cardiovascular diseases [13]. Brine-salting for fish processing may be a better method to reduce salt uptake and water loss and, thus, to reach a higher weight yield and better quality in salted fish compared to dry-salting [14]. Therefore, in recent years, brine- and light-salting milkfish has gained popularity with Taiwanese people. However, the quality of brined and dried fish is influenced by the brine concentrations and dry methods used for drying the fish [15].

There is no information of the occurrence of hygienic quality and histamine in brined and dried milkfish products, and the formation of histamine and the quality of brined and dried milkfish produced with different brine concentrations. Therefore, the main aim of this study was to monitor the bacteriological and chemical quality, including histamine content, in 20 brined and dried milkfish samples sold in retail stores in southern Taiwan. This work also aimed to examine the effects of different brine concentrations (0%, 3%, 6%, 9%, and 15%) on the bacteriological and chemical quality and histamine contents in brined and dried milkfish products during sun-drying for five days.

2. Materials and Methods

2.1. Materials

Twenty brined and dried milkfish products were collected from four retail stores in southern Taiwan, including store A (six samples), store B (five samples), store C (five samples), and store D (four samples). All brined and dried milkfish products were home-made by the farmer or manufacturer and delivered to the store for sale. Trackback information indicated that the samples collected from store A and D were processed using higher brine concentration (>10%) and longer sun-drying days (5–7 days); on the other hand, the samples of store B and C were processed using lower brine concentration (<6%) and shorter sun-drying days (3–5 days). In general, the processing of brined and dried milkfish include scaling, back-cutting, degutting, and brine-salting with 3–15% NaCl concentrations at room temperature for 1-2 h, followed by sun-drying for 3–7 days. After the samples were purchased, they were wrapped in aseptic bags, placed in an ice box, and instantly delivered to the laboratory for analysis within 6 h. The dorsal part of the commercial dried milkfish samples were cut and taken for microbiological and chemical determinations.

Sixty fresh milkfish (weights of 546 ± 11.6 g, lengths of 31.9 ± 1.2 cm) were purchased from the fish market of the city of Kaohsiung in Taiwan and transported to our laboratory within half an hour in an ice box. Once the fish samples arrived at the laboratory, they were manually scaled, back-cut, gutted, washed with clean water, and then drained.

2.2. Reagents

Histamine dihydrochloride, trichloroacetic acid, 2-thiobarbituric acid, and butylated hydroxytoluene were purchased from Sigma-Aldrich (St. Louis, MO, USA). Acetonitrile (LC grade) and dansyl chloride (GR grade) were purchased from E. Merck (Darmstadt, Germany).

2.3. Brine-Salting and Drying of Milkfish

The back-cut milkfish were brine-salted with concentrations of 3%, 6%, 9%, or 15% NaCl with a fish-to-brine ratio of 1:2 for 60 min at 20 °C, and unsalted milkfish were used as controls. After brine-salting, all milkfish samples were placed under sun light at 30–33 °C for seven hours each day for five days. Sampling analyses were conducted at days 1, 3, and 5 for sun-drying. The experiments were conducted in triplicate for each brine concentration and sampling time. The dorsal part of the fish samples was used for analysis.

2.4. Determination of pH Value, Moisture Content, Water Activity, and Salt Content

Ten grams of the samples was weighted and homogenized with a mixer (FastPrep-24, MP Biomedicals, Solon, OH, USA) for 2 min with 20 mL of deionized water to make a thick slurry. The pH of this slurry was determined using a digital pH meter (Mettler Toledo FE20/EL20, Schwerzenbach, Switzerland). The moisture of each sample (1–3 g) was measured using the oven-dry method at 105.0 ± 1.0 °C for drying, followed by the determination of the sample weight until a constant weight was achieved. Water activity was determined by an Aqualab 4TE (Decagon Devices, Pullman, WA, USA) at 25 °C. The salt (NaCl) content was determined using Mohr's titration method [16].

2.5. Determination of Total Volatile Basic Nitrogen (TVBN) and Thiobarbituric Acid (TBA)

The TVBN values were measured using Conway's dish method as described by Cobb et al. [17]. Five grams of the minced samples was homogenized with 45 mL of 6% trichloroacetic acid (TCA; Sigma-Aldrich, St. Louis, MO, USA). After the extract was filtered, saturated K_2CO_3 was added to the filters. The released TVBN was absorbed by boric acid and then titrated with 0.02 N HCl, while the TVBN value was expressed in milligrams per 100 g fish sample. The TBA values were determined by the modified method of Faustman et al. [18]. Briefly, 20 g of dried milkfish sample was added into a tube containing 180 mL of deionized water and then homogenized with a mixer for 3 min. Twelve milliliters of 0.1 M TBA reagent in 0.2% HCl and 0.15 mL of 0.2% butylated hydroxytoluene (BHT) in 95% ethanol were added into 2 mL of the homogenate and then mixed well. The mixtures were heated in a water bath at 90 °C for 20 min and then filtered, and the absorbance of the filtrates was detected using a spectrophotometer (UV-1201, Shimazu, Tokyo, Japan) at 532 nm. The TBA values in the fish samples are expressed in milligrams of malondialdehyde (MDA) per kilogram.

2.6. Microbiological Analysis

Twenty-five grams of the minced samples was homogenized with 225 mL of sterile 0.85% (w/v) physiological saline in a sterile blender at a 1200 rpm speed for 2 min. The homogenate was serially diluted with a sterile physiological saline for 1:10 (v/v) dilutions. With regard to spread plate counting, 0.1 mL of the dilutes was spread on aerobic plate count (APC) agar (Difco, BD, Sparks, MD, USA) with 0.5% NaCl and then incubated at 30 °C for 24–48 h. After the bacterial colonies grown on the plate were counted, the data were expressed as log_{10} colony forming units (CFUs) per gram. The levels of coliform and Escherichia coli in the milkfish samples were performed according to the three-tube most probable number (MPN) method as described by the FDA [19].

2.7. Histamine Analysis

Histamine dihydrochloride (82.8 mg) was dissolved in 50 mL of 0.1 M HCl and used as the working solution, and the final concentration of histamine (free base) was 1.0 mg/mL. Five grams of the ground milkfish samples were homogenized with 20 mL of 6% cold trichloroacetic acid (TCA) using a Polytron PT-MR 3100 homogenizer for 3 min. The homogenates were collected via centrifugation at 4500× g for 8 min at 7 °C and filtered through Advantec Toyo No. 2 filter paper. The filtrates were diluted up to 50 mL with a 6% TCA solution. For the derivatization reaction of histamine, 1 mL aliquots of the TCA extract of each sample and histamine standard solution were derivatized with dansyl chloride using the method of Chen et al. [3] with some modifications. Briefly, 0.2 mL of 2 M sodium hydroxide and 0.3 mL of saturated sodium bicarbonate were added to 1 mL aliquots of the TCA extract of each sample and the histamine standard solution. The solution was added to 2 mL of 1% dansyl chloride solution dissolved in acetone, mixed by a vortex mixer, and left to stand at 40 °C for 45 min. After the reaction, 100 µL of ammonia was added to terminate the derivatization reaction. Acetonitrile was added to a final volume of 5 mL and the solution was centrifuged (10,000× g, 5 min, 4 °C). After the supernatants were filtered through 0.22 µm membrane filters, 20 µL of the filtrates were injected into high-performance liquid chromatography (HPLC). The histamine levels in each milkfish sample were analyzed by HPLC (Hitachi, Tokyo, Japan) equipped with a LiChrospher 100 RP-18 reversed-phase column (5 µm, 125 × 4.6 mm, E. Merck, Damstadt, Germany) and a UV-Vis detector (Model L-4000, Hitachi, wavelength at 254 nm). The mobile phase consisted of eluent A (acetonitrile) and eluent B (water). At the beginning, eluents A and B at a ratio of 50:50 (v/v) were applied for 19 min, followed by a linear gradient with an increase of eluent A up to 90% during the next minute. In the final 10 min, the eluent A and B mix was set to a linear decrease to 50:50 (v/v). The flow rate was 1.0 mL/min. Validation of the histamine analysis method including inter- and intra-day repeatability (expressed as % and relative standard deviation, RSD) was determined by fortifying homogenized dried milkfish meats with 1.0, 5.0, and 10 mg/100 g of standard histamine. Each spiked amount was extracted and derivatized with dansyl chloride using the above procedure in triplicate, including a blank test to evaluate the average recovery.

2.8. Statistical Analysis

One-way analysis of variance (ANOVA) and Tukey's pairwise comparison tests were performed within the 95% confidence interval. Pearson correlation was carried out to determine relationships between pH, moisture, water activity, salt content, TVBN, APC, coliform, and histamine contents in the brined and dried milkfish samples. All statistical analyses were carried out using the Statistical Package for Social Sciences (SPSS) Version 16.0 for Windows (SPSS Inc., Chicago, Il, USA), and $p < 0.05$ was used to consider significant deviation.

3. Results and Discussion

3.1. Chemical and Bacteriological Quality of the Brined and Dried Milkfish Samples

For all 20 brined and dried milkfish samples collected from the four retail stores, the pH, moisture, water activity, salt content, TVBN, APC, coliform, and histamine ranged from 5.67 to 6.05, 38.27% to 69.78%, 0.89 to 0.98, 0.16% to 4.37%, 8.86 to 19.88 mg/100 g, 3.51 to 8.25 log CFU/g, <3 to >2400 MPN/g, and 0.16 to 301 mg/100 g, respectively (Table 1). *E. coli* was not detected in any milkfish samples. Store A samples had significantly lower ($p < 0.05$) mean water activity (0.94) than did samples collected from the other three stores, while the mean salt content (3.23%) in store A samples was higher ($p < 0.05$) than the others (Table 1). Moreover, the mean TVBN and APC values in store B samples (16.06 mg/100 g and 6.62 log CFU/g, respectively) and store D samples (16.82 mg/100 g and 6.33 log CFU/g, respectively) were markedly higher ($p < 0.05$) than those samples obtained from the other two stores, while the mean coliform level (356 MPN/g) in store D samples were higher than that of the other stores (Table 1). The highest mean histamine content of 79 mg/100 g was obtained from five samples from store B, followed by store D with a mean of 4.9 mg/100 g of histamine.

Table 1. pH, moisture, water activity, salt content, total volatile basic nitrogen (TVBN), aerobic plate count (APC), coliform, *Escherichia coli*, and histamine values in brined and dried milkfish products.

Sample Sources	Number of Samples	pH	Moisture (%)	Water Activity	Salt Content (%)	TVBN (mg/100 g)	APC (log CFU/g)	Coliform (MPN/g)	E. coli (MPN/g)	Histamine (mg/100 g)
A	6	5.74~5.83	38.27~65.37	0.89~0.97	2.47~4.37	8.86~17.36	3.51~7.65	<3~70	<3	0.34~4.9
		(5.78 ± 0.04) A	(51.78 ± 9.28) B	(0.94 ± 0.03) B	(3.23 ± 0.78) A	(12.79 ± 3.25) B	(5.50 ± 1.05) B	(42 ± 25) B		(1.3 ± 1.8) C
B	5	5.67~6.05	52.31~60.18	0.97~0.98	0.16~1.88	13.86~19.32	5.87~8.03	<3~40	<3	0.62~301
		(5.76 ± 0.17) A	(55.41 ± 3.23) AB	(0.98 ± 0.01) A	(0.80 ± 0.77) C	(16.06 ± 2.08) A	(6.62 ± 0.83) A	(25 ± 13) B		(79 ± 57) A
C	5	5.80~5.95	54.68~69.78	0.97~0.98	0.61~1.02	10.85~13.93	4.18~5.92	<3~240	<3	0.16~0.31
		(5.87 ± 0.07) A	(64.17 ± 5.95) A	(0.98 ± 0.01) A	(0.86 ± 0.18) C	(12.35 ± 1.39) B	(5.32 ± 0.66) B	(90 ± 130) B		(0.20 ± 0.11) C
D	4	5.75~5.82	51.07~54.12	0.95~0.96	1.47~1.80	15.40~19.88	5.80~8.25	20~>2400	<3	0.24~19
		(5.78 ± 0.03) A	(52.46 ± 1.26) B	(0.96 ± 0.01) AB	(1.68 ± 0.15) B	(16.82 ± 2.66) A	(6.33 ± 0.60) A	(356 ± 47) A		(4.9 ± 4.8) B

The values in parentheses represent the mean ± standard deviation (SD). Values in the same column with different letters are statistically different ($p < 0.05$). CFU, colony forming unit; MPN, most probable number.

In this study, the proportion of the 20 brined and dried milkfish samples that did not meet the 6.47 log CFU/g Taiwanese regulatory standard for APC was 35% (7/20). Therefore, brined and dried milkfish manufacturers may need to be more careful with hygienic handling or processing in their preparation of brined and dried milkfish products. The distribution of histamine contents in the brined and dried samples is shown in Table 2. Five samples (25%, 5/20) failed to meet the 5 mg/100 g level of histamine, the allowable limit by the USFDA for scombroid fish and/or products, while two (10%) had 69 and 301 mg/100 g of histamine, greater than the potential toxicity level (50 mg/100 g). According to information by Bartholomev et al. [20], which showed that fish with histamine levels >100 mg/100 g could result in illness and health hazards if ingested by humans, one sample with 301 mg/100 g of histamine could have caused disease symptoms if consumed (Table 2). In contrast, our previous research showed that 78.1% (25 samples) and 43.7% (14 samples) of 32 dry-salted and dried milkfish products contained more than 5 mg/100 g and 50 mg/100 g of histamine, respectively [21].

Table 2. Distribution of the histamine content in the 20 brined and dried milkfish products.

Content of Histamine (mg/100 g)	Brined and Dried Milkfish Products	
	Number of Samples	% of Samples
<4.9	15	75
5.0–49.9	3	15
50.0–99.9	1	5
>100	1	5
Total	20	100

High levels of histamine have been found in various types of milkfish implicated in HFP. Our research group detected 61.6 mg/100 g of histamine in dried milkfish products that were implicated in an incident of HFP [6]. Two fried milkfish sticks implicated in a poisoning incident contained 86.6 mg/100 g and 235.0 mg/100 g of histamine [10]. The high content of histamine (i.e., 91.0 mg/100 g) in a suspected milkfish surimi product could be the etiological factor for this fish-borne poisoning in Taiwan [11]. Therefore, it is also very important for people, especially those from the Indo-Pacific region, such as the Philippines, Indonesia, and Taiwan, to be aware that milkfish products could become a hazardous food item, causing histamine poisoning.

Pearson correlation was conducted to determine if there existed any relationship among the pH, moisture, water activity (a_w), salt content, TVBN, APC, coliform, and histamine contents of the tested 20 samples. In general, positive correlations existed between moisture and a_w (r, correlation coefficient = 0.81, $p < 0.05$), TVBN and APC ($r = 0.76$, $p < 0.05$), APC and histamine ($r = 0.71$, $p < 0.05$), and histamine and TVBN ($r = 0.76$, $p < 0.05$). However, negative correlations were noted between moisture and salt content ($r = -0.73$, $p < 0.05$), and a_w and salt content ($r = -0.76$, $p < 0.05$).

3.2. Effect of Brine Concentrations on the Quality of Brined and Dried Milkfish

Changes in the moisture and water activity (a_w) of the milkfish samples pre-immersed in different brine concentrations (i.e., 0%, 3%, 6%, 9%, and 15%) during a sun-drying period of five days are shown in Figure 1. The initial moisture of the fish samples was 70.3%, while the moisture of all fish samples rapidly decreased with increasing drying time. At the end of the drying period, the moisture content in all of the samples ranged from 44.2% to 46.9%, and no significant differences ($p > 0.05$) were observed among the samples of the various brine concentrations and control samples (Figure 1A). For all fish samples with an initial a_w value of 0.985, the a_w values gradually decreased with an increase in the drying time and reduced to 0.967 in the control sample, 0.959 in the 3% NaCl sample, 0.950 in the 6% NaCl sample, 0.945 in the 9% NaCl sample, and 0.942 in the 15% NaCl sample at the end of the sun-drying period (Figure 1B). It was found that the milkfish samples with higher brine concentrations had lower a_w values ($p < 0.05$).

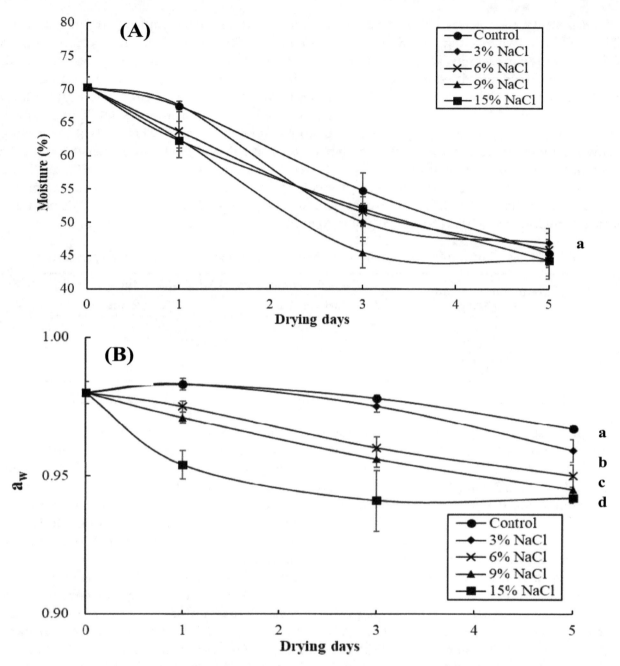

Figure 1. Changes in the moisture (**A**) and water activity (a_w) (**B**) in the milkfish samples as a result of brine-salting with 0% (control), 3%, 6%, 9%, and 15% NaCl during sun-drying. Each value represents the mean ± SD of three replications. Different lower letters indicate significant differences ($p < 0.05$) within the data at the end of the sun-drying period.

Changes in the pH and salt content of the milkfish samples pre-immersed in different brine concentrations (i.e., 0%, 3%, 6%, 9%, and 15%) over a sun-drying period of five days are presented in Figure 2. The pH values of the milkfish samples slightly increased from the initial reading of 5.41 to 5.69 for the control sample, 5.70 for the 3% and 6% NaCl samples, 5.87 for the 9% NaCl sample, and 5.89 for the 15% NaCl sample at the end of the sun-drying period. The increase in the pH for all of the group samples may be due to the formation of basic components, including ammonia, trimethylamine, and other amines by bacterial spoilage [22]. Moreover, the final pH values of the 9% and 15% NaCl samples were higher ($p < 0.05$) than those of the control and the 3% and 6% NaCl samples (Figure 2A). As shown in Figure 2B, the salt content in the fish sample slightly increased from 0.05% to 0.13% in the control sample, 0.20% to 0.70% in the 3% NaCl sample, 0.51% to 1.17% in the

6% NaCl sample, 0.85% to 2.24% in the 9% NaCl sample, and 1.62% to 2.87% in the 15% NaCl sample after give days of sun-drying. The results also show that the milkfish samples pre-immersed in a higher brine concentration had a higher salt content ($p < 0.05$).

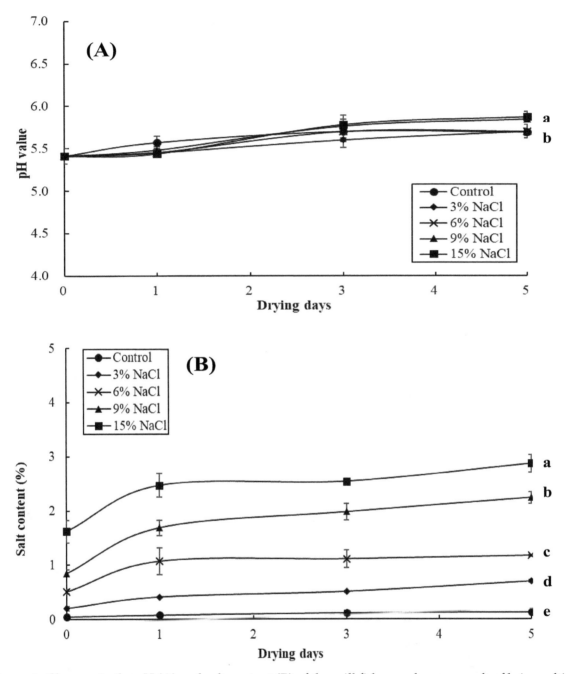

Figure 2. Changes in the pH (**A**) and salt content (**B**) of the milkfish samples as a result of brine-salting with 0% (control), 3%, 6%, 9%, and 15% NaCl during sun-drying. Each value represents the mean ± SD of three replications. Different lower letters indicate significant differences ($p < 0.05$) within the data at the end of the sun-drying period.

Figure 3 shows the changes in the TVBN and TBA values in the milkfish samples pre-immersed in different brine concentrations (i.e., 0%, 3%, 6%, 9%, and 15%) during a sun-drying period of five days. Initially, the milkfish samples had 13.7 mg/100 g of TVBN, and subsequently, the TVBN content in all fish samples increased gradually while drying, reaching 34.0 mg/100 g for the control sample, 30.5 mg/100 g for the 3% NaCl sample, 29.76 mg/100 g for the 6% NaCl sample, 27.0 mg/100 g for the 9% NaCl sample, and 26.9 mg/100 g for the 15% NaCl sample at the end of the sun-drying period. Thus, the highest TVBN level was detected in the control sample, followed by the 3% and 6% NaCl samples, and the lowest levels were observed for the 9% and 15% NaCl samples ($p < 0.05$) (Figure 3A). Connell [23] revealed that the increase in TVBN is due to the production of volatile basic compounds, including ammonia, trimethylamine and dimethylamine, via decomposition by autolytic enzymes and spoilage bacteria. Moreover, Nooralabettu [15] demonstrated that the addition of NaCl addition in Bombay duck can decrease autolytic enzyme activity in fish meat. An increase in salt content above 1% in fish can have an inhibitory effect on the bacteria associated with fish spoilage [24]. Consequently, the high content of TVBN in the unsalted samples (i.e., the control sample) obtained in this study was probably due to the increasing decomposition by enzymes and spoilage bacteria with the lack of salt's inhibitory effect.

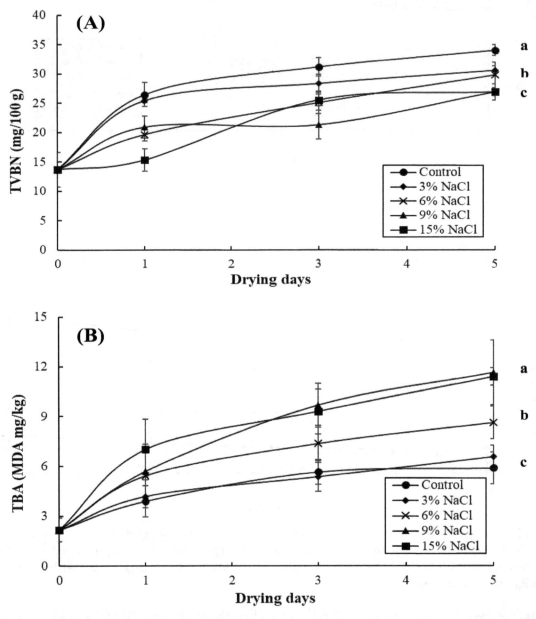

Figure 3. Changes in total volatile basic nitrogen (TVBN) (**A**) and thiobarbituric acid (TBA) (**B**) in the milkfish samples as a result of brine-salting with 0% (control), 3%, 6%, 9%, and 15% NaCl during sun-drying. Each value represents the mean ± SD of three replications. Different lower letters indicate significant differences ($p < 0.05$) within the data at the end of the sun-drying period.

Thiobarbituric acid (TBA), a measure of MDA as a secondary lipid oxidation product, is one of the most widely used indicators for the assessment of food lipid oxidation [25]. Initially, the TBA values for the control and brined samples were 2.18 MDA mg/kg. The value of TBA in all of the samples increased during the sun-drying period, reaching 5.9 MDA mg/kg for the control sample, 6.5 MDA mg/kg for the 3% NaCl sample, 8.6 MDA mg/kg for the 6% NaCl sample, 11.5 MDA mg/kg for the 9% NaCl sample, and 11.4 MDA mg/kg for the 15% NaCl sample at the end of the sun-drying period. In contrast to TVBN, the highest levels of TBA were observed in the 9% and 15% NaCl samples, followed by the 6% NaCl sample, and the lowest TBA level was detected in the control and 3% NaCl samples ($p < 0.05$) (Figure 3B). Yanar et al. [26] also reported that hot-smoked tilapia samples treated with a 15% brine concentration contained very high levels of TBA. Sodium chloride can promote lipid oxidation, while sodium ions may replace iron from myoglobin, thereby resulting in free iron ions for the catalysis of lipid oxidation [26,27].

Therefore, the results in this study reveal that the high TBA values in the samples prepared with 9% and 15% brine concentrations may be attributed to the addition of NaCl by accelerating the rate of lipid oxidation. In addition, when seafood is dried by exposure to sunlight, lipids can be oxidized and low molecular weight carbonyl components can be produced [28]. The results of this study are in agreement with a previous study reporting that the TBA values of dried yellow corvina increased rapidly during sun-drying [28].

Figure 4 shows the changes in APC and coliform bacteria in the milkfish samples pre-immersed in different brine concentrations (i.e., 0%, 3%, 6%, 9%, and 15%) over a sun-drying period of five days. The APC numbers of the milkfish sample gradually increased from the initial population of 3.21 to 6.88 log CFU/g for the control sample, 6.81 log CFU/g for the 3% NaCl sample, 6.15 log CFU/g for the 6% NaCl sample, 6.0 log CFU/g for the 9% NaCl sample, and 5.86 log CFU/g for the 15% NaCl sample at the end of the sun-drying period. Thus, the APC bacteria detected in the control and 3% NaCl samples were markedly higher ($p < 0.05$) than those of other brine concentration samples (Figure 4A) and exceeded the 6.47 log CFU/g Taiwanese regulatory standard. Similar to the APC population, the growth of coliform in this fish samples was considerably faster in the unsalted (control) sample than in the other brined samples ($p < 0.05$). The coliform counts in the control, 3%, 6%, 9%, and 15% NaCl samples increased to 3.51, 2.87, 2.75, 2.70, and 2.41 log MPN/g, respectively, at the end of the sun-drying period (Figure 4B). These results are in agreement with our previous report, in which the APC and coliform levels of dry-salted and sun-dried milkfish samples decreased with increasing salt concentrations [21].

A similar finding was also reported by Yang et al. [14], who found that higher brine-salting could inhibit the growth of bacteria in grass carp. Moreover, higher brine concentrations (>6%) in the milkfish samples obviously had a repressive action on microbiological growth in this study, indicating that salt content is able to inactivate or inhibit bacteria.

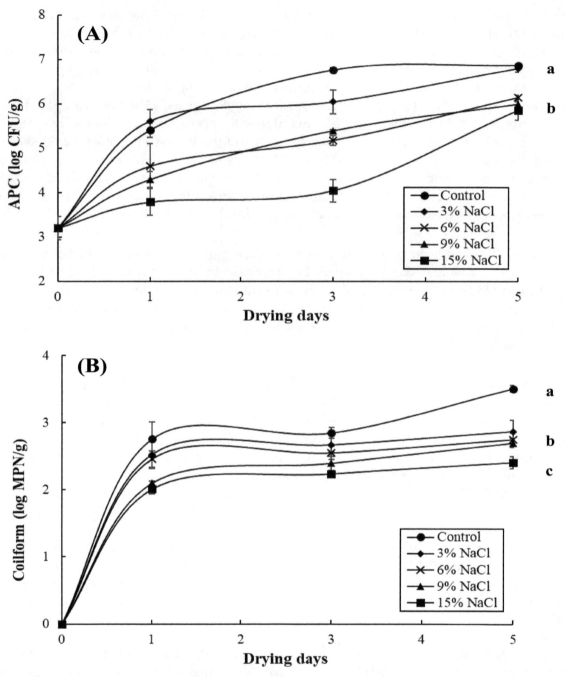

Figure 4. Changes in aerobic plate count (APC) (**A**) and coliform (**B**) in the milkfish samples as a result of brine-salting with 0% (control), 3%, 6%, 9%, and 15% NaCl during sun-drying. Each value represents the mean ± SD of three replications. Different lower letters indicate significant differences ($p < 0.05$) within the data at the end of the sun-drying period.

Figure 5 shows that the histamine content in the control sample increased gradually during the sun-drying period, reaching 4.8 mg/100 g by the end. On the other hand, the histamine contents in the 3%, 6%, 9%, and 15% NaCl samples only slightly increased during the sun-drying period, reaching 2.8, 2.0, 0.79, and 0.27 mg/100 g, respectively, by the end. In conclusion, the histamine content observed in the control sample was markedly higher ($p < 0.05$) than that of the other brine concentrations samples (Figure 5). These results agree with the previous research of Hwang et al. [21], where high contents of histamine at 67 mg/100 g were found in unsalted dried milkfish samples via sun-drying. The low levels of histamine (<2.8 mg/100 g) detected in the salted samples (>3% NaCl) in this study may be due to the growth reduction of histamine-forming bacteria by the preservative effect of salt, indicating that the addition of salt could be effective in reducing or inhibiting histamine accumulation. In our previous study, high levels of a_w, moisture, TVBN, APC, and histamine were detected in unsalted dried milkfish samples produced by sun-drying; therefore, dried milkfish producers should be aware that dried milkfish with low salt and sun-drying periods could become a vehicle for histamine poisoning [21]. Similarly, since high levels of TVBN (>30 mg/100g), APC (>6.81 log CFU/g), and histamine (>2.8 mg/100 g) were observed in the unsalted and 3% NaCl samples during the sun-drying period, brined and dried milkfish manufacturers should pay attention to the fact that dried milkfish brined with a low amount of salt (<3% NaCl) and a sun-drying period could lead to worse hygienic quality and potential hazards, such as food poisoning. However, the samples with higher brine concentrations (>9% NaCl) had higher TBA values (>11.4 MDA mg/kg) (Figure 3B). With regard to an assessment of APC, TBA, TVBN, and histamine, this study suggests that dried milkfish brined with a 6% NaCl addition has better chemical and bacteriological quality.

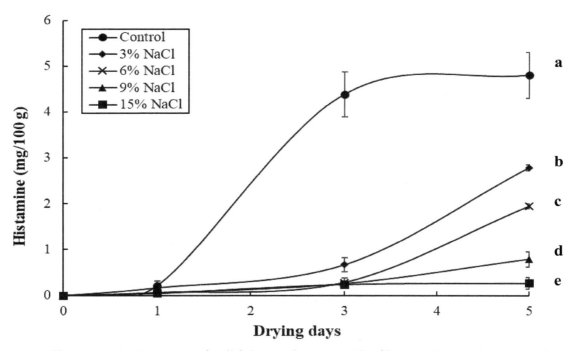

Figure 5. Changes in the histamine of milkfish samples as a result of brine-salting with 0% (control), 3%, 6%, 9%, and 15% NaCl during sun-drying. Each value represents the mean ± SD of three replications. Different lower letters indicate significant differences ($p < 0.05$) within the data at the end of the sun-drying period.

Pearson correlation was conducted to determine if there existed any relationship among the moisture, a_w, pH, salt content, TVBN, TBA, APC, coliform, and histamine contents of the samples at the end of the sun-drying period. In general, positive correlations existed between APC and a_w ($r = 0.95$, $p < 0.05$), APC and histamine ($r = 0.88$, $p < 0.05$), coliform and a_w ($r = 0.93$, $p < 0.05$), coliform and histamine ($r = 0.90$, $p < 0.05$), a_w and TVBN ($r = 0.88$, $p < 0.05$), salt content and TBA ($r = 0.85$, $p < 0.05$), a_w and histamine ($r = 0.89$, $p < 0.05$), and histamine and TVBN ($r=0.86$, $p < 0.05$). However, negative correlations were noted between salt content and APC ($r = -0.89$, $p < 0.05$), salt content and coliform ($r = -0.90$, $p < 0.05$), salt content and a_w ($r = -0.92$, $p < 0.05$), a_w and TBA ($r = -0.90$, $p < 0.05$), and salt content and histamine ($r = -0.95$, $p < 0.05$).

4. Conclusions

This study, aimed at investigating the hygienic quality of 20 brined and dried milkfish products, revealed that the APC numbers in seven samples (35%) exceeded the 6.47 log CFU/g Taiwanese regulatory standard. Moreover, 25% of the tested samples had histamine contents greater than the

5 mg/100 g recommended by the USFDA in their guideline levels, and 10% (2/20) of the fish samples had >50 mg/100 g of histamine. After the consumption of these samples, histamine fish poisoning could occur. In addition, the chemical and bacteriological quality of the brined and dried milkfish pre-immersed in various brine concentrations during a sun-drying period were observed in this study. Although the samples prepared with higher brine concentrations presented a retarded APC growth rate and a reduced formation of TVBN and histamine, as compared with the control sample, they produced higher TBA values. It is suggested that 6% NaCl for brined milkfish is the optimal condition for maintaining the quality of brined and dried milkfish. Our results could suggest that application of brine concentration information is effective in controlling quality and enhancing the safety of brined and dried milkfish products.

Author Contributions: Conceptualization, C.-C.H., Y.-C.L., and Y.-H.T.; methodology, C.-C.H., Y.-H.T., and H.-H.C.; analysis, C.-C.H., Y.-C.L., and H.-F.K.; data curation, Y.-C.L., H.-F.K., and C.-C.H.; writing—original draft preparation, C.-C.H. and Y.-H.T.; writing—review and editing, C.-C.H., Y.-H.T., and C.-Y.H.; supervision, Y.-C.L. and H.-F.K.; project administration, C.-C.H.; funding acquisition, C.-C.H. and Y.-H.T. All authors read and agreed to the published version of the manuscript.

Acknowledgments: The authors gratefully acknowledge Shinn-Lih Yeh, a director of Mariculture Research Center, Fisheries Research Institute with providing equipment, Su-Sing Liu with the operation and maintenance of the equipment, and partly financial support from higher education sprout project of National Kaohsiung University of Science and Technology.

References

1. Hugerford, J.M. Scombroid poisoning: A review. *Toxicon* **2010**, *56*, 231–243. [CrossRef] [PubMed]
2. Chen, H.C.; Kung, H.F.; Chen, W.C.; Lin, W.F.; Hwang, D.F.; Lee, Y.C.; Tsai, Y.H. Determination of histamine and histamine-forming bacteria in tuna dumpling implicated in a food-borne poisoning. *Food Chem.* **2008**, *106*, 612–618. [CrossRef]
3. Chen, H.C.; Huang, Y.R.; Hsu, H.H.; Lin, C.S.; Chen, W.C.; Lin, C.M.; Tsai, Y.H. Determination of histamine and biogenic amines in fish cubes (*Tetrapturus angustirostris*) implicated in a food-borne poisoning. *Food Control* **2010**, *21*, 13–18. [CrossRef]
4. Chang, S.C.; Kung, H.F.; Chen, H.C.; Lin, C.S.; Tsai, Y.H. Determination of histamine and bacterial isolation in swordfish fillets (*Xiphias gladius*) implicated in a food borne poisoning. *Food Control* **2008**, *19*, 16–21. [CrossRef]
5. Tsai, Y.H.; Kung, H.F.; Lee, T.M.; Chen, H.C.; Chou, S.S.; Wei, C.I.; Hwang, D.F. Determination of histamine in canned mackerel implicated in a food borne poisoning. *Food Control* **2005**, *16*, 579–585. [CrossRef]
6. Tsai, Y.H.; Kung, H.F.; Chen, H.C.; Chang, S.C.; Hsu, H.H.; Wei, C.I. Determination of histamine and histamine-forming bacteria in dried milkfish (*Chanos chanos*) implicated in a food-borne poisoning. *Food Chem.* **2007**, *105*, 1289–1296. [CrossRef]
7. Hsieh, S.L.; Chen, Y.N.; Kuo, C.M. Physiological responses, desaturase activity, and fatty acid composition in milkfish (*Chanos chanos*) under cold acclimation. *Aquaculture* **2003**, *220*, 903–918. [CrossRef]
8. Chiang, F.S.; Sun, C.H.; Yu, J.M. Technical efficiency analysis of milkfish (*Chanos chanos*) production in Taiwan-an application of the stochastic frontier production function. *Aquaculture* **2004**, *230*, 99–116. [CrossRef]
9. Chiou, T.K.; Shiau, C.Y.; Chai, T.J. Extractive nitrogenous components of cultured milkfish and tilapia. *Nippon Suisan Gakk.* **1990**, *56*, 1313–1317. [CrossRef]
10. Lee, Y.C.; Kung, H.F.; Wu, C.H.; Hsu, H.M.; Chen, H.C.; Huang, T.C.; Tsai, Y.H. Determination of histamine in milkfish stick implicated in a foodborne poisoning. *J. Food Drug Anal.* **2016**, *24*, 63–71. [CrossRef]
11. Hwang, C.C.; Kung, H.F.; Lee, Y.C.; Wen, S.Y.; Chen, P.Y.; Tsen, D.I.; Tsai, Y.H. Histamine fish poisoning and histamine production by *Raoultella ornithinolytica* in milkfish surimi. *J. Food Prot.* **2020**, *83*, 874–880. [CrossRef]
12. Hsu, H.H.; Chuang, T.C.; Lin, H.C.; Huang, Y.R.; Lin, C.M.; Kung, H.F.; Tsai, Y.H. Histamine content and histamine-forming bacteria in dried milkfish (*Chanos chanos*) products. *Food Chem.* **2009**, *114*, 933–938. [CrossRef]

13. Gallart-Jornet, L.; Rustad, T.; Barat, J.M.; Fito, P.; Escriche, I. Effect of superchilled storage on the freshness and salting behaviour of Atlantic salmon (*Salmo salar*) fillets. *Food Chem.* **2007**, *103*, 1268–1281. [CrossRef]

14. Yang, W.; Shi, W.; Qu, Y.; Wang, Z.; Shen, S.; Tu, L.; Huang, H.; Wu, H. Research on the quality changes of grass carp during brine salting. *Food Sci. Nutri.* **2020**, *8*, 2968–2983. [CrossRef]

15. Nooralabettu, K.P. Effect of sun drying and artificial drying of fresh, salted Bombay duck (*Harpodon neherius*) on the physical characteristics of the product. *J. Aquat. Food Prod. Technol.* **2008**, *17*, 99–116. [CrossRef]

16. AOAC. *Official Methods of Analysis of AOAC International*, 21st ed.; AOAC International: Arlington, VA, USA, 2019.

17. Cobb, B.F.; Alaniz, I.; Thompson, C.A. Biochemical and microbial studies on shrimp: Volatile nitrogen and amino nitrogen analysis. *J. Food Sci.* **1973**, *38*, 431–435. [CrossRef]

18. Faustman, C.; Spechtm, S.M.; Malkus, L.A.; Kinsman, D.M. Pigment oxidation in ground veal: Influence of lipid oxidation, iron and zinc. *Meat Sci.* **1992**, *31*, 351–362. [CrossRef]

19. FDA. *Bacteriological Analytical Manual*; AOAC International: Arlington, VA, USA, 1998.

20. Bartholomev, B.A.; Berry, P.R.; Rodhouse, J.C.; Gilhouse, R.J. Scombrotoxic fish poisoning in Britain: Features of over 250 suspected incidents from 1976 to 1986. *Epidemiol. Infect.* **1987**, *99*, 775–782. [CrossRef]

21. Hwang, C.C.; Lin, C.M.; Kung, H.F.; Huang, Y.L.; Hwang, D.F.; Su, Y.C.; Tsai, Y.H. Effect of salt concentrations and drying methods on the quality and formation of histamine in dried milkfish (*Chanos chanos*). *Food Chem.* **2012**, *135*, 839–844. [CrossRef]

22. Arulkumar, A.; Paramasiam, S.; Rameshthangam, P.; Rabie, M.A. Changes on biogenic, volatile amines and microbial quality of the blue swimmer crab (*Portunus pelagicus*) muscle during storage. *J. Food Sci. Technol.* **2017**, *54*, 2503–2511. [CrossRef]

23. Connell, J.J. Methods of assessing and selecting for quality. In *Control of Fish Quality*, 3rd ed.; Fishing News Books: Oxford, UK, 1990; pp. 122–150.

24. Mohamed, S.B.; Mendes, R.; Slama, R.B.; Oliveira, P.; Silva, H.A.; Bakhrouf, A. Changes in bacterial counts and biogenic amines during the ripening of salted anchovy (*Engraulis encrasicholus*). *J. Food Nutr. Res.* **2016**, *5*, 318–325.

25. Fernandez, J.; Perez-Alvarez, J.A.; Fernandez-Lopez, J.A. Thiobarbituric acid test for monitoring lipid oxidation in meat. *Food Chem.* **1997**, *59*, 343–353. [CrossRef]

26. Yanar, Y.; Celik, M.; Akamca, E. Effects of brine concentration on shelf-life of hot-smoked tilapia (*Oreochromis niloticus*) stored at 4 °C. *Food Chem.* **2006**, *97*, 244–247. [CrossRef]

27. Kanner, J.; Harel, S.; Jaffe, R. Lipid peroxidation in muscle food as affected by NaCl. *J. Agri. Food Chem.* **1991**, *39*, 1017–1021. [CrossRef]

28. Gwak, H.; Eun, J.B. Changes in the chemical characteristics of *Gulbi*, salted and dried yellow Corvenia, during drying at different temperatures. *J. Aquat. Food Prod. Technol.* **2010**, *19*, 274–283. [CrossRef]

13

Screening Method to Evaluate Amino Acid-Decarboxylase Activity of Bacteria Present in Spanish Artisanal Ripened Cheeses

Diana Espinosa-Pesqueira, Artur X. Roig-Sagués * and M. Manuela Hernández-Herrero

CIRTTA—Departament de Ciència Animal i dels Aliments, Universitat Autònoma de Barcelona, Travessera dels Turons S/N, 08193 Barcelona, Spain; diespe@gmail.com (D.E.-P.); manuela.hernandez@gmail.com (M.M.H.-H.)
* Correspondence: arturxavier.roig@uab.cat

Abstract: A qualitative microplate screening method, using both low nitrogen (LND) and low glucose (LGD) decarboxylase broths, was used to evaluate the biogenic amine (BA) forming capacity of bacteria present in two types of Spanish ripened cheeses, some of them treated by high hydrostatic pressure. BA formation in decarboxylase broths was later confirmed by High Performance Liquid Chromatography (HPLC). An optimal cut off between 10–25 mg/L with a sensitivity of 84% and a specificity of 92% was obtained when detecting putrescine (PU), tyramine (TY) and cadaverine (CA) formation capability, although these broths showed less capacity detecting histamine forming bacteria. TY forming bacteria were the most frequent among the isolated BA forming strains showing a strong production capability (exceeding 100 mg/L), followed by CA and PU formers. *Lactococcus, Lactobacillus, Enterococcus* and *Leuconostoc* groups were found as the main TY producers, and some strains were also able to produce diamines at a level above 100 mg/L, and probably ruled the BA formation during ripening. *Enterobacteriaceae* and *Staphylococcus* spp., as well as some *Bacillus* spp. were also identified among the BA forming bacteria isolated.

Keywords: biogenic amines; decarboxylase activity; screening method; artisanal cheese; high hydrostatic pressure

1. Introduction

Cheese is, after fish, the food product that most usually causes poisoning due to the presence of high amounts of biogenic amines (BA), compounds with psychoactive and vasoactive properties that can be formed in foodstuffs due to the microbial decarboxylation of amino acids [1–6]. Amino acid decarboxylase activity has been described for several groups of microorganisms, such as *Enterobacteriaceae, Pseudomonas* spp, *Enteroccocus, Micrococcus* and Lactic Acid Bacteria (LAB). These BA-producing organisms may be part of the microbiota of the raw materials or may be introduced by contamination during or after processing of foodstuffs [4,7–14]. The specificity of the amino acid decarboxylases is strain dependent [11,15]. Lactic acid bacteria (LAB) have an important role in cheese elaboration and they are also the most important bacterial group that may build-up biogenic amine (BA), especially tyramine (TY) and putrescine (PU), but also cadaverine (CA) and histamine (HI) [13,16–18]. Sumner et al. [19] isolated a strain of *Lactobacillus buchneri* (strain St2A) from a Swiss cheese involved in an outbreak of HI poisoning occurred in the USA in 1980 that was able to form high amounts of HI. This LAB, later classified al *L. parabuchneri*, is able to grow and produce histamine at refrigeration temperatures [20]. *Enterobacteriaceae, Staphylococcus* spp. or *Bacillus* spp. have also been related to the accumulation of diamines in foods, including cheese, but also TY and/or HI [8,17,21,22].

Diverse qualitative and quantitative methods have been described in the literature to evaluate the amino acid decarboxylase activity of microorganisms isolated in food products. Different culture media have been proposed to be used as screening qualitative procedures, the most being formulated as a basal medium that include sources of carbon (glucose), nitrogen (peptone, yeast or meat extract), vitamins, salt, a relative high amount of one (or several) precursor amino acids and a pH indicator (e.g., bromocresol purple). Decarboxylase activity is then detected by the pH shift that changes the color of the medium when the carboxylic group is released from the amino acid(s) leaving in the medium the more alkaline BA(s) [23,24]. False-positive results have been described probably due to the formation of other alkaline compounds [9,22,25], but also false-negative responses are possible as a result of the fermentative activity of some bacteria, such as LAB, which produce acid that neutralize the alkalinity of BA [23,26].

In a previous work, the formation of BA in two artisanal varieties of Spanish ripened cheese, one made of ewe's raw milk and other of goat's raw milk, was presented. The effect of high hydrostatic pressure (HHP) treatments on both the levels of BA formed and on the main microbial groups present was also evaluated [27]. The aim of the present work has been to develop a fast, reliable and easy to perform screening method to evaluate the bacterial formation capacity of a wide range of BA, and evaluate the BA formation capability of the microbiota present in these two varieties of cheese to understand why HHP treatments reduce the formation of BA.

2. Materials and Methods

2.1. Cheese Manufacturing

Two types of artisan ripened cheeses elaborated in Spain were studied in this survey, both made of enzymatic curd and pressed paste. The first one was produced from goat's raw milk in the region of Catalonia, northeast of Spain, and the second was made from ewe's raw milk in Castilla y León, central Spain. The procedure of sampling as well as the HHP treatment applied have been described in a previous work [27]. Three independent batches of each type of cheese were produced following the usual manufacturing procedures used by the manufacturers. Cheese samples were separated in three batches: samples not HHP treated (Control samples); samples HHP treated before the 5th day of ripening (HHP1) and samples treated after 15 days of ripening (HHP15, only for ewe's milk cheeses). HHP treatments were performed at 400 MPa for 10 min at a temperature of 2 °C using an Alstom HHP equipment (Alstom, Nantes, France) with a 2 L pressure chamber.

2.2. Strain Isolation

Ten grams of each cheese sample were homogenized in 90 mL of sterile Buffered Peptone Water (Oxoid, Basingstoke, Hampshire, UK) with a BagMixer 400 paddle blender (Interscience, St Nom la Bretèche, France) and plated on M-17 agar (Oxoid) supplemented with a bacteriological grade lactose solution (5 g/L, Oxoid) and incubated at 30 °C, 48 h to isolate *Lactococcus* spp.; on de Man Rogosa Sharpe agar (MRS, Oxoid) incubated at 30 °C for 48 h to isolate Lactobacilli; on Kenner Fecal *Streptococcus* Agar (KF, Oxoid) supplemented with 2,3,5-triphenyltetrazolium chloride solution 1% (Oxoid) and incubated at 37 °C for 48 h to isolate Enterococci; Violet Red Bile Glucose Agar (VRBG, Oxoid) incubated at 37 °C for 24 h to isolate *Enterobactericeae* and Baird Parker Agar (BPA, BioMérieux, Marcy L'Etoile, France) incubated at 37 °C for 24–48 h to isolate *Staphylococcus* strains.

A total of 688 isolates were randomly picked out from the different selective media. The purification of each isolated was made by streaking single colonies on Petri plates with Tryptone Soy Agar (Oxoid) and incubating at 30 °C for 24–48 h. Two TY producing strains of *Lactobacillus brevis* and *Lactobacillus casei* and an HI producing strain of *Staphylococcus epidermidis*, isolated from previous surveys were used as positive controls [8,9]. These cultures were recovered in 10 mL of Tryptone Soy Broth (Oxoid) and incubated at 30 °C for 24 h. The purity of each culture was verified by subculturing the *Lactobacillus brevis* and *Lactobacillus casei* strains onto MRS agar (Oxoid), incubated at 30 °C for 24 h, and the *Staphylococcus epidermidis* strain on BPA (BioMérieux) incubated at 37 °C for 24 h. Before performing the decarboxylase assay, each strain was suspended in a tube with physiological solution of NaCl 0.85% (Panreac, Barcelona, Spain) until reaching a turbidity of about 0.5 in the McFarland scale.

2.3. Preparation of Decarboxylase Media

Table 1 shows the composition of the two synthetic media formulated to determine the ability to form the most toxic BA (HI and TY) and their enhancers (PU and CA): Low Nitrogen Broth (LND), prepared with the objective to decrease the incidence of false positive results of bacteria with a strong peptidase (or deaminase) activity; and the Low Glucose Broth (LGD) developed with the aim to decrease the incidence of false negative responses of bacteria with a great fermentative activity. Before performing the tests both base broth media were supplemented with the precursor amino acids (L-Lysine monohydrate (Merck, Darmstadt, Germany), L-Ornithine monohydrate (Sigma-Aldrich, Steinheim, Germany), L-Histidine monohydrochloride (Merck) and L-Tyrosine disodium salt (Sigma-Aldrich), individually, or adding a mixture of all them (described in the next section as total amino acid broth). The base broth without amino acids added was used as negative control. All media were adjusted to the pH values indicated in Table 1 and autoclaved at 120 °C during 5 min.

Table 1. Composition of broth media (g/100 mL)) used to evaluate decarboxylase ability of strains isolated from cheeses.

Reagent	Low Nitrogen Decarboxylase Broth (LND)	Low Glucose Decarboxylase Broth (LGD)	Adjusted pH
Tryptone	0.125	0.25	
Yeast extract	0.125	0.25	
NaCl	0.25	0.25	
$CaCO_3$	0.01	0.01	
Pyridoxal-5-phosphate	0.03	0.03	
Glucose	0.05	0.001	
Bromocresol purple	0.01	0.01	
All amino acids	1.0	1.0	5.5
L-Lysine	1.0	1.0	5.0
L-Ornithine	1.0	1.0	5.5
L-Histidine	1.0	1.0	5.7
L-Tyrosine	0.25	0.25	5.5

2.4. Assessment of Amino Acid Decarboxylase Activity

In order to detect the capacity of the isolated strains to form BA and to determine which of the two decarboxylase broths (LND and LGD) show the best results in each one, a screening test was performed on a 96-wells flat bottom Microtiter plate. Aliquots of 200 μL of total amino acid broth (TAB) and 20 μL saline solution were added into 6 wells of a 96 well (decarboxylase control assay: DCA); 200 μL of TAB and 20 μL of bacterial suspension were added into another 6 wells (positive decarboxylase assay: PDA); and 200 μL of broth base without amino acids with 20 μL of bacterial suspension were added into another 6 wells (negative decarboxylase assay: NDA). Microplates were incubated at 30 °C for 24 h. A positive result was considered in PDA wells when a purple color appeared due to an increase of alkalinity (Figure 1a). In LGD broth positive results were also considered when no color changes were observed in PDA wells and yellow color appeared in NDA wells, because a high acidification was produced due to the bacterial growth (Figure 1b). Negative results were considered when no color changes were observed in PDA wells (Figure 1a,b), or when a purple color appeared in NDA wells due to another alkaline compounds different than BA (Figure 1a).

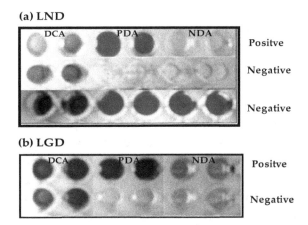

(a) LND

Positve

Negative

Negative

(b) LGD

Positve

Negative

Figure 1. Example of negative and positive responses to amino acid decarboxylase activity in Low Nitrogen Decarboxylase (LND) and Low Glucose Decarboxylase (LGD) media. DCA: decarboxylase control assay; PDA: positive decarboxylase assay; NDA: negative decarboxylase assay.

2.5. Confirmation of Amino Acid Decarboxylase Activity by HPLC

Decarboxylase activity of strains was confirmed by the quantitative analysis of BA produced in the decarboxylase broths by means of reverse-phase High Performance Liquid Chromatography (HPLC), using an automated HPLC system (HPLC P680, Dionex, Sunnyvale, CA, USA) equipped with an Ultra Violet (UV) detector Dionex UVD170U (Thermo-Fisher Scientific, Waltham, MA, USA). Briefly: one mL of each bacterial suspension (0.5 McFarland) was inoculated into a tube containing 4 mL of the TAB version of LND or LGD broths (depending on the previous results for each strain). After 4 days of incubation at 30 °C, the media was centrifuged ($9000\times g$, 10 min, 20 °C) and 3 mL of the supernatant was extracted with 2 mL of 0.4 M $HClO_4$ (Panreac). Determination of BA was carried out according to the RP-HPLC method described by Eerola et al. [28] and modified by Roig Sagués et al. [8] using dansyl chloride reagent (Sigma-Aldrich Chemical) to derivate the sample. The separation was performed on a Waters Spherisorb S5 ODS 2 45 × 150 mm column (Waters Corporation, Milford, MA, USA). All reagents were of analytical grade and all solvents involved in derivatization and in the separation process were of HPLC grade. The BA standards: putrescine (PU), cadaverine (CA), histamine (HI), tyrosine (TY), and the internal standard 1,7-diaminoheptane, were all purchased from Sigma-Aldrich Chemical.

2.6. Analytical Validation of the Qualitative Microplate Method of Amino Acid Decarboxylase Activity

The sensitivity, specificity, and the positive and negative predictive values were obtained to determine the diagnostic properties of the qualitative method [29–31] and were calculated by the following equations:

$$\text{Sensitivity} = \frac{\text{TP}}{\text{TP} + \text{FN}} \times 100$$

where TP is the truly positive amino acid decarboxylating isolates, correctly identified by the screening test and FN is the false negative responses obtained.

$$\text{Specifity} = \frac{\text{TN}}{\text{TN} + \text{FP}} \times 100$$

where TN is the truly negative (TN) amino acid decarboxylating isolates, correctly identified by the screening test and FP is the false positive responses obtained.

$$\text{PPV} = \frac{\text{TP}}{\text{TP} + \text{FP}} \times 100$$

where PPV is the positive predictive value and reflects the proportion of truly positive isolates confirmed by HPLC among all positive isolates evaluated by Microtiter plate screening.

$$NPV = \frac{TN}{TN + FN} \times 100$$

where negative predictive value (NPV) is reflects the proportion of truly negative isolates confirmed by HPLC among all negative isolates evaluated by Microtiter plate screening.

The Receiver Operating Characteristic (ROC) curves were assessed using the MedCalc statistical software, version 11.2.1 (MedCalc, Ostend, Belgium), to know the discriminative power of the qualitative method referred to the HPLC method with its 95% confidence interval. In a ROC curve the true positive rate (Sensitivity) is plotted in function of the false positive rate (100-Specificity). A test with perfect discrimination (no overlap in the two distributions) has a ROC curve that passes through the upper left corner (100% sensitivity, 100% specificity). Therefore, the closer the ROC curve is to the upper left corner, the higher the overall accuracy of the test [32]. The area under the ROC curve (AUC) is a measure of how well a parameter can distinguish between two groups (isolates with amino acid decarboxylase activity/isolates without this capacity). The better overall diagnostic performance of the test is when the AUC value is closer to 1 and the practical lower limit for the AUC of a diagnostic test is 0.5 [31,33]. A classification of diagnostic accuracy for the qualitative method is given according to AUC value: AUC 0.90–1.0 excellent, 0.80–0.90 good, 0.70–0.80 fair, 0.60–0.70 poor, 0.50–0.60 deficient and 0.50 null [34].

The point of intersection of the ROC curve with the diagonal line drawn from 100% sensibility to 100% 1-specificity was chosen as the best discriminator value. The optimal cut-off value showed the highest accuracy, the lowest false negative (FN) and the highest false positive (FP) results.

2.7. Identification of Strains with Decarboxylase Activity

Confirmed decarboxylase-positive strains were identified based on Gram stain and catalase and citochromooxidase activity [35]. Further identification to the species level was carried out by a variety of biochemical tests using API 20-E, API 20-Strep, API-Staph and API 50-CH strips (BioMérieux, Marcy l'Etoile, France).

3. Results and Discussion

3.1. Validation of the Qualitative Microplate Method of Amino Acid Decarboxylase Activity

ROC curve analysis was used to determine the discriminative power and the cuts-off of the amino acid decarboxylase screening method with both media (LND and LGD) to evaluate the specific amino acid decarboxylase activity (Table 2).

Tyrosine decarboxylase test showed an area under the ROC curve (AUC) around 0.98, with an optimal cut-off value at 25 mg/L and 20 mg/L of TY on LND and LGD broths, respectively. This means that the microplate screening method could discriminate the isolates with tyrosine decarboxylase activity the 98% of the time at optimal cut-off. The sensitivity and specificity values obtained with both broths were higher than 92%, reflecting that the number of false negative (FN) and false positive (FP) responses obtained by the qualitative method were generally low. However, the negative predictive value (NPV) was considered low (<66%). AUC for the lysine decarboxylase test displayed was greater than 0.930 with an optimal cut-off concentration of 15 mg/L and 10 mg/L for LND and LGD broths, respectively. In this case, the sensitivity and specificity values using LND broth were about 98% and 93%, respectively, while for LGD broth values were about 88% and 98%, respectively. The assay to detect ornithine decarboxylase with the LND broth showed the highest diagnostic values (over 98%) with the lowest cut-off concentration (10 mg/L) and an AUC higher than 0.995. On the other hand, for the same test using LGD broth an 84% of sensitivity and a 97.5% of specificity were reached at a cut-off value of 15 mg/L with an AUC of 0.907.

Histidine decarboxylase test showed the lowest sensitivity values (below 60%) using both broths with the highest optimal cut-off concentration set at 50 mg/L with specificity values up to 90%. Likewise, the AUC value was the lowest, possibly due to a 16.5% of FP and 24% of FN reactions observed at the optimal cut-off using LND broth, whereas a 9.9% and 25% of FP and FN were obtained, respectively, in LGD broth.

Table 2. Receiver Operating Characteristic (ROC) curve analysis of qualitative method to predict specific amino acid decarboxylase activity in isolates using Low Nitrogen Decarboxylase (LND) and Low Glucose Decarboxylase (LGD) broths.

Broth	Data of ROC Analysis	Ornithine	Lysine	Histidine	Tyrosine
LND	AUC	0.999 (0.974–1.00)	0.992 (0.961–1.00)	0.737 (0.659–0.806)	0.980 (0.943–0.996)
	Optimal cut-off (mg/L)	10	15	50	25
	Sensitivity at optimal cut-off (%)	98.53	98.68	65.38	92.25
	Specificity at optimal cut-off (%)	100	93.15	91.75	100
	PPV (%)	100	83.7	81	100
	NPV (%)	99.0	98.6	83.2	66.7
	FN at cut-off	3	6	11	1
	FP at cut-off	1	1	17	6
LGD	AUC	0.907 (0.849–0.948)	0.935 (0.883–0.969)	0.592 (0.509–0.672)	0.989 (0.956–0.999)
	Optimal cut-off (mg/L)	15	10	50	20
	Sensitivity at optimal cut-off (%)	84.37	88	30	93.06
	Specificity at optimal cut-off (%)	97.5	97.6	100	100
	PPV (%)	90	88	100	100
	NPV (%)	95.8	97.6	90.3	37.5
	FN at cut-off	5	5	2	0
	FP at cut-off	5	3	14	8

AUC: Area under ROC curve and 95% confidence interval; PPV: Positive predictive value; NPV: Negative predictive value; FN: Number of false negative: FP: Number of false positive.

In many occasions it has been reported that qualitative screening decarboxylase methods have some limitations in terms of sensitivity in detecting BA production. The presence of FP and FN reactions reported has not been insignificant. Hernández-Herrero et al. [9] observed that 96.5% of the suspected histamine formers detected by Niven decarboxylase media were finally considered as FP. Likewise, Roig-Sagués et al. [36] found that only a 15.8% of the total presumptively histamine-formers obtained in Joosten and Northolt media [37] were confirmed. Similar results were observed when tyramine decarboxylase capacity was tested in the same media, where only 8.4% of the suspected isolates with tyrosine decarboxylase activity were confirmed. The FP results were attributed to the production of other substances able to alkalinize the media [25]. Similarly, Moreno-Arribas et al. [38] used the Maijala modified decarboxylase media and noticed a high number of FP reactions to PU and agmatine production, but less than were found in the tyrosine decarboxylase activity test. On the contrary, de las Rivas et al. [15] did not find any correlation between the positive responses in the decarboxylase activity media and the BA detected by HPLC. They suggested that the screening Maijala modified decarboxylase media underestimates the number of BA-producing strains. On the contrary, Bover-Cid and Holzapfel [23], in their improved screening media tested on LAB, did not observe FP reactions and only 3 strains showed a negative response with the screening procedure. They justified these FN results due to the low amount of tyramine formed that did not neutralize the acid production of LAB. Although these authors proposed their improved decarboxylase medium as a rapid preliminary method to select strains with low decarboxylase activity, the optimal cut-off value was around 300 mg/L. Torracca et al. [17] reported, using the same decarboxylase medium described by Bover-Cid and Holzapfel [23], an optimal cut-off value of 631 mg/L and 810 mg/L for PUT and TY, respectively.

3.2. Amino Acid Decarboxylase Activity of the Control Strains

Table 3 shows the results after testing the control strains in the microplate screening method and the result of the confirmation by HPLC. Lactobacillus brevis and Lactobacillus casei showed tyrosine decarboxylase activity in LGD broth, and Staphylococcus epidermidis histidine decarboxylase capacity in LND broth. All strains were also able to produce low amounts of PU (around 1 mg/L) but were only detected by HPLC.

Table 3. Biogenic amine production by positive control bacteria strains in the amino acid decarboxylase microplate assay (DMA) and HPLC analysis (mg/L).

Strain	Broth	PU		CA		HI		TY	
		DMA	HPLC	DMA	HPLC	DMA	HPLC	DMA	HPLC
L. brevis	LGD	(−)	1.11	(−)	ND	ND	ND	(+)	109.8
L. casei	LGD	(−)	0.78	(−)	ND	ND	ND	(+)	77.14
S. epidermidis	LND	(−)	0.85	(−)	ND	(+)	46.48	(−)	ND

PU: putrescine; CA: cadaverine; HI: histamine; TY: tyramine; LND: Low Nitrogen Decarboxylase; LGD: Low Glucose Decarboxylase; DMA (detection on microplate assay): (+) Positive; (−) Negative; ND: not detected.

3.3. Biogenic Amine Production by Isolates from Goat's and Ewe's Milk Cheeses

3.3.1. Total Amino-Acid Decarboxylase Activity of the Isolated Strains

A total of 688 strains were obtained from the different culture media and a 43.02% of them gave a positive response in the microplate assay with TAB, being subsequently confirmed by HPLC. A 37.7% of the bacteria isolated from goat's milk cheeses and a 47% of the strains picked up from ewe's milk cheeses were BA-formers. The number of isolates obtained from VRBG and BPA media was much lower since these two groups of microorganisms are a minority among the microbiota of ripened cheeses, and their counts are usually low [27], but the percentage of decarboxylase positive results among these isolates was higher (87.5% and 92% on VRBG and BPA, respectively) than in KF, M-17 and MRS media (49.7, 36.8, and 35% respectively). Several studies found that the decarboxylase activity is more frequent in Enterobacteriaceae strains (from 80 to 95%) and in less extension among LAB strains (from 9.5 to 65%) [8,21,36,38,39]. Nevertheless, Enterobacteriaceae and Staphylococcus usually do not achieve high counts in ripened cheeses made under good hygienic practices and normally become undetectable after few days of ripening, reason why decarboxylase positive LAB is usually considered the main responsible for the formation of high concentrations of BA in cured cheeses [17,21].

3.3.2. Specific Amino Acid Decarboxylase Activity of the Isolated Strains

The assessment of the specific amino acid decarboxylase activity was done with the strains that gave positive responses in the screening assay with TAB. Up to 150 of these isolates were recovered and tested in LND and LGD media, respectively. In general, the capability to decarboxylate tyrosine was the most frequent activity detected (91.6% of the total isolates tested) in the specific amino acid decarboxylase screening assay, followed by the ability to decarboxylate lysine and ornithine (around 33.5%). In these cases, the 95%, 96% and 94% were confirmed by the HPLC analysis, respectively. However, histidine decarboxylase activity was detected in only a 24% of the isolates tested, a 76% of them confirmed by HPLC.

Table 4 shows the number of strains with HPLC-confirmed BA-producing capability obtained from goat´s and ewe's milk cheeses according to the culture media of origin. These results are shown as a whole without considering the HHP treatment to which cheese samples were subjected. Strains were grouped in four categories according to Aymerich et al. [38]: medium amine formers (25–50 mg/L), good amine formers (50–100 mg/L), strong amine formers (100–1000 mg/L) and prolific amine formers (>1000 mg/L). In general, strong amine formers were more frequent among strains with TY-producing capability, followed by those able to form CA and PU. Prolific amine production capacity was only observed in some diamine formers, while the formation of HI in amounts above 100 mg/L was a rare event (Table 4).

Table 4. Biogenic amine forming capacity of bacteria isolated from goat's and ewe's milk cheeses according to the culture media. High hydrostatic pressure (HHP) treatments to which cheese samples were subjected are not considered.

Medium	BAP	PU				CA				HI				TY			
		±	+	++	+++	±	+	++	+++	±	+	++	+++	±	+	++	+++
VRBG	25	1	2	17	4	0	0	17	7	8	10	4	0	3	6	9	0
BPA	15	1	1	3	3	0	0	4	3	4	4	1	0	2	2	7	0
KF	89	5	3	22	4	13	0	22	9	19	21	4	0	4	8	71	0
M-17	98	7	2	5	0	7	1	2	1	14	6	2	0	5	6	69	0
MRS	72	2	2	2	3	0	0	2	3	7	0	0	0	5	8	50	0
Total	299	16	10	49	14	20	1	47	23	52	41	11	0	19	30	206	0

BAP: Number of positive BA producers; PU: putrescine; CA: cadaverine; Hi: histamine; TY: tyramine; Number of BA-forming isolates detected depending on their production (in mg L^{-1}): (±) 25–50, medium; (+) 50–100, good; (++) 100–1000, strong; and (+++), >1000, prolific.

The isolates obtained from VRBG medium showed the highest frequency of PU and CA forming activity. Lysine and ornithine decarboxylases are very common among enterobacteria, and their detection is commonly used for the biochemical identification of *Enterobacteriaceae* species. A 100% of the isolates picked up from goat's and ewe's milk cheeses showed strong activity (>100 mg/L) for CA, and 87.5% and 85.35% for PU, respectively. Strong TY forming capacity was detected in only one isolate obtained from goat's milk cheeses and in 8 from ewe's milk cheese. The isolates with histidine decarboxylase activity showed a weak production and only one isolate obtained from goat's milk cheese and three from ewe's milk cheese presented the ability to produce more than 100 mg/L. *Enterobacteriaceae* are known to decarboxylate several amino acids, specially arginine, lysine and ornithine [11,39–41] and histidine [17,25,42]. The number of isolates with amino acid decarboxylase activity obtained from BPA culture medium was low. Between 75–100% of these isolates displayed a strong PU, CA and TY forming ability. Whereas histamine accumulation was detected especially in a range of 25–100 mg/L in the 89% of the cases. Little information is available about the production of BA by *Staphylococcus* spp. in cheese. However, some species of this group have been related to a variable formation of TY, PU, CA and/or HI in meat and fish fermented products. Martin et al. [43] found in fermented sausages that TY was the main amine produced by this group and some strains also were able to produce PU, CA and HI. Hernández-Herrero et al. [9] reported that the main HI-formers detected in salted anchovies belonged to this genus and de las Rivas et al. [15] reported some strains isolated from "Chorizo", a Spanish ripened sausage, as TY-formers. Most of the isolates obtained from the KF medium were strong TY formers (76% from goat's milk cheese and 81% from ewe's milk cheese) and some of them were also able to produce above 100 mg/L of PU and CA. The ability to form TY was also frequent in the strains isolated from M-17 medium in both kind of cheeses (79% from goat's and 65% from ewe's milk cheeses, respectively) in amounts above 100 mg/L. In that case, the ability to form diamines was less frequent (around 5% of the isolates) in both type of cheeses. Similar results were found in the isolates obtained from MRS medium where around 69% of them were considered strong TY-producers, but no PU or CA formation was detected in any of the isolates obtained from goat's milk cheese and only a 10% of those obtained from ewe's milk cheese were able to from up to 100 mg/L.

3.3.3. Identification of BA-Producing Strains

The result of the identifications of the BA-forming strains isolated from cheese samples, as well as their BA forming capacity expressed as mg/L, are shown in Table 5 (goat's milk cheese) and Table 6 (ewe's milk cheese). It was not possible to establish a clear effect of the HHP treatments to which some of the cheese samples were submitted on the type of strain and its amine-forming capacity. Consequently, results are shown globally, without considering the type of treatment to which the samples were submitted. In both types of cheese, most of the strains with decarboxylase activity were Gram positive. In the case of the goat's milk cheese only 5 Gram negative strains showed decarboxylase activity, four of them identified as *Hafnia alvei*, all of them with a strong PU and CA forming capacity. One of these strains also showed a strong capacity to form TY, but much lower than diamines. Nevertheless, the maximum capacity to form these amines (and also HI) was shown by a strain that could not be precisely identified. In the case of the ewe's milk cheese, 11 Gram negative strains showed decarboxylase activity, most of them (5) also identified as Hafnia alvei and one of them showing the maximum capability to form CA, PU and HI. The maximum TY-forming capacity was shown by a strain of *Citrobacter freundii*. Strains of *Klebsiella oxytoca* and *Escherichia coli* with strong diamine production capacity were identified from ewe's cheese samples. However only one strain of each specie and other identified as *C. freundii* were able to produce CA and TY in a considerable amount (above 100 mg/L). *H. alvei*, *K. oxytoca* and *E. coli* have been previously associated with the formation of PU, CA and/or HI in foods [8,9,11,17,21,25,36,41]. Also, some strains of these species have been reported to possess the ability to decarboxylate tyrosine [21]. The formation of high amounts of PU and CA, as well as of TY, has been previously reported by *C. freundii* [21,40], indicating that this specie is more prolific forming PU than CA. Enterobacteria is usually a minor group among the microbiota present on fermented products. In the cheese object of this work, enterobacteria counts were usually below 3 Log_{10} after 60 days of ripening, but their counts at the beginning of the process were above 6 Log_{10} [27]. When unhygienic manufacturing practices allow for achieving high counts of enterobacteria at the beginning of the process, the fact that their counts would be later reduced during ripening does not necessarily imply the inhibition of their decarboxylase activity and consequently may contribute to BA formed in the final product [36].

Among strains identified as Gram positive, 50 that showed decarboxylase activity were obtained from the goat's milk cheese and 55 from the ewe's milk cheese, and in general this activity was much higher than the Gram negative strains. Among the positive decarboxylase bacteria isolated from the ewe's milk cheeses one strain of *Staphylococcus chromogenes* showed to be a prolific diamine former and strong TY former. Likewise, strains of *Staphylococcus xylosus* and *Staphylococcus aureus* with strong TY and HI production, respectively, were also found. On the other hand, the most frequent strains with high BA forming capacity obtained from goat's milk cheeses were identified as *Staphylococcus hominis*, that were capable to produce high levels of PU, CA and TY, and *Staphylococcus warneri*, which showed to be a strong diamine producer.

S. chromogenes was previously reported as a prolific PU former, good CA and strong TY and HI-forming bacteria in Spanish salted anchovies [44]. Masson et al. [45] detected a weak TY-production capacity in strains of *S. xylosus* isolated from fermented sausages and Martin et al. [43] found in slightly fermented sausages some strains of *S. xylosus* capable to produce strong and prolific amounts of TY, PU and/or HI, and in a lesser extent of CA. Silla-Santos [46] observed HI-production in the 76% of *S. xylosus* strains isolated from Spanish sausages. Strains of *S. warneri* have also been reported to possess tyrosine decarboxylase, but with great variability of production [15,43,45]. CA and PU formation, in medium-good and strong concentration, have also been described [39]. Drosinos et al. [47] isolated one strain of *S. hominis novobiosepticus* in traditional fermented sausages with lysine and tyrosine decarboxylase activity. As enterobacteria, *Staphylococcus* spp. are associated to the contamination of food during unhygienic handling, and consequently it is important to follow always good manufacturing practices to avoid the proliferation of this kind of bacteria in the product. Nevertheless, in the cheeses from where the studied strains were obtained *Staphylococcus* spp. counts were always low (below 3 Log_{10} at the beginning of the ripening), and their counts reduced during ripening until being undetectable after 60 days in most cases.

Enterococcus faecalis, *Enterococcus faecium* and *Enterococcus durans* were the most frequent amine producing bacteria identified from both types of cheese, with a varied production capacity of TY,

PU and CA (Tables 5 and 6). In addition, one strain of *E. faecium* and two of *E. durans* showed a strong HI formation capacity. Enteroccoci are commonly associated with unhygienic conditions during the production and processing of dairy, although they can play an important role for developing the aroma and flavor of certain type cheeses, especially traditional cheeses produced in the Mediterranean area [48]. Several authors have described *E. faecalis*, *E. faecium* and *E. durans* as the most frequently TY formers in food [12,17,21,23,25,49–52] and also some strains of *E. faecalis* and *E. faecium* have been registered as capable of producing amounts up to 100 mg/L of PU [12,17,21,53] and CA [17,21,50] and/or HI [17,21,54]. Tyrosine is a relevant amino acid for the formation of BA in cheese as it can be an inducer of PU production in *E. faecalis* and would be received by the enterococci cells as a signal to growth, what would lead in an increment in the number of BA-producing cells increasing the risk of accumulating TY and PU in cheese [55]. No references concerning histidine decarboxylase activity of *E. durans* have been found in the literature.

Several BA producing strains isolated from M-17 and MRS media of goat's milk cheeses were identified as *Lactococcus lactis* subsp. *lactis*, followed by Lactobacillus brevis, *Lactobacillus plantarum*, and *Leuconostoc* spp. All of them showed strong TY forming ability and *Leuconostoc* spp. also showed a strong-prolific PU formation. In the case of ewe's milk cheeses, *L. lactis* subsp. *lactis*, *L. lactis* subsp. *cremoris*, *Pediococcus pentosaceus*, *Lactobacillus paracasei* subsp. *paracasei*, *L. plantarum* and *Leuconostoc* spp. were often associated with a strong TY-forming capability. Moreover, two strains identified as *L. lactis* subsp. *lactis* showed strong PU and CA forming ability, respectively, while two strains of *P. pentosaseus* species were strong PU and prolific CA producers, respectively.

Table 5. Identification of strains obtained from goat's raw milk cheeses with decarboxylase activity and their BA production in decarboxylase broths (mg/mL). Results are shown without considering the HHP treatments to which some cheese samples were subjected.

Identification	N		PU		CA		HI		TY
Gram Negative									
Hafnia alvei	4	4	537.7–889.54	2	641–1001.30	4	30.43–95.68	4	22.44–151.54
Enterobacteriaceae	1	1	1037.52	1	1173.24	1	111.43	1	73.41
Gram Positive									
Staphylococcus cohni subsp. *cohni*	1	-		1	1.50	1	1.57	1	25.78
Staphylococcus warneri	2	2	69.29–753.55	2	240.94–694.22	1	88.63	1	65.50
Staphylococcus capitis	1	1	310.30	1	246.37	1	95.13	-	
Staphylococcus lentus	2	1	23.44	2	5.41–19	-		2	39.58–191.69
Staphylococcus hominis	2	1	890.96	1	998.20	1	91.63	2	102.42–245.22
Enterococcus faecalis	8	8	39.66–884.27	8	32.44–972	6	25.62–92.35	8	327.5–477.20
Enterococcus durans	2	-		-		-		2	337.40–357.44
Enterococcus avium	1	-		1	26.52	1	56.51	1	24.88
Enterococcus faecium	6	3	19.68–1113.80	3	1.32–1281.50	3	20.22–111.38	6	9.91–366.47
Lactococcus lactis subsp. *lactis*	10	1	7.25	1	7.50	2	21.24–33.38	10	198.77–450.77
Pediococcus pentosaceus	1	-		-		1	33.44	1	55.71
Lactobacillus brevis	5	-		-		1	22.51	5	212.58–519.52
Lactobacillus plantarum	3	1	22.41	-		1	24.5	3	307.62–528.45
Lactobacillus paracasei subps. *paracasei*	1	-		-		-		1	307.03
Leuconostoc spp.	3	2	252.88–749.05	2	14.47–28.83	2	22.74–25.75	3	330.01–417
Bacillus macerans	1	-		-		-		1	418.10
Bacillus licheniformis	1	-		-		-		1	403.07

N: number of strains identified; PU: putrescine; CA: cadaverine; HI: histamine; TY: tyramine.

Several studies have reported different species of LAB able to form BA, especially TY [8,11,16–18,21,23,25,50,54–57], but also HI [58,59]. Within the species of LAB that may occur in food some strains of *L. brevis* and *L. plantarum* were reported to possess the potential to form TY, PU and/or HI [15,18,21,23,25,60,61]. Strains of *P. pentosaceus* isolated from commercial starters [26] and ripened sausages [25] were reported to form TY. Although *L. lactis* subsp. *lactis*, *L. lactis* subsp. *cremoris* and *L. paracasei* subsp. *paracasei* are species usually used as starter cultures or probiotic strains and usually reported as non-decarboxylating strains [16–18,62], some strains of *L. lactis* and *L. paracasei* have been reported with the ability to form TY, HI, PU and/or CA in amounts up to 1000 mg/L [8,17,18,21,23,25,51,58] and *L. lactis* subsp. *cremoris* has been also described as a PU-producer [12]. *Leuconostoc* spp. has been frequently described among the microbiota present in several Spanish farm house cheeses [3,63,64], and has also been described to possess tyrosine, lysine and/or histidine decarboxylase activity [39]. Likewise, González de Llano et al. [63] reported the production of TY in a range of 100–1000 mg/L by strains of *Leuconostoc*. Pircher et al. [21] found that strains of this genus could form TY, PU, CA and /or HI in amounts up to 100 mg/L.

Strains belonging to the *Bacillus* genus isolated from ripened salted anchovies, cheese and raw sausages have previously been described as BA-formers [9,22,36]. One of them was *Bacillus macerans* isolated from Italian cheese, which was capable to from prolific amounts of HI. Formation of CA and PU was also observed in this strain [22].

Table 6. Identification of strains obtained from ewe's raw milk cheeses with decarboxylase activity and their BA production in decarboxylase broths (mg/mL). Results are shown without considering the HHP treatments to which some cheese samples were subjected.

Identification	N		PU		CA		HI		TY
Gram Negative									
Escherichia coli	2	2	746–857.48	2	762.35–983.93	2	41.37–74.12	2	21.7–281.48
Hafnia alvei	5	5	738.50–1049.51	5	787.20–1180.12	4	43.52–185.54	5	48.83–184.79
Klebsiella oxytoca	2	2	30.63–458.50	2	493.41–866.96	2	16.04–27.22	2	5.01–167.24
Citrobacter freundii	1	1	67.1	1	1095.3	1	45.62	1	372.38
Enterobacteriaceae	1	1	832.76	1	846.95	1	83.95	1	69.42
Gram Positive									
Staphylococcus xylosus	2	-		-		2	42.31–68.5	2	92.09–475.35
Staphylococcus chromogenes	1	1	1142.92	1	1760.06	1	31.12	1	441.40
Staphylococcus aureus	1	-		-		1	100.54	-	
Enterococcus faecalis	5	3	860.29–978.9	4	35.43–1394.87	5	20.7–92.98	5	338.01–461.50
Enterococcus durans	9	9	12.63–1160.44	8	13.32–1773.03	9	30.8–179.32	9	46.09–747.33
Enterococcus faecium	2	2	24.89–847.05	2	877.24–941.69	2	28.98–29.97	2	100.54–149.41
Enterococcus hirae	2	2	552.4–579.25	2	600.67–615.75	2	32.36–45.95	2	108.01–187.58
Enterococcus avium	2	-		1	15.17	1	2.18	2	19.93–434.25
Streptococcus salivarius	1	-		-		-		1	742.1
Lactococcus lactis subsp. *lactis*	9	3	34.39–795.26	2	14.95–844.13	2	16.38–30.72	9	211.93–566.2
Lactococcus lactis subsp. *cremoris*	4	3	4.78–37.63	3	5.23–22.94	2	11.15–19.03	4	229.47–406.87
Pediococcus pentosaceus	4	3	9.91–897.67	3	10.25–1018.4	3	27.06–88.78	4	34.58–411.66
Lactobacillus paracasei subsp. *paracasei*	6	2	17.42–45.05	1	17.18	1	51.93	6	365.79–575.73
Lactobacillus plantarum	4	1	8.58	1	17.58	1	26.26	4	45.1–353.31
Lactobacillus brevis	1	-		-		-		1	240.08
Lactobacillus pentosus	1	1	50.79	-		-		1	422.75
Leuconostoc spp.	5	2	11.13–1162.76	2	22.52–1781.26	3	29.63–33.82	5	392.01–626.42

N: number of strains identified; PU: putrescine; CA: cadaverine; HI: histamine; TY: tyramine.

3.4. Consequences of HHP Treatments on the Formation of BA

The effect of the HHP treatments on the microbial counts, the proteolytic activity and the formation of BA in the cheeses from where we obtained the studied strains are described in the previous work of Espinosa-Pesqueira et al. [27]. TY and PU were the main BA formed in control (untreated) cheeses, whereas in ewe's milk chesses the level of CA was also relevant. However, in cheeses that were pressurized on the 5th day of ripening (HHP1) the amounts TY formed at the end of the ripening (60th day) were about 93% and 88% lower than in control goat's and ewe's raw milk cheeses, respectively, and similar was the result for PU. The application of an HHP treatment on the 15th day of ripening (HHP15) showed to be less efficient reducing the formation of BA. HHP1 treatments caused a significant decrease on microbiological counts, specially of LAB, enteroccocci and enterobacteria, and also reduced the proteolytic activity, showing a reduction of about 34% and 49% of the free amino-acid content in goat's and ewe's cheeses, respectively. Although HHP15 samples also reduced the microbial counts, this did not affect the proteolysis, and consequently the release of amino acids. Several authors mentioned that the specificity of the amino acid decarboxylases is specially strain dependent [11,15,39,60] and a great variability in BA production by different groups and species of bacteria, either in type or amount, was found in this survey. LAB and enterococci were among the most efficient TY producers found in this work. LAB ruled the ripening process of cheeses from the beginning, especially *Lactococcus*, becoming the most important group of microorganisms in the ripened cheese.

The application of the HHP treatments in the early stages of ripening (HHP1) caused a significant reduction of LAB counts, although they recovered along the ripening process. It should be considered that the BA forming rate is greater during the first 15–30 days of ripening, and consequently, the reduction on the counts of the most prolific BA forming groups at these stages could have contributed to reduce the decarboxylase potential, as well as the availability of amino acids, reducing the final amounts of BA. Enterococci were also affected by HHP1 treatments and could not recover their initial counts during the rest of the ripening, affecting their contribution to the formation of BA.

The role of other minority bacterial groups, such as enterobacteria, on the BA formation is not so clear. In the cheeses object of this work their counts were above 6 Log_{10} at the beginning of the ripening, although practically disappeared after both HHP1 and HHP15 treatments samples. Different enterobacteria have been identified as prolific PU, CA and even TY formers, and consequently its elimination the first days of ripening could have reduced their contribution to the amino acid decarboxylase activity. When treatments are applied after 15 days of maturation, the decarboxylase capacity of these microorganisms in the early stages of maturation was still present and consequently their elimination from the 15th day of ripening reduced the consequences on the BA formation.

4. Conclusions

The microplate screening method allows for a rapid preliminary selection of strains with low decarboxylase activity, with a detection limit estimated around 50 mg/L. Moreover, the use of Microtiter plates allows for processing a large numbers of samples, reducing the volume of material and culture media needed. The data indicates that, in general, the specific amino acid decarboxylase assay with LND and LGD broths have satisfactory diagnostic parameters to discriminate bacterial isolates with ornithine, lysine and tyrosine decarboxylases. Moreover, the sensitivity and specificity values for ornithine, lysine and tyrosine decarboxylase test with both types of media were acceptable with low numbers of FP and FN responses. Generally, FN responses were due to weak BA producers. The detection of histidine decarboxylase activity in bacterial isolates using LND and LGD broths have a low sensitivity, but a high specificity value. About a 43% of the strains isolated from cheeses showed decarboxylase activity on one or more amino acids and most of them were later confirmed, especially of TY, PU and CA that were the most important present in cheeses after ripening. The application of the HHP treatments, especially in the early stages of ripening, caused a significant reduction among the most prolific BA forming groups including LAB. Considering that the BA forming rate is greater during the first 15–30 days of ripening, this effect on the microbiota reduces the decarboxylase potential, as well as the availability of amino acids, reducing the final amounts of BA formed. Most of the strains were LAB, including some species that are important for the development of the typical cheese characteristics, such as *L. lactis*, that can be used as starter culture, or *Lactobacillus casei/paracasei*, *Lactobacillus plantarum* and *Lactobacillus curvatus*, of which there is increasing interest to be employed in dairy products with 'protected geographic indication' [18]. Consequently, the formation of BA should always be considered among the selecting criteria for strains considered as suitable to be used as starter cultures.

Author Contributions: Conceived and designed the experiments: D.E.-P., M.M.H.-H. and A.X.R.-S.; performed the experiments: D.E.-P.; analysed the data; D.E.-P. and M.M.H.-H.; wrote the paper: D.E.-P. and A.X.R.-S.

References

1. Karovicova, J.; Kohajdova, Z. Biogenic amines in food. *ChemInform* **2005**, *36*, 70–79. [CrossRef]
2. Collins, J.D.; Noerrung, B.; Budka, H.; Andreoletti, O.; Buncic, S.; Griffin, J.; Hald, T.; Havelaar, A.; Hope, J.; Klein, G.; et al. Scientific Opinion on risk based control of biogenic amine formation in fermented foods. *EFSA J.* **2011**, *9*, 2393. [CrossRef]
3. Costa, M.P.; Rodrigues, B.L.; Frasao, B.S.; Conte-junior, C.A. *Chemical Risk for Human Consumption*; Elsevier Inc.: Amsterdam, The Netherlands, 2018; ISBN 9780128114421.
4. Benkerroum, N. Biogenic amines in dairy products: Origin, incidence, and control means. *Compr. Rev. Food Sci. Food Saf.* **2016**, *15*, 801–826. [CrossRef]
5. Naila, A.; Flint, S.; Fletcher, G.; Bremer, P.; Meerdink, G. Control of biogenic amines in food-existing and emerging approaches. *J. Food Sci.* **2010**, *75*, R139–R150. [CrossRef] [PubMed]
6. Novella-Rodriguez, S.; Veciana-Nogues, M.T.; Izquierdo-Pulido, M.; Vidal-Carou, M.C. Distribution of biogenic amines and polyamines in cheese. *J. Food Sci.* **2003**, *68*, 750–756. [CrossRef]
7. Roig-Sagués, A.X.; Ruiz-Capillas, C.; Espinosa, D.; Hernández, M. The decarboxylating bacteria present in foodstuffs and the effect of emerging technologies on their formation. In *Biological Aspects of Biogenic Amines, Polyamines and Conjugates*; Dandrifosse, G., Ed.; Transworld Research Network: Trivandrum, India, 2009; pp. 201–230.
8. Roig-Sagués, A.; Molina, A.; Hernández-Herrero, M. Histamine and tyramine-forming microorganisms in Spanish traditional cheeses. *Eur. Food Res. Technol.* **2002**, *215*, 96–100. [CrossRef]
9. Hernández-Herrero, M.M.; Roig-Sagués, A.X.; Rodríguez-Jerez, J.J.; Mora-Ventura, M.T. Halotolerant and halophilic histamine-forming bacteria isolated during the ripening of salted anchovies (*Engraulis encrasicholus*). *J. Food Prot.* **1999**, *62*, 509–514. [CrossRef] [PubMed]
10. Kalac, P.; Abreu Gloria, M.B. Biogenic amine in cheeses, wines, beers and sauerkraut. In *Biological Aspects of Biogenic Amines, Poliamines and Conjugates*; Dandrifosse, G., Ed.; Transworld Research Network: Kerala, India, 2009; pp. 267–285, ISBN 9788178952499.
11. Bover-Cid, S.; Hugas, M.; Izquierdo-Pulido, M.; Vidal-Carou, M.C. Amino acid-decarboxylase activity of bacteria isolated from fermented pork sausages. *Int. J. Food Microbiol.* **2001**, *66*, 185–189. [CrossRef]
12. Ladero, V.; Fernández, M.; Calles-Enríquez, M.; Sánchez-Llana, E.; Cañedo, E.; Martín, M.C.; Alvarez, M.A. Is the production of the biogenic amines tyramine and putrescine a species-level trait in enterococci? *Food Microbiol.* **2012**, *30*, 132–138. [CrossRef] [PubMed]
13. Linares, D.M.; Martín, M.; Ladero, V.; Alvarez, M.A.; Fernández, M. Biogenic amines in dairy products. *Crit. Rev. Food Sci. Nutr.* **2011**, *51*, 691–703. [CrossRef] [PubMed]
14. Latorre-Moratalla, M.L.; Bover-Cid, S.; Veciana-Nogués, M.T.; Vidal-Carou, M.C. Control of biogenic amines in fermented sausages: Role of starter cultures. *Front. Microbiol.* **2012**, *3*, 169. [CrossRef] [PubMed]
15. De las Rivas, B.; Ruiz-Capillas, C.; Carrascosa, A.V.; Curiel, J.A.; Jiménez-Colmenero, F.; Muñoz, R. Biogenic amine production by Gram-positive bacteria isolated from Spanish dry-cured "chorizo" sausage treated with high pressure and kept in chilled storage. *Meat Sci.* **2008**, *80*, 272–277. [CrossRef] [PubMed]
16. Novella-Rodríguez, S.; Veciana-Nogués, M.T.; Roig-Sagués, A.X.; Trujillo-Mesa, A.J.; Vidal-Carou, M.C. Influence of starter and nonstarter on the formation of biogenic amine in goat cheese during ripening. *J. Dairy Sci.* **2002**, *85*, 2471–2478. [CrossRef]
17. Torracca, B.; Pedonese, F.; Turchi, B.; Fratini, F.; Nuvoloni, R. Qualitative and quantitative evaluation of biogenic amines in vitro production by bacteria isolated from ewes' milk cheeses. *Eur. Food Res. Technol.* **2018**, *244*, 721–728. [CrossRef]
18. Ladero, V.; Martín, M.C.; Redruello, B.; Mayo, B.; Flórez, A.B.; Fernández, M.; Alvarez, M.A. Genetic and functional analysis of biogenic amine production capacity among starter and non-starter lactic acid bacteria isolated from artisanal cheeses. *Eur. Food Res. Technol.* **2015**, *241*, 377–383. [CrossRef]
19. Sumner, S.S.; Speckhard, M.W.; Somers, E.B.; Taylor, S.L. Isolation of histamine-producing *Lactobacillus buchneri* from Swiss cheese implicated in a food poisoning outbreak. *Appl. Environ. Microbiol.* **1985**, *50*, 1094–1096. [PubMed]
20. Diaz, M.; del Rio, B.; Sanchez-Llana, E.; Ladero, V.; Redruello, B.; Fernández, M.; Martin, M.C.; Alvarez, M.A. *Lactobacillus parabuchneri* produces histamine in refrigerated cheese at a temperature-dependent rate. *Int. J. Food Sci. Technol.* **2018**, *53*, 2342–2348. [CrossRef]

21. Pircher, A.; Bauer, F.; Paulsen, P. Formation of cadaverine, histamine, putrescine and tyramine by bacteria isolated from meat, fermented sausages and cheeses. *Eur. Food Res. Technol.* **2007**, *226*, 225–231. [CrossRef]

22. Rodriguez-Jerez, J.J.; Giaccone, V.; Colavita, G.; Parisi, E. *Bacillus macerans*—A new potent histamine producing micro-organism isolated from Italian cheese. *Food Microbiol.* **1994**, *11*, 409–415. [CrossRef]

23. Bover-Cid, S.; Holzapfel, W.H. Improved screening procedure for biogenic amine production by lactic acid bacteria. *Int. J. Food Microbiol.* **1999**, *53*, 33–41. [CrossRef]

24. Marcobal, Á.; de las Rivas, B.; Moreno-Arribas, M.V.; Muñoz, R. Multiplex PCR method for the simultaneous detection of histamine-, tyramine-, and putrescine-producing lactic acid bacteria in foods. *J. Food Prot.* **2005**, *68*, 874–878. [CrossRef] [PubMed]

25. Roig-Sagués, A.X.; Hernàndez-Herrero, M.M.; López-Sabater, E.I.; Rodríguez-Jerez, J.J.; Mora-Ventura, M.T. Evaluation of three decarboxylating agar media to detect histamine and tyramine-producing bacteria in ripened sausages. *Lett. Appl. Microbiol.* **1997**, *25*, 309–312. [CrossRef] [PubMed]

26. Maijala, R.L. Formation of histamine and tyramine by some lactic acid bacteria in MRS-broth and modified decarboxylation agar. *Lett. Appl. Microbiol.* **1993**, *17*, 40–43. [CrossRef]

27. Espinosa-Pesqueira, D.; Hernández-Herrero, M.; Roig-Sagués, A.; Espinosa-Pesqueira, D.; Hernández-Herrero, M.M.; Roig-Sagués, A.X. High Hydrostatic Pressure as a tool to reduce formation of biogenic amines in artisanal Spanish cheeses. *Foods* **2018**, *7*, 137. [CrossRef] [PubMed]

28. Eerola, S.; Hinkkanen, R.; Lindfors, E.; Hirvi, T. Liquid chromatographic determination of biogenic amines in dry sausages. *J. AOAC Int.* **1993**, *76*, 575–577. [PubMed]

29. Altman, D.G.; Bland, J.M. Statistics Notes: Diagnostic tests 1: Sensitivity and specificity. *BMJ* **1994**, *308*, 1552. [CrossRef] [PubMed]

30. Altman, D.G.; Bland, J.M. Statistics Notes: Diagnostic tests 2: Predictive values. *BMJ* **1994**, *309*, 102. [CrossRef] [PubMed]

31. Altman, D.G.; Bland, J.M. Statistics Notes: Diagnostic tests 3: Receiver operating characteristic plots. *BMJ* **1994**, *309*, 188. [CrossRef] [PubMed]

32. Zweig, M.H.; Campbell, G. Receiver-operating characteristic (ROC) plots: A fundamental evaluation tool in clinical medicine. *Clin. Chem.* **1993**, *39*, 561–577. [PubMed]

33. Park, S.H.; Goo, J.M.; Jo, C.-H. Receiver Operating Characteristic (ROC) Curve: Practical review for radiologists. *Korean J. Radiol.* **2004**, *5*, 11–18. [CrossRef] [PubMed]

34. Hanley, J.A.; McNeil, B.J. The meaning and use of the area under a receiver operating characteristic (ROC) curve. *Radiology* **1982**, *143*, 29–36. [CrossRef] [PubMed]

35. Harrigan, W.F. *Laboratory Methods in Food Microbiology*; Academic Press: Cambridge, MA, USA, 1998; ISBN 0123260434.

36. Roig-Sagues, A.X.; Hernandez-Herrero, M.; Lopez-Sabater, E.I.; Rodriguez-Jerez, J.J.; Mora-Ventura, M.T. Histidine decarboxylase activity of bacteria isolated from raw and ripened salchichón, a Spanish cured sausage. *J. Food Prot.* **1996**, *59*, 516–520. [CrossRef]

37. Joosten, H.M.; Northolt, M.D. Detection, growth, and amine-producing capacity of lactobacilli in cheese. *Appl. Environ. Microbiol.* **1989**, *55*, 2356–2359. [PubMed]

38. Moreno-Arribas, M.V.; Polo, M.C.; Jorganes, F.; Muñoz, R. Screening of biogenic amine production by lactic acid bacteria isolated from grape must and wine. *Int. J. Food Microbiol.* **2003**, *84*, 117–123. [CrossRef]

39. Marino, M.; Maifreni, M.; Bartolomeoli, I.; Rondinini, G. Evaluation of amino acid-decarboxylative microbiota throughout the ripening of an Italian PDO cheese produced using different manufacturing practices. *J. Appl. Microbiol.* **2008**, *105*, 540–549. [CrossRef] [PubMed]

40. Suzzi, G.; Gardini, F. Biogenic amines in dry fermented sausages: A review. *Int. J. Food Microbiol.* **2003**, *88*, 41–54. [CrossRef]

41. Marino, M.; Maifreni, M.; Moret, S.; Rondinini, G. The capacity of *Enterobacteriaceae* species to produce biogenic amines in cheese. *Lett. Appl. Microbiol.* **2000**, *31*, 169–173. [CrossRef] [PubMed]

42. Halász, A.; Baráth, Á.; Simon-Sarkadi, L.; Holzapfel, W. Biogenic amines and their production by microorganisms in food. *Trends Food Sci. Technol.* **1994**, *5*, 42–49. [CrossRef]

43. Martín, B.; Garriga, M.; Hugas, M.; Bover-Cid, S.; Veciana-Nogués, M.T.; Aymerich, T. Molecular, technological and safety characterization of Gram-positive catalase-positive cocci from slightly fermented sausages. *Int. J. Food Microbiol.* **2006**, *107*, 148–158. [CrossRef] [PubMed]

44. Pons-Sánchez-Cascado, S.; Vidal-Carou, M.C.; Mariné-Font, A.; Veciana-Nogués, M.T. Influence of the Freshness grade of raw fish on the formation of volatile and biogenic amines during the manufacture and storage of vinegar-marinated anchovies. *J. Agric. Food Chem.* **2005**, *53*, 8586–8592. [CrossRef] [PubMed]

45. Masson, F.; Talon, R.; Montel, M.C. Histamine and tyramine production by bacteria from meat products. *Int. J. Food Microbiol.* **1996**, *32*, 199–207. [CrossRef]

46. Santos, M.H. Amino acid decarboxylase capability of microorganisms isolated in Spanish fermented meat products. *Int. J. Food Microbiol.* **1998**, *39*, 227–230. [CrossRef]

47. Drosinos, E.H.; Paramithiotis, S.; Kolovos, G.; Tsikouras, I.; Metaxopoulos, I. Phenotypic and technological diversity of lactic acid bacteria and staphylococci isolated from traditionally fermented sausages in Southern Greece. *Food Microbiol.* **2007**, *24*, 260–270. [CrossRef] [PubMed]

48. Foulquié Moreno, M.R.; Sarantinopoulos, P.; Tsakalidou, E.; De Vuyst, L. The role and application of enterococci in food and health. *Int. J. Food Microbiol.* **2006**, *106*, 1–24. [CrossRef] [PubMed]

49. Leuschner, R.G.K.; Kurihara, R.; Hammes, W.P. Formation of biogenic amines by proteolytic enterococci during cheese ripening. *J. Sci. Food Agric.* **1999**, *79*, 1141–1144. [CrossRef]

50. Galgano, F.; Suzzi, G.; Favati, F.; Caruso, M.; Martuscelli, M.; Gardini, F.; Salzano, G. Biogenic amines during ripening in 'Semicotto Caprino' cheese: Role of enterococci. *Int. J. Food Sci. Technol.* **2001**, *36*, 153–160. [CrossRef]

51. Landete, J.M.; Pardo, I.; Ferrer, S. Tyramine and phenylethylamine production among lactic acid bacteria isolated from wine. *Int. J. Food Microbiol.* **2007**, *115*, 364–368. [CrossRef] [PubMed]

52. Burdychova, R.; Komprda, T. Biogenic amine-forming microbial communities in cheese. *FEMS Microbiol. Lett.* **2007**, *276*, 149–155. [CrossRef] [PubMed]

53. Martuscelli, M.; Gardini, F.; Torriani, S.; Mastrocola, D.; Serio, A.; Chaves-López, C.; Schirone, M.; Suzzi, G. Production of biogenic amines during the ripening of Pecorino Abruzzese cheese. *Int. Dairy J.* **2005**, *15*, 571–578. [CrossRef]

54. Tham, W.; Karp, G.; Danielsson-Tham, M.L. Histamine formation by enterococci in goat cheese. *Int. J. Food Microbiol.* **1990**, *11*, 225–229. [CrossRef]

55. Perez, M.; Ladero, V.; del Rio, B.; Redruello, B.; de Jong, A.; Kuipers, O.; Kok, J.; Martin, M.C.; Fernandez, M.; Alvarez, M.A. The relationship among tyrosine decarboxylase and agmatine deiminase pathways in *Enterococcus faecalis*. *Front. Microbiol.* **2017**, *8*, 2107. [CrossRef] [PubMed]

56. Novella-Rodriguez, S.; Veciana-Nogues, M.T.; Trujillo-Mesa, A.J.; Vidal-Carou, M.C. Profile of biogenic amines in goat cheese made from pasteurized and pressurized milks. *J. Food Sci.* **2002**, *67*, 2940–2944. [CrossRef]

57. Fernández-García, E.; Tomillo, J.; Núñez, M. Effect of added proteinases and level of starter culture on the formation of biogenic amines in raw milk Manchego cheese. *Int. J. Food Microbiol.* **1999**, *52*, 189–196. [CrossRef]

58. Sumner, S.S.; Taylor, S.L. Detection method for histamine-producing, dairy-related bacteria using diamine oxidase and leucocrystal violet. *J. Food Prot.* **1989**, *52*, 105–108. [CrossRef]

59. Ascone, P.; Maurer, J.; Haldemann, J.; Irmler, S.; Berthoud, H.; Portmann, R.; Fröhlich-Wyder, M.-T.; Wechsler, D. Prevalence and diversity of histamine-forming *Lactobacillus parabuchneri* strains in raw milk and cheese—A case study. *Int. Dairy J.* **2017**, *70*, 26–33. [CrossRef]

60. Kung, H.F.; Tsai, Y.H.; Hwang, C.C.; Lee, Y.H.; Hwang, J.H.; Wei, C.I.; Hwang, D.F. Hygienic quality and incidence of histamine-forming *Lactobacillus* species in natural and processed cheese in Taiwan. *J. Food Drug Anal.* **2005**, *13*, 51–56.

61. Leuschner, R.G.; Heidel, M.; Hammes, W.P. Histamine and tyramine degradation by food fermenting microorganisms. *Int. J. Food Microbiol.* **1998**, *39*, 1–10. [CrossRef]

62. Straub, B.W.; Kicherer, M.; Schilcher, S.M.; Hammes, W.P. The formation of biogenic amines by fermentation organisms. *Z. Lebensm. Unters. Forsch.* **1995**, *201*, 79–82. [CrossRef] [PubMed]

63. González de Llano, D.; Ramos, M.; Rodriguez, A.; Montilla, A.; Juarez, M. Microbiological and physicochemical characteristics of Gamonedo blue cheese during ripening. *Int. Dairy J.* **1992**, *2*, 121–135. [CrossRef]

64. Fontecha, J.; Peláez, C.; Juárez, M.; Requena, T.; Gómez, C.; Ramos, M. Biochemical and microbiological characteristics of artisanal hard goat's cheese. *J. Dairy Sci.* **1990**, *73*, 1150–1157. [CrossRef]

PERMISSIONS

LIST OF CONTRIBUTORS

Ricarda Torre, Henri P. A. Nouws and Cristina Delerue-Matos
REQUIMTE/LAQV, Instituto Superior de Engenharia do Porto, Instituto Politécnico do Porto, Dr. António Bernardino de Almeida 431, 4200-072 Porto, Portugal

Estefanía Costa-Rama
REQUIMTE/LAQV, Instituto Superior de Engenharia do Porto, Instituto Politécnico do Porto, Dr. António Bernardino de Almeida 431, 4200-072 Porto, Portugal Departamento de Química Física y Analítica, Universidad de Oviedo, Av. Julián Clavería 8, 33006 Oviedo, Spain

Young Hun Jin, Jae Hoan Lee, Young Kyung Park, Jun-Hee Lee and Jae-Hyung Mah
Department of Food and Biotechnology, Korea University, 2511 Sejong-ro, Sejong 30019, Korea

Hana Buchtova
Department of Meat Hygiene and Technology, Faculty of Veterinary Hygiene and Technology, University of Veterinary and Pharmaceutical Sciences Brno, 61242 Brno, Czech Republic

Dani Dordevic
Department of Plant Origin Foodstuffs Hygiene and Technology, Faculty of Veterinary Hygiene and Technology, University of Veterinary and Pharmaceutical Sciences Brno, 61242 Brno, Czech Republic Department of Technology and Organization of Public Catering, South Ural State University, Lenin prospect 76, 454080 Chelyabinsk, Russia

Iwona Duda and Piotr Kulawik
Department of Animal Product Technology, Faculty of Food Technology, University of Agriculture, 31-120 Krakow, Poland

Alena Honzlova
Department of Chemistry, State Veterinary Institute Jihlava, 58601, Jihlava, Czech Republic

Maria Manuela Hernández-Herrero
CIRTTA-Departament de Ciència Animal i dels Aliments, Universitat Autònoma de Barcelona, Travessera dels Turons S/N, 08193 Barcelona, Spain

Umile Gianfranco Spizzirri and Donatella Restuccia
Department of Pharmacy, Health and Nutritional Sciences, University of Calabria, I-87036 Rende (CS), Italy

Francesca Ieri, Margherita Campo and Annalisa Romani
Department of Statistic, Informatics and Applications "G. Parenti" (DiSIA) — University of Florence, Phytolab Laboratory, via Ugo Schiff 6, 50019 Sesto Fiorentino (FI), Italy

Donatella Paolino
Department of Experimental and Clinical Medicine, University of Catanzaro "Magna Græcia", 88100 Catanzaro, Italy

Claudia Ruiz-Capillas and Ana M. Herrero
Department of Products, Institute of Food Science, Technology and Nutrition, ICTAN-CSIC, Ciudad Universitaria, 28040 Madrid, Spain

So Hee Yoon and BoKyung Moon
Department of Food and Nutrition, Chung-Ang University, Gyeonggi-do 17546, Korea

Eunmi Koh and Bogyoung Choi
Major of Food & Nutrition, Division of Applied Food System, Seoul Women's University, Seoul 01797, Korea

Federica Barbieri and Chiara Montanari
Interdepartmental Center for Industrial Agri-Food Research, University of Bologna, 47521 Cesena, Italy

Fausto Gardini and Giulia Tabanelli
Interdepartmental Center for Industrial Agri-Food Research, University of Bologna, 47521 Cesena, Italy Department of Agricultural and Food Sciences, University of Bologna, 40126 Bologna, Italy

Tianjiao Niu
School of Chemistry and Chemical Engineering, Harbin Institute of Technology, Harbin 150090, China Mengniu Hi-tech Dairy (Beijing) Co., Ltd., Beijing 101107, China

Xing Li and Ying Ma
School of Chemistry and Chemical Engineering, Harbin Institute of Technology, Harbin 150090, China

Yongjie Guo
Mengniu Hi-tech Dairy (Beijing) Co., Ltd., Beijing 101107, China

Martin Grootveld and Benita C. Percival
Leicester School of Pharmacy, De Montfort University, The Gateway, Leicester LE1 9BH, UK

Jie Zhang
Green Pasture Products, 416 E. Fremont Street, O'Neill, NE 68763, USA

Mehdi Triki
Ministry of Public Health, Doha, Qatar

Ana M. Herrero, Francisco Jiménez-Colmenero and Claudia Ruiz-Capillas
Department of Products, Institute of Food Science, Technology and Nutrition, ICTAN-CSIC, Ciudad Universitaria, 28040 Madrid, Spain

Chiu-Chu Hwang
Department of Hospitality Management, Yu Da University of Science and Technology, Miaoli 361027, Taiwan

Yi-Chen Lee, Chung-Yung Huang and Yung-Hsiang Tsai
Department of Seafood Science, National Kaohsiung University of Science and Technology, Kaohsiung 811213, Taiwan

Hsien-Feng Kung
Department of Pharmacy, Tajen University, Pingtung 907391, Taiwan

Hung-Hui Cheng
Mariculture Research Center, Fisheries Research Institute, Council of Agriculture, Tainan 724028, Taiwan

Diana Espinosa-Pesqueira, Artur X. Roig-Sagués and M. Manuela Hernández-Herrero
CIRTTA—Departament de Ciència Animal i dels Aliments, Universitat Autònoma de Barcelona, Travessera dels Turons S/N, 08193 Barcelona, Spain

Index

Printed in the USA
CPSIA information can be obtained
at www.ICGtesting.com
JSHW051625061123
51533JS00005B/105